Applied Forest Tree Improvement

Applied Forest Tree Improvement

Bruce Zobel
John Talbert
North Carolina State University

WAVELAND
PRESS, INC.
Prospect Heights, Illinois

For information about this book, write or call:

Waveland Press, Inc.
P.O. Box 400
Prospect Heights, Illinois 60070
(708) 634-0081

This book is dedicated to Dr. C. Syrach-Larsen.

Among the numerous pioneers in tree improvement, one of the most outstanding was C. Syrach-Larsen. His vision and "missionary zeal" gave the needed impetus to have tree improvement accepted as a useful tool in operational forestry. (Photo of Dr. Syrach-Larsen in his younger years through courtesy of Dr. Bent Søegaard, Director of the Arboretum at Hørsholm, Denmark).

Shown are stages in a tree improvement program. To the left is a 34-year-old selected loblolly pine tree that has many of the desired characteristics of a parent tree. To the right (above) is a mature grafted seed orchard of pine, now in full production. The results from a tree improvement program are shown in the 16-year-old progeny test (lower right) in which growth, tree form, disease resistance, wood qualities, and adaptability have been improved.

Preface

There has been a tremendous increase of interest recently in tree improvement and the use of genetics in forest management. Applied tree improvement activities began in earnest during the 1950s. There is now a considerable amount of information and research data available that makes efficient planning and operation of tree improvement programs possible.

The objective of our book is to consolidate and summarize the concepts that are necessary for useful and efficient operational tree improvement programs. This book will concentrate on the biological and the practical, rather than on the more theoretical and statistical aspects, even though basic statistical concepts that are vital to breeding and selection programs will be presented. The book is based on more than 30 years' experience with large applied tree improvement programs. It will emphasize why and how certain things should or should not be done. Much of the information presented comes from background that was obtained from the large cooperative tree improvement programs in the southeastern United States, and many of the examples used in the book will involve pines, although the hardwoods will receive considerable coverage. Because the authors have had experience in programs in countries in South America and Central America, the West Coast of the United States and Canada, the northeastern United States, Australia, New Zealand, and Europe, the book will have a global coverage. Because of the rapidly expanding use of exotic species in intensive forest management, there is much emphasis in the book on exotics in the tropical and subtropical regions as well as in the temperate zones.

The book has an emphasis on young tree improvement programs. It does not deal with specifics or minutiae but covers the general concepts and principles necessary to manipulate and use forest tree populations in operational tree improvement programs. The concepts discussed are applicable to most forest tree species throughout the world; when feasible, specific examples are used to clarify the concepts. Because the greatest gains from the use of genetics in forestry will be from plantation programs, artificial regeneration will be stressed, but not to the exclusion of programs using natural regeneration.

The book has been written for three different audiences.

1. **Students** This book will be of special value to new and beginning students who wish to know how to develop a tree improvement program. It will also be of particular interest to advanced students who have forgotten, or who never knew, the basic biological concepts and economic considerations that are necessary for success in applied tree improvement. Introductory courses in genetics and statistics would be helpful but are not mandatory for a basic understanding of the concepts presented in the book.

2. **Personnel of tree improvement programs** Many programs for improving forest trees are carried out by persons with limited formal training in genetics. Although many of them are doing an excellent job, both their interest and efficiency will be greatly improved by an understanding of

genetic principles and a knowledge of how to manipulate genetic variation to achieve the greatest gains. Although the book is written in a format that is useful to students, it will also be readily understandable to persons without formal training in genetics.

3.　**Forest management personnel**　Although managerial foresters deal with the whole range of activities, from regeneration through harvesting, tree improvement is often a vital and large part of their activities. Management foresters do not want to become specialists in tree improvement, but they need to know enough to use tree improvement wisely. Some major errors have been made by forest managers who were ignorant of, or who overlooked, tree improvement principles. This book will enable them to obtain an overview of important concepts in tree improvement quickly that will help them coordinate their operations.

In rapidly developing fields such as tree improvement and forest genetics, a textbook is often out of date before it is published. We have tried to avoid this problem by emphasizing the biological basis of concepts that are fundamental to forest tree populations and their manipulation, while avoiding a concentration on the rapidly evolving and changing techniques and methodologies. The book emphasizes both practical and applied aspects; thus, economic considerations are either stated briefly or implied throughout most of the presentation. The book will be useful to anyone interested in improving forest tree growth, quality, adaptability, and pest resistance. It is of value for all stages of tree improvement programs, but it is of special importance to those who are just beginning this area of endeavor.

Pertinent references will be used throughout the text, but there will be no attempt to cite all available literature; to do so would make the book unwieldy and difficult to read.

Bruce Zobel
John Talbert

Acknowledgments

The authors greatly appreciate the extensive help obtained in preparing our book *Applied Forest Tree Improvement*. Many persons edited various chapters, made suggestions, and supplied photographs. To each of these we owe a debt of gratitude. Those who have contributed to the success of the book are listed alphabetically. It is unfortunate that the contribution of each cannot be personally recognized.

However, one reviewer deserves special mention. Dr. Ted Miller, professor emeritus of the School of Forest Resources at North Carolina State University, reviewed every chapter. He knows several languages, and he especially helped with the spelling, accent marks, and the use of foreign titles that are listed. Dr. Miller's help was invaluable.

Following is an alphabetical listing of many of those who contributed to the development of the book.

NAME	ORGANIZATION	LOCATION
Jim Barker	International Paper Company	Tuxedo Park, New York
Walt Beineke	Purdue University	Lafayette, Indiana
Floyd Bridgwater	U.S. Forest Service	Raleigh, North Carolina
Arno Brune	Federal University of Minas Gerais	Viçosa, Minas Gerais, Brazil
Rollie Burdon	Forest Research Institute	Rotorua, New Zealand
Jeff Burley	Oxford University	Oxford, England
Cheryl Busby	Graduate student, North Carolina State University	Raleigh, North Carolina
Roger Blair	Potlach Corp.	Lewiston, Idaho
Mike Carson	Forest Research Institute	Rotorua, New Zealand
Susan Carson	Forest Research Institute	Rotorua, New Zealand
Donald Cole	Consultant	Raleigh, North Carolina
Ellis Cowling	School of Forest Resources, North Carolina State University	Raleigh, North Carolina
Bill Critchfield	U.S. Forest Service	Berkeley, California
Gary DeBarr	U.S. Forest Service	Athens, Georgia

NAME	ORGANIZATION	LOCATION
Wade Dorsey	Graduate student, North Carolina State University	Raleigh, North Carolina
Keith Dorman	U.S. Forest Service, retired	Asheville, North Carolina
Jack Duffield	Professor emeritus, North Carolina State University	Shelton, Washington
Dean Einspahr	Institute of Paper Chemistry	Appleton, Wisconsin
Carlyle Franklin	School of Forest Resources, North Carolina State University	Raleigh, North Carolina
Carlos Gallegos	A.I.D. Washington, D.C.	Washington, D.C.
Bill Gladstone	Weyerhaeuser Timber Company	Tacoma, Washington
Max Hagman	Forest Research Institute	Marsala, Finland
J. B. Jett	School of Forest Resources, North Carolina State University	Raleigh, North Carolina
Hyun Kang	U.S. Forest Service	Rhinelander, Wisconsin
Lauri Karki	Foundation for Forest Tree Breeding	Helsinki, Finland
Bob Kellison	School of Forest Resources, North Carolina State University	Raleigh, North Carolina
Bill Ladrach	Cartón de Colombia	Cali, Colombia
Gladys Ladrach	Translator	Cali, Colombia
Clem Lambeth	Weyerhaeuser Company	Hot Springs, Arkansas
Bill Libby	University of California	Berkeley, California
Bill Lowe	Texas Forest Service	College Station, Texas

NAME	ORGANIZATION	LOCATION
Steve McKeand	School of Forest Resources, North Carolina State University	Raleigh, North Carolina
Kris Morgenstern	University of New Brunswick	Fredericton, New Brunswick, Canada
Garth Nikles	Queensland Forest Service	Brisbane, Queensland, Australia
Ron Pearson	School of Forest Resources, North Carolina State University	Raleigh, North Carolina
B. Phillion	Ministry of Natural Resources	Orono, Canada
Dick Porterfield	Champion Papers	Stamford, Connecticut
Harry Powers	U.S. Forest Service	Macon, Georgia
Lindsay Pryor	Consultant	Canberra, Australia
Pamela Puryear	Librarian, North Carolina State University	Raleigh, North Carolina
Diane Roddy	Prince Albert Pulp Company	Prince Albert, Saskatchewan, Canada
Marie Rauter	Ministry of Natural Resources	Toronto, Canada
Leroy Saylor	School of Forest Resources, North Carolina State University	Raleigh, North Carolina
Earl Sluder	U.S. Forest Service	Macon, Georgia
Dan Struve	Ohio State University	Columbus, Ohio
Richard Sniezko	Graduate Student, North Carolina State University	Raleigh, North Carolina
Per Stahl	Swedish Forest Service	Falun, Sweden
Bent Søegaard	Arboretum	Hørsholm, Denmark
Oscar Sziklai	University of British Columbia	Vancouver, British Columbia, Canada

NAME	ORGANIZATION	LOCATION
Hans van Buijtenen	Texas Forest Service	College Station, Texas
Bob Weir	School of Forest Resources, North Carolina State University	Raleigh, North Carolina
Ozzie Wells	U.S. Forest Service	Gulfport, Mississippi
Tim White	International Paper Company	Lebanon, Oregon
Peter Wood	Commonwealth Forest Institute	Oxford, England
Marvin Zoerb	Union Camp Corporation	Savannah, Georgia

B. Z.
J. T.

Contents

Applied Forest
Tree Improvement

CHAPTER 1

General Concepts
of Tree Improvement

Historically, foresters generally did not view trees as typical plants having systems of heredity similar to all other living organisms. Genetic variability was ignored, and it was somehow felt that a tree's development depended only upon the environment in which it was grown. It has only been in relatively recent years that there has been a general recognition that *forest tree parentage is important* and that changes and improvements in tree growth and quality can be brought about through breeding and parental control. Forest tree improvement activities were undertaken seriously on an operational scale only after this was recognized.

A whole book could be written about the fascinating steps in the development of tree improvement and about the contributions made by the early pioneers. Several publications deal with the history of forest genetics, both overall and locally. Examples are Ohba (1979), Toda (1980), and Wright (1981). No attempt will be made in this book to completely cover the historical development or even to mention all of the pioneers, who had much faith and foresight in doing such a radical thing as applying genetic principles to improve forest trees. Even though most forest genetics research is rather recent, some was done centuries ago. Perhaps the most striking early activities were in Japan. These have been abstracted in two volumes by Toda (1970, 1974). His references relating to tree improvement date to the seventeenth century. While helping Toda edit these books into acceptable English, it was evident to the senior author that foresters are today rediscovering a large number of concepts that were known or suspected hundreds of years ago.

A few of the pioneers in forest genetics are listed in Table 1.1, which is a portion of a table originally entitled "Chronology of Forest Genetics" that was prepared by Sziklai (1981).

Some excellent ideas on the use of genetics in forestry were suggested by the earlier workers, although many of the publications contained mostly generalized concepts and ideas. Examples of some early writings in the United States are Austin (1927), Leopold (1929), Schreiner (1935), and Righter (1946). These, along with publications such as the one by Richens (1945), contributed to tree improvement by making foresters cognizant of the fact that parentage is important in forest trees and can be manipulated to help in forest management.

A major contribution by the early workers was that they observed and cataloged patterns of variation in commercially important tree species. In this way, they achieved familiarity with the various species, which is an absolute essential to the success of any breeding program. The early activities resulted in only limited amounts of firm evidence regarding the possible improvements, but by drawing on methods employed in plant and animal breeding the early workers were able to make some very good estimates about potential gains in forest trees. Therefore, when the forest industry finally became interested in developing large programs of applied genetics in the early 1950s, it had little proof of the returns that would be obtained to justify the investments that had been made, but it did have some reasonably useful projections on which plans could be based.

The success of the first large programs was due primarily to good intuition about how the important forest tree characteristics would respond to genetic

TABLE 1.1.
Some of the Early Forest Geneticists and Their Areas of Interest[a]

1717 Bradley (England)	Importance of seed origin
1760 Duhamel de Monceau (France)	Inheritance: oak
1761 Koehlreuter (Germany)	Hybridization
1787 Bursdorf (Germany)	Plantation for seed production
1840 Marrier de Boisdhyver (France)	Vegetative propagation
1840 de Vilmorin (France)	Fir hybrids
1845 Klotzsch (Berlin)	Intraspecific hybrids: oak, elm, and alder
1904 Cieslar (Austria)	Provenances: larch and oak
1905 Engler (Switzerland)	Elevation differences of species: fir, pine, spruce, larch and maple
1905 Dengler (Germany)	Provenance tests: fir and spruce
1906 Andersson (Sweden)	Vegetative propagation
1907 Sudworth, Pinchot (U.S.A.)	Breeding nut and other forest trees
1908 Oppermann (Denmark)	Straightness: beech and oak
1909 Johannsen (Sweden)	Elite stands
1909 Sylven (Sweden)	Self-pollination: Norway spruce
1912 Zederbauer (Austria)	Crown form: Austrian pine
1918 Sylven (Sweden)	Seed orchards
1922 Fabricius (Austria)	Plantation for seed production
1923 Oppermann (Denmark)	Seedling seed orchards
1924 Schreiner (U.S.A.)	Poplar breeding
1928 Burger (Switzerland)	Pine selection
1928 Bates (U.S.A.)	Seed orchards
1930 Larsen S. (Denmark)	Controlled pollination: larch
1930 Heikinheimo and M. Larsen (Finland)	Curly grain: birch
1930 Nilsson-Ehle, Sylven, Johnsson, Linquist (Sweden)	Pine and aspen breeding
1935 Nilsson-Ehle (Sweden)	Triploid aspen

[a]After Sziklai (1981).

manipulation. This point can be illustrated by the results that were achieved by including wood properties in some initial breeding programs of several pine species in the southeastern United States. Essentially nothing was known about the inheritance of wood specific gravity, the most important wood quality. However, the patterns of natural variation of specific gravity were well known (Figure 1.1). These indicated that significant gains might be made through a selection and breeding program, although there was no available proof. Based on these patterns, wood-specific gravity was included as a major characteristic in some tree improvement programs. It is now clear that including that trait in the breeding program was helpful both biologically and economically because specific gravity, which greatly influences the yield and quality of wood products, has proved to be controlled strongly enough by genetics so that it will respond well to a

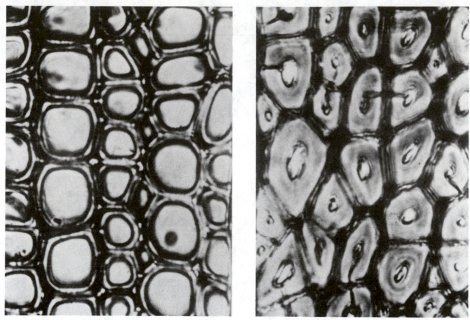

FIGURE 1.1

Variability among trees of the same age growing adjacent on the same site is shown by wall thickness of two *Eucalyptus grandis* trees. Wall thickness affects specific gravity, which is reasonably strongly inherited, enabling the development of trees with either high or low-wood specific gravities.

selection program. If characteristics of importance but of unknown inheritance, such as wood specific gravity, had not responded well to breeding, the entire program of genetic manipulation of forest trees on a large scale probably would have faltered.

A major emphasis on *applied* tree improvement developed rapidly in the early 1950s in a number of countries throughout the world. There are dozens of references that could be cited. However, for the sake of simplicity a few are indicated in Table 1.2 to give the reader an idea about the international scope of activities in tree improvement during this period. There was a literal explosion of activities, including both applied and fundamental aspects of forest genetics, and many new programs were established. Note that even though tree improvement started in many countries throughout the world about the same time, there have been great differences from country to country about how vigorously the programs were followed up.

It is of interest how three different incidents triggered the establishment of large, well-organized, and adequately financed tree improvement programs in the southeastern United States over 30 years ago. The first was the publication in 1948 of a book on forest genetics by Bertil Lindquist of the Göteborg Botanical Garden in Sweden. The book was translated into English and was circulated widely among

TABLE 1.2.
A Sampling of Authors Who Published on Applied Tree Improvement and Forest Genetic Activities During the 1950s

Author	Date	Country	Type of Article
Barner	1952	Denmark	General tree breeding
Bouvarel	1957	France	General tree improvement
Buchholz	1953	Russia (Germany)	Review of Soviet activities
Duffield	1956	U.S.A.	Breeding approaches
Fielding	1953	Australia	Variation studies
Fischer	1954	Germany	General tree breeding
Greeley	1952	U.S.A.	History of Institute of Forest Genetics
Haley	1957	Australia	Status of tree breeding in Queensland
Heimburger	1958	Canada	General tree breeding
Hellinga	1958	Indonesia	General tree improvement
Hyun	1958	Korea	General tree breeding
Johnsson	1949	Sweden	Results of breeding
Langner	1954	Germany	General tree breeding
Larsen	1951	World (Denmark)	General tree breeding
Matthews	1953	Great Britain	General tree breeding
Pauley	1954	U.S.A.	General tree breeding
Perry and Wang	1958	U.S.A.	Value of genetically improved seed
Rao	1951	India	General tree improvement
Schreiner	1950	U.S.A.	General tree breeding
Thulin	1957	New Zealand	General tree breeding
Toyama	1954	Japan	General tree breeding
Wright	1953	U.S.A.	General tree breeding
Zobel	1952	U.S.A.	Improving wood quality

Note. See Literature Cited for complete references.

foresters in the southern United States. It was written in a manner that caught the interest of foresters, and from it many obtained their first insights into the use of genetics in forest trees. In his book *Forest Tree Breeding in the World,* Toda (1974) makes reference to Lindquist's lectures in Japan in 1952 and their strong influence on the senior foresters in that country.

Another major influence was a series of lectures on crop breeding given in Texas by Åke Gustafsson of the Royal College of Forestry in Stockholm. In one lecture on forest genetics he predicted that trees would respond to genetic manipulation and urged that tree breeding be incorporated into silviculture. The third influence was a series of articles on tree breeding that were published by a newspaper in Texas as a result of Gustafsson's lectures. The articles resulted in a public campaign to raise money for a program of tree improvement. On behalf of the Texas Forest Service and with the participation of 14 forest industries, Bruce Zobel organized such a program in 1951. The working territory included Texas,

Louisiana, and Arkansas. The major initial task was to convince foresters that the interaction of the environment and genetics and not the environment alone, determine the growth, form, and adaptability of a tree.

After a slow beginning, enthusiasm for forest genetics grew to such an extent in certain areas that some people considered it a cure for most of forestry's ills. Since then, tree improvement has gained its proper perspective as a powerful tool of the forest manager, and it is successful in relation to the degree that it is employed in conjunction with good forest management practices.

WHAT IS TREE IMPROVEMENT?

In order to understand what tree improvement is, it is necessary to know three terms and their development and relationship to each other. Although some authors, such as Toda (1974), equate the three terms, most scientists differentiate between *forest tree breeding, forest genetics,* and *forest tree improvement.* Activities that are restricted to genetic studies of forest trees are termed *forest genetics;* here, the objective is to determine the genetic relationships among trees and species. An example of a forest genetic activity is the attempt to determine crossability patterns among species within a genus. The crosses are made to determine relationships, but otherwise they have no special breeding objective. The next term is *forest tree breeding,* in which activities are geared to solve some specific problem or to produce a specially desired product. An example of such directed breeding is the development of pest-resistant strains of trees or breeding trees that possess specially desired wood. The third term, *forest tree improvement,* is applied when control of parentage is combined with other forest management activities, such as site preparation or fertilization, to improve the overall yields and quality of products from forestlands.

Tree improvement is effective only when it consists of the combination of *all* silvicultural and tree-breeding skills of the forester to grow the most valuable forest products as quickly as possible and as inexpensively as possible. It consists of a "marriage" of silviculture and tree parentage to obtain the greatest overall returns (Figure 1.2). Stated simply, tree improvement is an additional tool of silviculture that deals with the kind and genetic makeup of the trees used in forest operations.

It has taken foresters a long time—much too long—to recognize that intensive forest management activities, such as site preparation or fertilization, never will yield maximum returns unless the genetically best trees are also used. Conversely, in recent years, foresters have learned from bitter experience that no matter how excellent trees may be genetically, maximum production cannot be achieved unless good forest management practices are used along with the improved plants. This concept of combining forest management with parentage is now quite widely accepted as tree improvement, but unfortunately there still are some who do not recognize this most critical relationship.

Tree improvement activities have sometimes developed without a suitable knowledge of the necessary genetic principles. Basic information is now becoming

FIGURE 1.2

Tree improvement can only be successful when combined with intensive silviculture, such as the bedding prior to planting shown in southern Brazil. The combination of good culture and good genetic stock makes possible optimum timber production.

available more rapidly than in the past, but there still is a major gap that needs to be filled. An inspiration for those who have been involved with tree improvement from the start is that in most forested areas in the world, tree improvement is currently included as an essential part of forest management operations. No longer is tree improvement considered an impractical academic exercise that requires special treatment and financing but contributes little to the income from forest land. Most forestry organizations now handle tree improvement as a regular part of silviculture. After a slow start and years of intensive propaganda to make tree improvement accepted by the forestry profession, a reversal in attitude took place, and for a time in the late 1950s and 60s tree improvement specialists found themselves in the unusual position of having to deemphasize the genetic approach because some people were "carried away" by its supposed potential and value. For a time, tree improvement was treated as a utopian activity that would solve all forestry problems. Luckily, tree improvement is now realistically considered to be an essential tool of silviculture.

There is nothing mysterious or difficult about tree improvement. It consists mostly of the use of common sense in forest management, looked at from the "point of view" of the basic biological organism—the tree. It is concerned with how trees vary and how this variation can be utilized to improve forest productivity. Although some people have the impression that a good tree improver must have some special training and possess certain mystical powers, this is not the case.

All beginning tree improvement programs rely on and consist of the following:

1. A determination of the species, or geographic sources within a species, that should be used in a given area.
2. A determination of the amount, kind, and causes of variability within the species.
3. A packaging of the desired qualities into improved individuals, such as to develop trees with combinations of desired characteristics (Figure 1.3).
4. Mass producing improved individuals for reforestation purposes.
5. Developing and maintaining a genetic base population broad enough for needs in advanced generations.

In some instances, step 1 may have been completed before intensive tree improvement programs are initiated. In others, considerable time, money, and skill may be required to complete step 1. This first step must be done well before steps 2 to 5 can be really effective. There is a danger, however, that an organization may become so involved with testing species and sources that all available resources are used in this phase of a tree improvement program. In our opinion, an overemphasis on species and provenance testing by some organizations has become a most serious stumbling block to their making maximum progress in the total tree improvement activity.

The five steps essentially outline what is needed for the development of any tree improvement program, and although they may make the scope of activities clear, they do not indicate the time and effort required to develop each of them properly. The outline may appear to be simple, but getting the job done is not easy, and it requires a great deal of skill, common sense, time, and money. For example, with the southern pines (where the first step on species determination was not a major activity), the senior author has spent over 30 years of intensive effort trying to satisfy steps 2 through 5.

Nature has created the variation needed for use in a tree improvement program (Figure 1.4). The tree improver's major job is to be able to recognize the variability, isolate it, package it in a desired tree, and multiply it. As results are obtained and advanced generations of improved trees are developed, a much more sophisticated and scientifically based approach than that used initially will

FIGURE 1.3

The yellow poplar tree shown represents a "package" of good characteristics in a single individual. The tree has desirable limbs, good form, fast growth, and desirable wood.

FIGURE 1.4

Natural stands of forest trees show great variability as is indicated by differences in branch habit among eastern white pines (*Pinus strobus*). The tree improver must first recognize and then use the variability related to the genetic makeup of the tree.

be necessary to maintain and increase variability and to take full advantage of the natural variation found in trees growing in unimproved forests.

In tree improvement programs using hybridization and/or vegetative propagation, the principles previously listed will still hold, although their order of importance may vary. In truth, all methods of tree improvement require that all five steps must be followed. The special advantage with a vegetative propagation program is that once a suitable "package" has been located, or developed through breeding, it can be reproduced rapidly many times, and the propagules are essentially the same genetically as the desired parent tree (Figure 1.5). Vegetative propagation allows quick and large gains because all types of genetic variation can be captured. (This will be described in later chapters.) When seed reproduction is used, only a portion of the genetic variation of the trees used as parents will be passed on to the progeny. Hybridization has an advantage because it enables the creation of something entirely different by recombining the variability produced in nature into a new "package," the hybrid tree. Thus, it may be possible through hybridization to create a plant with characteristics for difficult environments, pest resistance, or specially desired products.

One aspect of a tree improvement operation that is of great importance, but that is not usually mentioned in a textbook, is a vague thing called *intuition* or *feel*. Many of the earlier phases of a program were very often intuitive. The user could not always explain exactly why certain things were done or why they worked. Such an intuitive approach cannot be taught, but it does begin with a working knowledge of the species on which the improvement is to be done. Many of the younger and more highly technically trained scientists scoff at the importance of an intuitive approach, but it is valuable and successful. Any really new, large-scale

FIGURE 1.5
A vegetatively reproduced stand of radiata pine (*P. radiata*) in Australia is shown. Note the great similarity among the trees, all of which came from the same "donor" tree. There is a strong trend toward more use of vegetative propagation in forestry that will yield great gains and good uniformity.

operational program must initially rely on this vague "green thumb" approach as part of its methods if it is to quickly achieve its goals; this is true because usually only a limited amount of fundamental data are available. Some of the more successful tree improvers state that they "think or feel like the tree," and the resultant intuitive actions, often aided by experience, frequently prove to be the correct ones.

WHERE AND WHEN SHOULD TREE IMPROVEMENT BE USED?

The contributions that tree improvement can make to growth, quality, pest resistance, and adaptability of forest stands are greater under some conditions than under others. Obviously, large-scale, aggressive planting programs are most conducive to the use of genetic improvement. Tree improvement is more difficult to justify economically when forests are regenerated naturally. The basic fact remains, though, that all forest management activities can profit from using tree improvement concepts. If this is not done, forest managers can only partially achieve their objectives. Forest trees are plants with responses controlled by the environment and by genetics, like all other organisms. The result obtained from any forest management operation will be determined both by the genetic makeup of the tree and by its interaction with the environment in which it grows. Tree improvement should play a significant role in forest management any time the production of high volumes of good quality timber is the principal management objective. The most intensive tree improvement efforts will be made where stands are regenerated artificially at least once every few rotations. It is in these cases that the greatest gains can be obtained from tree improvement.

The *objective* of *intensive tree farming* can be stated simply as the *production of the desired-quality timber in maximum amounts in the shortest period of time at a reasonable cost*. This objective is simpler to state than it is to achieve, of course, but in the past three decades tree improvement has played an increasingly important role in helping to achieve it by increasing forest productivity and reducing the time needed to harvest. Tree improvement can be used to help accomplish many management objectives to overcome problems, but this will not be done without some adverse reactions. For example, the harvest age of trees can be reduced through genetic selection for growth rate, but the reduction may lead to significant changes in wood quality, harvesting costs, and regeneration costs. A major job of the tree breeder is also to help overcome problems that arise from intensive forest management. There are essentially three lines of attack that the tree improver can use to increase timber production: (1) Breeding can be accomplished for improved yields and quality on the more productive forested areas; (2) trees can be developed that will grow satisfactorily on land that is currently submarginal and noneconomic for timber production; and (3) strains of forest trees can be developed that are more suitable for specialized products or uses. The first approach has been widely utilized for several tree species, and dramatic results have been obtained. Many persons only associate tree improvement and the use of genetics with yield and quality improvement.

The development of trees especially suitable for marginal sites is long term in nature, but it will result in substantial benefits as pressures for forest land use intensify. Competition for land is increasing, which is forcing forestry operations from the more productive sites to areas that were previously considered to be marginal or useless for timber production. As a result, large amounts of genetically improved seeds are needed quickly that are specifically developed to grow on the vast forest areas that are currently marginal or submarginal for economic forest or agricultural production. The current emphasis in tree improvement is toward breeding for adaptability to marginal sites, in addition to improving trees for better products or better growth on forest sites that are suitable for forest production. Potential gains from breeding for adaptability to marginal sites are great, and huge areas of such land are available for forestry use. However, forest managers and tree breeders must be constantly alert to the basic biological constraints to the productive potential on a given area of land. High production will not be obtained from deficient soils, no matter how good the trees are genetically.

One of the more serious errors being made by some forest geneticists is to predict ever-increasing gains from the use of genetics without suitable consideration of the absolutely essential accompanying better forest management practices. As do all organisms, trees respond to the "law of limiting factors." Today, the most important limiting factor to increased forest yields often is the genetic potential of the plants being used. However, as genetically improved stock becomes more generally available, other factors necessary for good tree growth, such as limited moisture, excess moisture, or limited nutrients, will still limit tree growth no matter how excellent the genetic potential of the trees may be. It is not possible to obtain continued large gains from genetics without correcting whatever factor or factors in the environment may presently be, or may become, limiting. Thus, predictions of gains of several hundred percent from the use of genetics are misleading if the commensurate improvement of forest management methodology is not undertaken concurrently. The idea held by some persons that improved trees can be planted in the grass, brush, or briars with little care and still grow outstandingly well is totally unrealistic.

Tree improvement without commensurate intensive forestry is generally of marginal value, and there must be a union between the two if maximum gains from either are to be achieved. This combination of genetics and culture has been shown many times to be essential in agriculture. For example, the old open-pollinated strains of corn did not take full advantage of intensive culture and fertilization. Conversely, the highly genetically improved corn varieties will not produce anywhere close to their potential without intensive management. Trees react in the same way; the total genetic potential can only be exploited if the trees are grown in the best environments.

There are essentially two ways to ameliorate conditions limiting forest productivity. First, the forester can help to reduce the limiting factors of the environment through use of better forest management techniques and silviculture. This has been, and for a long time will continue to be, the easiest and most common method used to increase forest productivity. However, as forestry operations are driven from the better lands that are needed for agriculture, the second option of developing strains of trees to overcome severely limiting environmental factors will become increasingly important. This type of breeding has already made some forestry operations profitable on land that had been considered marginal or submarginal for economical forest production.

It is frequently possible to overcome a limiting factor by breeding. For example, drought-tolerant (Brix, 1959; Bey, 1974), water-tolerant (Zobel, 1957; Hosner and Boyce, 1962; Heth and Kramer, 1975), or cold-tolerant (Dietrichson, 1961; Schönbach, 1961; Parker, 1963; Sakai and Okada, 1971; and many others) strains can be developed to overcome moisture imbalance or excess cold. It is possible to breed trees that can tolerate low levels of nutrients, thus partially overcoming the limitation of nutrient deficiency (Lacaze, 1963; Goddard et al., 1976; Roberds et al., 1976; McCormick and Steiner, 1978). In the northern coniferous forests where a deficiency in nitrogen is severe, improving growth through fertilization is generally accepted as a better option than breeding for trees that can grow with a lesser nutrient supply. A large part of the decision about whether to breed tolerant trees or take corrective action by supplying nutrients depends on the costs and availability of fertilizers. Occasional trees grow well on soils deficient in micronutrients such as boron (Figure 1.6). A number of appar-

FIGURE 1.6

Shown is a radiata pine (*P. radiata*) growing on a boron-deficient site in southcentral Chile. Note the health and vigor of the tree when compared to the veritable bushes surrounding it that result from a boron deficiency. Sometimes it is possible to use such trees to develop a new strain that is more tolerant to some deficiency in the environment.

ently boron-deficiency-tolerant radiata pine trees have been found indicating the possibility of developing a special strain to grow on the nutrient-deficient soils. Success will depend on the extent and kind of genetic variation in boron tolerance as well as the relative returns and costs of fertilizing versus developing a nutrient-tolerant strain of trees. Breeding for adaptability and pest resistance has proven to be feasible, and soon millions of hectares will be converted to productive forests on lands formerly considered to be nonproductive (Batzer, 1961; Goddard et al., 1975; Zobel and Zoerb, 1977; and numerous others).

It is essential, therefore, to remember that when production has not been hampered by limiting factors of the environment, greater yields of better-quality products will result from tree improvement. For many situations where the environment is limiting, such as drought or cold, the most effective approach is to develop trees that have greater tolerance to the limiting factors when it is not possible economically to overcome the deficiencies through forest management actions.

ESSENTIALS OF A TREE IMPROVEMENT PROGRAM

There are two aspects to any successful tree improvement program. The first relates to obtaining an immediate gain of desired products as rapidly and as efficiently as possible. This is achieved by intensively applying genetic principles to operational forestry programs that will result in better-quality, better-adapted, and higher-yielding tree crops. Maximum gains are achieved by the use of a few of the very best genetic parents to supply planting stock for operational programs. A benefit of a tree improvement program that is often not recognized is the production of large and regular seed crops that are suitable for the forest operations. Lack of suitable seed is one of the greatest deterrents to forestry.

The second aspect of a tree improvement program is concerned with the long-term need to provide the broad genetic base that is essential for continued progress over many generations. Although not emphasized in some current programs, the long-term aspect of tree improvement is of great importance.

All tree improvement programs must have an *operational* (production) and a *developmental* (research) phase. The two are closely linked, yet they require different approaches and philosophies. The two phases are roughly outlined in Figure 1.7; the solid boxes indicate operational activities for large-scale planting programs. The developmental, or research, phases are necessary for a successful long-range program. As the programs mature, the operational activities become increasingly dependent on continued progress in the developmental area. Therefore, to be successful, a tree improvement program must have the *developmental aspects initiated at an early stage* in the program, along with the operational activities. Too often, the developmental activities are not begun until later years because of the press of work in operations and, as a result, a gap develops in plant materials and information necessary for advanced breeding.

The operational phase produces quick economic gains from tree improvement and is the most easily understood and obvious to the general public and to forest managers. This phase consists of obtaining improved planting stock as quickly and efficiently as possible and with as much genetic improvement as possible. *Time* is of the essence, and often initial gains are only partially as large as the biological potential. Modest gains are often accepted in order to obtain improvement quickly. Although there is an inevitable time lag between the investment and returns on the investment in the operational phase, this lag is small compared to the time required for the payoff from the developmental phase.

The main objective of the developmental phase is to obtain and retain a broad genetic base and to combine desired characteristics into suitable trees that will be valuable for future generations. No program can be better than the base of genetic material upon which it is founded, and although the developmental phase takes considerable time to yield useful results, the provision of the skill and money required to maintain it is mandatory. A large number of tree improvement

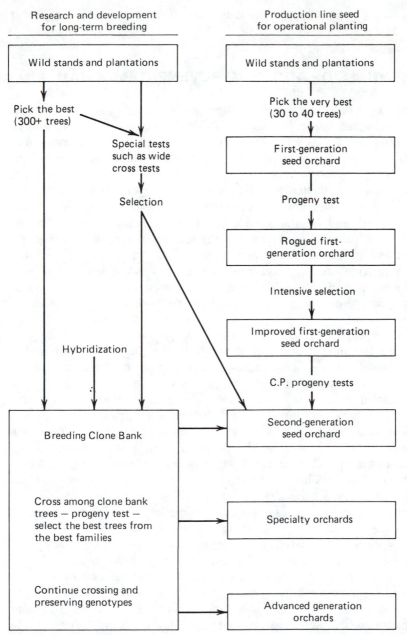

FIGURE 1.7

Tree improvement programs consist of two main lines of endeavor, the operational phase and developmental phase, as indicated. The quality of the breeding clone bank will determine long-term success; gains from the production seed orchards determine immediate success.

programs ignore the developmental or research aspects. Such programs will eventually come to a dead end. Because of the complexity and cost, the research or developmental work is ideally suited to a cooperative approach where several organizations jointly fund the work and share the results.

Some laymen, and even some foresters, are not aware of the need for the continuing long-range aspect of tree improvement and the necessity for active support of the developmental phase. Tree improvement work is never completed, and actions and success 10, 20, or 30 years from now are determined by the quality of the developmental phase that is established early. Many persons think that all tree improvement consists of is the location of trees with good phenotypes and manipulating them to produce large quantities of seeds. Nothing could be further from the truth. Continual improvement is always required. The operational portion of the program really touches only the surface of the total skill, cost, and energy required for a successful long-range tree improvement program. Problems and needs that arise in keeping a well-balanced program have been emphasized by Namkoong et al. (1980).

It is sad but true that the forest industry has been one of the most backward of the major industries in supporting biological research on forest trees. There seems to be an absolute distrust and suspicion of the words *fundamental* or *basic research,* and activity tagged in this way is often not encouraged and not funded. We have found, however, that substitution of the words *supportive research* has done wonders for acceptance by administrators in the forest industry. The idea that certain actions are necessary to obtain the basic information that will enable the operational program to continue to progress is usually received with no objections when it is called *supportive* research.

No matter what words are used—*fundamental, basic,* or *supportive*—information is needed if tree improvement programs are to be successful. Perhaps more than in any other discipline, the tree improver often does things on an operational scale before there is proof that they will work, or how much gain will be obtained. In a field like forestry, one cannot wait for all the answers, or even a good part of them, before taking action to make forestland more productive. Although this magnifies the danger of mistakes, the success of such empirical actions has been quite evident in newly begun tree improvement programs and has resulted in a tremendous improvement in forest productivity. A knowledge of the species and its variation in addition to a good bit of intuition enables rapid progress to be made with minimal basic knowledge. However, as tree improvement moves into advanced generations, much more scientific information will be needed for orderly and rapid development. Thus, the emphasis on supportive research needs to be increased.

THE IMPORTANCE OF TIME

One of the most important considerations in an active, ongoing tree improvement program is TIME (Zobel, 1978). It takes years to do the necessary selection, breeding, and testing to obtain the desired improvements. When an organization is planting large areas annually, each year that unimproved, rather than improved, stock is used represents a loss in future revenues. Therefore, operational programs are under great pressures to produce early returns. For example, members of the North Carolina State University–Industry Tree Improvement Cooperative were planting about 500,000 acres (200,000 ha) of pine each year when the tree improvement program was started. The urgency to engender as much improvement as possible as quickly as possible was great. Pressures for early improvements meant that all efforts had to be taken to obtain gain quickly, even if it required the use of some shortcuts.

In all applied tree improvement programs the time needs must be balanced against possible gains in both the short and long term; the key objective is to *obtain maximim gains per unit time*. For most forest trees, the generation time is long because of long-rotation ages and the 10 to 20 years required before abundant flowering occurs. To accrue the greatest gains in the shortest time through selective breeding, there must be a quick turnover of generations that are combined with some meaningful selection pressures in each generation. A most critical factor is the necessity for an assessment at a young age of such important characteristics as growth rate, adaptability, pest resistance, and tree form. Currently, for most species it generally is not satisfactory to assess growth parameters before about half-rotation (harvest) age, although certain form, pest resistance, wood, and adaptability characteristics can be accurately assessed earlier. In operational forestry, the rotation age can vary from as short as 6 to 8 years for the eucalypts, to 80 years or more for western U.S., northern U.S., Canadian, and northern European species. The southern pines, which are usually considered to be fast growing, are usually harvested at 20 to 35 years of age; therefore, 10- to 15-year assessments are feasible. On the surface, it would appear that a meaningful tree improvement program that will yield useful gains in a reasonable time would not be feasible, considering the restrictions caused by a lack of early flowering and an early assessment of growth. However, good progress has been made.

Although long-generation times are a major disadvantage in forest tree breeding, certain shortcuts have been developed. A good example of a most successful shortcut is illustrated by the seed orchard programs. Without the pressures and urgency of time, the best scientific method to establish a seed orchard that will yield maximum gains is to evaluate thoroughly the offspring of the select parents through progeny testing followed by the establishment of the production seed orchard from which improved seed will be obtained. Although eventually giving the greatest gain, this method entails several years' delay during the period required to make the tests, which can be up to one-half rotation age for some characteristics. Such a large loss of time before improved trees are established in orchards and seed are available cannot be accepted when seed are needed immediately for large-scale planting programs. To partially offset such a time loss, the shortcut usually taken in seed orchard programs is one in which good phenotypes are selected and are immediately established in the orchard before their genotypic worth has been determined. The grafted trees are established at a closer spacing than is ultimately desired, with the knowledge that some of the phenotypes will, in fact, not be good genotypes and therefore must ultimately be removed from the orchard. At the time (or shortly thereafter) of orchard establishment with the best phenotypes, progeny tests are initiated to determine the genetic worth of the parents used in the initial seed orchard. After progeny data are available, the orchard can then be rogued of the undesirable parents. This results in an orchard as good genetically as if it had been established after the parent trees were progeny tested. During the 10- to 20-year testing period, seed for planting have been obtained from the orchard that are considerably improved over wild seed, although they are not as good as those from proven parents. However, the added value of the partially improved seed to the forest management operation is large.

The concept of the value of time is indicated in Figure 1.8. Note that smaller but earlier gains are obtained when the orchard is established with parents chosen only on their appearance than when the parental genetic quality is known. These gains will increase as progeny test data are obtained, and roguing is done in the seed orchard. If the testing is properly designed, the genetic gain from both types of programs will be about the same following progeny assessment. This is illustrated in Figure 1.8. It is the shaded area in the first rectangle that indicates gain that is additional to that at some later time following progeny testing, which is shown by the nonshaded area. In programs in which there are small pressures for

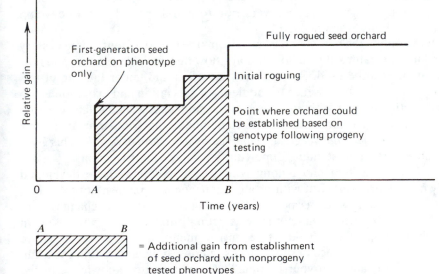

FIGURE 1.8

Time is all important in an operational tree improvement program associated with a large planting program. Shortcuts, such as establishing untested phenotypes in the seed orchard, may be taken to give early gains (AB). These are in addition to greater gains that are obtained after progeny test data are available.

immediate planting on a large scale, the prospective parents should be tested with respect to their genetic worth before the production orchards are established.

Although covered in detail elsewhere, a current major feature of tree improvement programs is to produce the *same amount and quality of product in less time at a reasonable cost* rather than to *produce more at a given time*. This objective of reducing time to harvest is now of prime importance in many tree improvement programs, and it will result in large monetary gains to the forest manager.

BREEDING OBJECTIVES

Although many aspects of tree improvement that are related to gains and genetic control will be covered in detail in later chapters, it is nevertheless necessary at this point to have some concept of those issues that will determine which approach to a tree improvement program should be used. Tree improvement specialists strive to improve the forest both for better yields and better quality (Zobel, 1974). Along with better forest management practices and the reduction of time to achieve specific goals, gains in tree improvement are determined by the intensity of inheritance and how well manipulated is the variation that is present in the population with which one is working.

There is little that the tree improver can do to change the inheritance pattern for a given characteristic; it is generally well established. The tree improver can, however, capture more of the genetic variation that is present in a population by suitable manipulation of the environment. The tree improver's best tool to increase gains is to use the existing variation to its fullest and to help develop additional variability when needed.

In order to obtain the best possible gains from tree improvement, it is necessary to understand the nature of wild populations, how they have developed, and how their variability can be used. A complete discussion and understanding of this subject encompassing the essence of the fields of speciation and evolution is too advanced for this volume, although some of the simpler concepts will be covered in several chapters. The important item to remember is that *forest trees* are mostly *wild populations that are not yet greatly changed by the action of people*. This gives the tree improver an outstanding opportunity to make improvements.

The differences within and among wild tree populations have developed naturally over many eons. But with proper management, intensity of selection, and suitable breeding systems people can bring about desired changes very rapidly. Prior to 1950, the senior author was strongly urged by a very well-known silviculturist not to pursue the field of tree improvement. His reason was "if nature has not been able to produce desirable trees in eons, how do you think you can make any useful changes in your lifetime?" He had no real knowledge of the forces that shape populations or why and how these work. What is most important, he had no concept of the pressures that people can apply to forest populations and individuals to cause rapid changes and to alter them in the desired direction. Although not totally intended, one of our major effects on changing forest trees, especially in areas like Europe and Japan, has resulted from widespread movement and the ultimate mixing of different geographic sources within a species.

Methods that can be used to change trees and the realized improvement arising from them vary with different characteristics. Because of the nature of the genetic patterns that influence the inheritance for most characteristics of forest trees, only part of the genetic variation that exists within a population can be utilized, especially when regeneration using seed is employed. Thus, progenies obtained for operational planting from a tree improvement program are only partially as good on the average as the combination of the parents from which they came. When vegetative propagation is feasible, gains that are realized may be greater, but even then, some of the desired parental characteristics are not obtained in the progenies because of the interaction of genotypes with the environment. There is always the desired goal of perfection, but this is rarely achieved. The practical approach is to come as close to the ideal as possible with a reasonable and justifiable expenditure of time and energy.

An understanding of the practical versus ideal goals is of special importance for pest *resistance,* the term generally used when breeding to reduce diseases or insects. Complete resistance or immunity to pests seldom can be obtained. What is hoped for is enough resistance or tolerance to enable growing a profitable crop of trees. A good example is the very bad fusiform rust disease on southern pines (*Cronartium quercuum f. sp. fusiforme*) in the southeastern United States. Tree improvement organizations will be quite satisfied if the disease incidence can be reduced from the very high level now found in some operational plantations to about 15 or 20% infection. This amount of rust infection will have only a slight effect on the yield and quality of products from the forest. The cost, time, and effort to reduce infection further are not warranted economically. Emphasis in the program can then be directed to improving characteristics that will give a greater response.

A similar situation applies to such things as straightness of tree bole or to essentially all adaptability characteristics. We have been able to improve tree straightness in loblolly pine (*Pinus taeda*) enough to produce high-quality products in one generation of intensive selection. Even though a small additional improvement in straightness can be obtained with continued breeding, it is not worth the additional time and effort to place intensive selection pressures on this characteristic.

When breeding for adaptability characteristics, the objective is to gain maximum tolerance to adverse environmental factors. One will never find trees that are totally resistant to drought, cold, or excess moisture. However, one should strive to obtain trees that can better tolerate severe droughts, or colder weather, or that will grow reasonably well in other adverse environments.

It is necessary for the reader to remember always that the objective of a tree improvement program is to obtain trees that will be closer to the desired state than those that are currently available. The situation of diminishing returns is encountered as increased efforts come to the point where they yield a lesser amount of improvement per "unit effort" (Figure 1.9). A well-designed tree improvement program will have as its achievement objective the production of the greatest

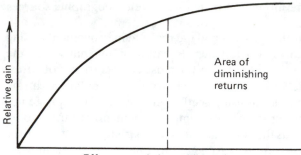

FIGURE 1.9

The law of diminishing returns is fundamental in tree improvement. As one gets closer and closer to the ideal, the cost and effort becomes greater for each added unit of gain. There comes the time when breeding for additional gains is not warranted by the time and effort involved (see Chapter 14 on economics).

amount of improvement that will still yield profitable returns. The problem is to determine when returns on the investment diminish to an extent that additional efforts for improvement are not warranted.

The most *important* and *exciting part of tree improvement* is that *it is usually possible to breed for improved economic characteristics while at the same time to maintain or broaden the genetic base for adaptability and pest resistance.* This is also true for changing product requirements. This is possible because very few of the important characteristics of forest trees are strongly correlated genetically. For example, one can develop disease resistance in either straight trees or crooked trees. Drought-resistant trees can either have high- or low-specific-gravity wood.

ADVANTAGES AND LIMITATIONS OF TREE IMPROVEMENT

Although this book will deal with the advantages and limitations of tree improvement in depth in later chapters, it is worthwhile to list here a few of the more important items in order to help the understanding of people interested in tree improvement.

One major advantage of genetic improvements in forest trees is that once a change is obtained, it can be kept over a number of generations. When improvements are made through silviculture (fertilization, for example), they will need to be done one or more times at each rotation. This quality of the "permanence" of genetic improvements over several rotations makes them very appealing economically, even though the initial cost of developing desired trees may be considerable. The advantage will be somewhat reduced when environments, pests, or markets change, so that the genetic material is no longer suitable.

Another major advantage of tree management is that the genetic material that is developed can be kept essentially intact for an indefinite time through methods of vegetative propagation. Many outcrossing annuals require continued replacement through producing new seeds because seeds lose viability after a few years in storage. Every new crop will differ somewhat from that of the parents because of genetic recombinations. Also, the cost and effort of producing new seed crops every few years is large. However, for forest trees, the desired genotype can be kept indefinitely in the form of grafts or cuttings.

Some difficulties associated with working with large long-lived plants are obvious. The size of the trees creates problems in measurement, crossing, and especially in seed collection. Another item related to size is finding suitable areas necessary for "storage" of desired genetic material and for testing. Many trees do not flower at a young age, thus making a quick breeding program difficult. To add to the problem of slow generation time, juvenile–mature correlations, especially for growth characteristics, often are not satisfactory. This requires the maintenance of tests for a considerable number of years before the growth rate potential of a given genotype can be assessed accurately. The poor juvenile-to-mature correlations for growth result from the fact that different trees posess different growth curves. Some are fast starters that mature quickly, followed by a growth slowdown, whereas others are slow starters that grow at a constant rate over a long period of time and eventually overtake the fast starters.

Availability of seed with the known or desired genetic background is a frequent problem. Even when sufficient seed is available, there is a very difficult situation with respect to the size and configuration of test plots. Because of great environmental variability within a short distance in forestland, rather small plots are desired for uniformity. However, when plots are too small, not enough trees are represented to obtain reliable results because most trees are quite variable, one from another, even within closely related groups. Such variability necessitates the use of a number of individuals to accurately categorize a genetic grouping, thus requiring a considerable area in which to make the tests.

A lack of knowledge about what will be desired in the future can be a major deterrent to a tree improvement program. For example, will high- or low-specific-gravity wood be desired 30 years from now? Decisions must be made early with respect to future requirements, and once made, the type of tree produced must be used even though it may no longer be the ideal. This dictates a cautious or conservative approach to making decisions in tree improvement. Fads must be avoided, but if gains are to be obtained, early decisions must be made about long-term forestry objectives. Uniformity of products and the characteristics of tolerance to pests and adverse environments are uniformly desired, but other tree properties of form and wood qualities are not so easy to predict relative to future needs.

One problem that was severe in the early years, which is much improved now but nevertheless is still with us, is the attitude of foresters themselves. Training in forestry has tended to emphasize that the reasons trees differ one from another are due to the differing environments in which they grow. There was little recognition that some of the differences that occur among individuals in a forest are the result of differing parentages. The idea that trees are plants and therefore will respond to genetic manipulation like other plants was not understood at all by early foresters. This past block about the use of genetics in forestry now seems to be rapidly disappearing.

A major problem in the development of tree improvement programs is the permanence of organizations and movements of people. Tree improvement is a long-term process and as such, it must have good permanent records that are handed on from person to person. This often has not been the case. Records are sometimes poorly kept or not kept at all, and valuable information moves on with the personnel who leave. Often, the new person is a highly trained specialist who wants to do "his thing"; therefore, the old studies that were inherited suffer because they are only of passing interest to the new person. The loss and wastage caused by poor records and constant personnel changes cannot be overemphasized. Although not based on available data, the senior author estimates that at least 50% of the forest research that has been started never comes to the desired fruition because of a lack of care, loss of records, or the movements of key persons. This is even more serious in the developing countries that have few trained local staff members and where heavy use is made of short-term visiting experts.

As stated previously, one advantage of forest tree improvement, often not thoroughly recognized, is that most forest stands have great genetic variability and have not been greatly changed by the actions of people. There are exceptions, of course, but generally poor or dysgenic selection has not been intensive enough, or has gone on long enough to cause a permanent degradation in the genetic constitution of a population. It is true that the ratio of desired trees may have been reduced, but the good genes and gene complexes are still present in the forest so that the tree improver can use them if sufficient effort is made to locate them. The problem sometimes encountered is that the tree characteristics desired by the silviculturist may differ somewhat from those developed over time through natural selection in wild forests. But generally, the tree improver has enough variation from wild stands to enable good gains to be achieved. This is especially true for adaptability characteristics.

GENOTYPE AND PHENOTYPE

Although they are widely used by many persons in publications and conversations, two important terms that must be clearly understood are *phenotype* and *genotype*. All concepts in tree improvement are dependent on knowing what a genotype and phenotype is.

The *phenotype is the tree we see*. It is influenced by the genetic potential of the tree and by the environment in which the tree is growing, including the managerial history of the site. The phenotype is often indicated by the simple formula $P = G + E$ (phenotype = genotype + environment). The pheonotype of the tree is what we measure and what we work with.

The *genotype is the genetic potential of the tree*. It cannot be seen directly, and it can only be determined through well-designed tests. The genotype is determined by the genes that reside in chromosomes in the nucleus of every cell in the tree. *The sum total of nongenetic factors that affect the growth and reproduction of trees is called the environment.* It is a very general and catchall term applied to soils, moisture, weather, and often also to the influence of pests and sometimes to the interference of people.

The *basic fact is that one cannot say anything definite about the genetic worth of a tree just by looking at it,* that is, from its phenotype (Figure 1.3). One is never sure whether the characteristics observed are primarily determined by the environment in which the tree is growing or by the genetic control from the genotype of the tree. Both will interact to affect the tree's phenotype. The objective of the tree improver is to package the better genes into improved genotypes and then to manipulate the environment so that this genotype will react in a positive way to produce the most desirable phenotype.

If we can say nothing definite about the genetics of a tree from its phenotype, how then does one proceed to determine genetic superiority in a tree improvement program? Because the genotype is a component of the phenotype, it is related to the genetic potential of the tree; therefore, if good phenotypes are selected, improved genotypes will often result. Once the general relationship between the phenotype and genotype is known, it will then be possible to make some assumptions about the genetic value of a tree, just from observing its phenotype. For example, it is known that the straightness of the tree (phenotype) and its genetic potential for straightness are correlated, so that predictions can be made with some confidence that straighter trees will be obtained if straight trees are used as parents.

The concepts of *genotype* and *phenotype* are so simple that there may be doubt about the need for the preceding discussion. Yet these concepts are widely misunderstood by many persons who, for example, feel that if a tall tree at a given age is used as a parent, then its progenies should all be as tall as the parent at the same age. Many people do not separate the genetic and environmental causes that contribute to tree height. Progeny of the tall trees will grow to a size that is dependent upon the environment and the genetic potential for height passed from parent to offspring.

TERMS COMMONLY USED IN TREE IMPROVEMENT

In this book, there will be no special glossary or definitions for terms used in tree improvement work. The most important terms will be defined when they are encountered in the specific subject matter area to which they belong. If exact definitions are required, they can be obtained from any one of a number of biological texts or glossaries, such as those by Richens (1945), King (1968), Ford-Robertson (1971), Snyder (1972), and Lamontagne and Corriveau (1973).

There are a few general terms that will constantly be encountered throughout the text that must be understood right from the beginning. These are defined very simply as follows:

Progeny   The trees produced from the seed of a parent tree are called its progeny. *Progeny tests* are established to determine the genetic worth of the parent trees or for determination of other genetic characteristics. Sometimes, a test of the vegetative propagules from a given donor is referred to as a progeny test, but usually it is called a *clonal test*.

Population   This term is very loosely used by workers in the field of tree improvement. In this text, the term *population* will be used in a general way to designate a community of interbreeding individuals. No degree of relationship is assumed. The word *stand* is often used synonymously with population; at other times, *stand* refers to a group of trees of special interest within a population.

Race   Groups of populations that generally interbreed with one another and that intergrade more or loss continuously are referred to as *races*. Many kinds of races are recognized, such as edaphic, climatic, elevational, and so forth.

Family   Individuals that are more closely related to each other than to other individuals in a population are called a *family*. Generally, the term *family* is used to denote groups of individuals who have one or both parents in common. As used in this book, the term *family* does not refer to a taxonomic category except when it is specifically indicated.

Siblings   A group of individuals within a family are referred to as *siblings;* the group of related individuals when only one parent is common is called a *half-sib family;* when both parents are common, they constitute a *full-sib family*. An *open-pollinated family* is one in which one parent is common and the other parent(s) are unknown.

Rotation age   The age at which a stand of trees is to be harvested is called the *rotation age*.

Seed orchard   An area where superior phenotypes or genotypes are established and managed intensively and entirely for seed production is referred to as a *seed orchard*.

REFERENCES AND PUBLICATIONS

There are a large number of publications that deal directly with, or are related to tree improvement. Only a few of the most important of these can be listed in this book. The authors have over 10,000 references in their personal libraries, and these are certainly not complete. Contrary to the practice in most books, some of the older references will be listed along with the new ones, because in some areas the most interesting and definitive work was done some years ago.

In any chapter, there will not be space to list all pertinent references to the subject covered. The interested reader will need to pursue the subject to the depth he or she desires. To aid in this activity, the best and most comprehensive references available have been listed. A reader who wishes to go further into a subject should look at the references listed and follow up on other pertinent references, especially those in the newest publications. Commonly, students and foresters ask, "How do I proceed to become better informed in subject X?" One answer is to use references from new publications to find older ones that may be pertinent. Although most foresters do not develop exhaustive libraries, they must keep lists of key references. Every personal library should include the most important periodicals, especially for those working in isolated areas. *Expenses for books and magazines are as legitimate to the forester as are costs of field equipment, tools, and vehicles.*

As computers become more widely used, the literature can be accessed through bibliographic data bases, many of which are the computer-readable counterparts of printed indexes. These data bases are termed *bibliographic* because searches result in a list of references, or a bibliography. Among the files useful for searching tree improvement topics are the FORESTRY ABSTRACTS subfile of the CAB data base, which is the computer equivalent of the printed FORESTRY ABSTRACTS, produced by the Commonwealth Agricultural Bureaux (CAB) in England; AGRICOLA, the computer counterpart of the BIBLIOGRAPHY OF AGRICULTURE, produced by the U.S. National Agriculture Library; and BIOSIS, the computer-readable equivalent of BIOLOGICAL ABSTRACTS, BIOLOGICAL ABSTRACTS/REPORTS, REVIEWS, MEETINGS, and BIORESEARCH INDEX, which is supplied by BioSciences Information Service in Philadelphia. Files containing information on soil science, chemistry, hydrology, tree improvement, and other topics pertinent to forestry are available. Coverage in many data bases is international, and it usually includes journal articles, books, translations, and technical reports. Occasionally, these and patents may also be included. Generally, computerized bibliographic files index the printed literature from the 1960s to the present.

Data base searches can be tailored to the individual user's needs, using genus and species, language of publication, date, and geography. Computerized searching also offers the capacity for scanning article titles and abstracts for relevant terms, a procedure that is very time consuming when done by hand. Computerized searches are particularly useful for topics that are too comprehensive or complex to be found easily when searched for manually. They are helpful when speed is desirable, and they are valuable in locations where printed indexes may be inaccessible.

Computerized search of the literature is available in most countries of the world. Although it is possible for the "end user" to conduct the search, users often obtain the service through intermediaries, such as libraries, information centers, or private information brokers. These agencies maintain the equipment needed for searching, they hire staff members trained to manipulate the files effectively, and, if they do not mount the data bases themselves, they will establish accounts with data base vendors. Those who are uncertain of the identity of agencies offering computerized searching in their areas can obtain this information from the national libraries in their respective countries.

Listed here are the addresses of the major American commercial data base vendors, all of whom offer service outside the United States. These corporations are not search intermediaries; however, the addresses are given for the benefit of institutions interested in establishing accounts.

LOCKHEED INFORMATION SERVICES, INC.
3460 Hillview Avenue
Palo Alto, California 94304
(405) 858–3785
(800) 982–5838 within California
(800) 227–1927 within the continental United States
TELEX 334499 (DIALOG)
TWX 910/339–9221

SDC SEARCH SERVICE
2500 Colorado Avenue
Santa Monica, California 90406
(800) 352–6689 within California
(800) 421–7229 within the continental United States
TELEX 65–2358
TWX 910/343–6443

BIBLIOGRAPHIC RETRIEVAL SERVICES, INC. (BRS)
1200 Route 7
Latham, New York 12110
(800) 833–4707 within the continental United States
(518) 783–7251 within New York state and Canada; call collect
TWX 710/444–4965

Although long lists of references have not been cited in the text, the most pertinent references have been available to the reader who is interested in pursuing a specific subject. Some good references have undoubtedly been missed, and we apologize for their omission; most notably, some publications in languages other than English may have been overlooked. The reference material in the book will be handled in four different ways:

1. Literature that is directly cited in a chapter will be listed alphabetically at the end of each chapter as "Literature Cited."

2. The following section contains a list of books on tree improvement or on closely related subjects. General textbooks, such as those on botany or dendrology, have not been listed, with the exception of a few that are of specific, direct interest to tree improvement professionals.

3. A list of journals that carry articles of interest in tree improvement appears in another of the following sections. Addresses are shown for readers who wish to subscribe to some specific journal; those that are particularly suitable for tropical areas are indicated with a footnote.

4. Another section contains an incomplete list of some important proceedings or symposia. Addresses are shown, when it is feasible to do so. The addresses of periodic meetings, such as tree improvement conferences, change with the place where the meeting was held. Therefore, no addresses will be shown for this type of publication. Some proceedings and symposia are not numbered, and therefore they are most difficult to index. However, outstanding information is available from them and they should be consulted whenever possible. Of key importance is the fact that many publications do not fall within the category of *refereed publications*. Therefore, they may not be listed in abstracts or computer reference lists.

In an effort to make the references in the following sections as comprehensive as possible, the initial lists of publications were sent to other workers in the area of tree improvement and to libraries, asking them to supply missing information. The good response is greatly appreciated.

SOME BOOKS OF SPECIAL INTEREST TO TREE IMPROVERS

Burley, J., and Styles, B. T. 1976. *Tropical Trees—Variation, Breeding and Conservation*. Commonwealth Forestry Institute. Academic Press, New York.

Burley, J., and Wood, P. J. 1976. *A Manual on Species and Provenance Research with Particular Reference to the Tropics*. Department of Forestry, CFI, University of Oxford, Oxford, England.

Dobzhansky, T. 1951. *Genetics and the Origin of Species*. Columbia University Press, New York.

Dorman, K. W. 1976. *The Genetics and Breeding of Southern Pines,* Agriculture Handbook No. 471. USDA, U.S. Forest Service, Washington, D.C.

Enescu, V. 1972. *Ameliorarea arborilor Partea Generala*. Editura "Ceres," Bucharest, Rumania.

Enescu, V. 1975. *Ameliorarea Principal elor Specii Forestiere*. Editura "Ceres," Bucharest, Rumania.

Falconer, D. S. 1960. *Introduction to Quantitative Genetics*. Ronald Press, New York.

Faulkner, R. 1975. "Seed Orchards." Forestry Commission Bulletin No. 54, Her Majesty's Stationary Office, London.

Goldschmitt, R. B. 1952. *Understanding Heredity: An Introduction to Genetics*. John Wiley & Sons, New York.

Hedlin, A. F., Yates, H. O., Lovar, D. C., Ebel, B. H., Koerber, T. W., and Merkel, E. P. 1980. *Cone and Seed Insects of North American Conifers*. USDA, U.S. Forest Service, Washington, D.C. (also Universidad Autónoma de Chapingo, Chapingo, Mexico—Environment Canada, Canadian Forestry Service, Ottawa, Canada).

Khosla, P. K. (Ed.). 1981 *Advances in Forest Genetics*. Ambika Publications, New Delhi, India.

Mettler, L. E., and Gregg, S. G. 1969. *Population Genetics and Evolution*. Prentice-Hall, Inc., Englewood Cliffs, N.J.

Namkoong, G. 1979. "Introduction to Quantitative Genetics in Forestry." Technical Bulletin No. 1588., U.S. Forest Service, Washington, D.C.

Snedecor, G. 1981. *Statistical Methods*. Iowa State College Press, Ames.

Srb, A. M., Owen, R. D., and Edgar, R. S. 1965. *General Genetics*. W. H. Freeman and Co., San Francisco/London.

Stebbins, G. L. 1950. *Variation and Evolution in Plants*. Columbia University Press, New York.

Stern, K., and Roche, L. 1974. *Genetics of Forest Ecosystems*. Springer-Verlag, New York.

Stonecypher, R. W., Zobel, B. J., and Blair, R. L. 1973. "Inheritance Patterns of Loblolly Pine from a Nonselected Natural Population." Agricultural Experimental Station Technical Bulletin No. 220, North Carolina State University, Raleigh.

Sziklai, O., and Katompa, T. 1981. *Erdészeti Növény-Nemesítés* [*Forest Tree Improvement*]. Mezögazdasagi Kiado, Budapest, Hungary.

Thielges, B. A. (Ed.). 1975. "Forest Tree Improvement—The Third Decade." 24th Annual Forestry Symposium, Louisiana State University, Baton Rouge.

Toda, R. (Ed.). 1974. *Forest Tree Breeding in the World*. Government Forest Experiment Station, Meguro, Tokyo, Japan (27 authors).

Toda, R. 1979. *Forest Genetics Up-to-Date*. Noorin Syuppan Co., Ltd., Tokyo Japan.

Wright, J. W. 1976. *Introduction to Forest Genetics*. Academic Press, New York.

Magazines That Include Articles
of Special Interest to Tree Improvers

Magazine	Address
AFOCEL	Association Foret-Cellulose, 164 Boulevard Haussman, 75008, Paris, France
Australian Forest Research	CSIRO (Commonwealth Scientific and Industrial Research Organization), P.O. Box 89, East Melbourne, Victoria 3002, Australia

(*continued*)

Magazine	Address
Biotropica[a]	Association for Tropical Biology, Inc., Washington State University Press, Pullman, Washington 99163 ORDERS TO: Clifford Evans, Secretary-Treasurer, c/o Anthropology-MNH 368, Smithsonian Institution, Washington, D.C. 20560
Bois et Forets des Tropiques[a]	Centre Technique Forestier Tropical, 45 bis, Ave. de la Belle Gabrielle, 94130 Nogent-sur-Marne, France
Boletim de Pesquisa Florestal[a]	EMBRAPA, Unidade Regional da Pesquisa Florestal Centro-Sul, Caixa Postal 3319, 80,000 Curitiba, Brazil
Brazil Florestal[a]	IBDF. Ministerio da Agricultura, Institute Brasileiro de Desenvolvimento Florestal, Brasilia, Brazil
Bulletin of Forestry and Forest Products Research Institute	Forestry and Forest Products Research Institute, P.O. Box 16, Tsukuba Norin Kenkyu, Danchi-nal, Ibavaki, 305, Japan
Bulletin Recherche Agronomie, Gembloux	Station de recherches des Eaux et Forets, Groenendaal—Hoeilaart, Belgium
Canadian Journal of Forest Research	National Research Council of Canada, Ottawa K1AOR6, Canada
Commonwealth Forestry Review[a]	The Commonwealth Forestry Association, c/o CFI, South Parks Road, Oxford OX1 3RB, England
Forest Ecology and Management[a]	Forest Ecology and Management, P.O. Box 330, 1000 AH, Amsterdam, The Netherlands
Forest Farmer	P.O. Box 95385, 4 Executive Park East, N.E., Atlanta, Georgia 30347
Forest Products Journal	2801 Marshall Court, Madison, Wisconsin 53705
Forest Science	Society of American Foresters, 5400 Grosvenor Lane, Washington, D.C. 20014
Forestry Abstracts[a]	Commonwealth Agricultural Bureaux, Farnham House, Farnham Royal, Slough SL2 3BN, England

(*continued*)

Magazine	Address
Indian Forester[a]	Forest Research Institute and Colleges, P.O. New Forest, Dehra Dun, India
IPEF (Institute de Pesquisas e Estudos Florestais, Brasil)[a]	Instituto de Pesquisas e Estudos Florestais, Caixa Postal 9, Escola Superior de Agricultura, Piracicaba, São Paulo, Brazil
Journal of Forestry	Society of American Foresters, 5400 Grosvenor Lane, Washington, D.C. 20014
New Zealand Journal of Forestry Science[a]	Forest Research Institute, Private Bag, Rotorua, New Zealand
Research Report of the Institute of Forest Genetics	Institute of Forest Genetics, Suwon, Korea
Rapporter, Institutionen för Skogsproduktion	Sveriges Lantbruksuniversitet, Department of Forest Yield Research, S-770 73 Garpenberg, Sweden
Rapporter och uppsatser	Sveriges Lantbruksuniversitet, Department of Forest Genetics, S-770 73 Garpenberg, Sweden
Revista Arvore[a]	Comissão Editorial da *Revista Arvore*, Sociedade de Investigações Florestais, Universidade Federal de Viçosa, 36.570 Viçosa, Minas Gerais, Brazil
Silvae Genetica[a]	Institut für Forstgenetik und Forstpflanzenzüchtung, Grosshandsdorf 2, Schmalenbeck, Federal Republic of Germany
South African Forestry Journal [*Suid-afrikaanse Bosboutydskrif*][a]	South African Forestry Association, 62 Lugan Road, Johannesburg 2193, South Africa
Southern Journal of Applied Forestry	Society of American Foresters, 5400 Grosvenor Lane, Washington, D.C. 20014
Studia Forestalia Suecica	Sveriges Lantbruksuniversitet, Ultunabiblioteket, S-750 07 Uppsala, Sweden
Sveriges Skogsvårdsförbunds Tidskrift	Sveriges Skogsvordsförbund, Box 273, S-18252 Djursholm, Sweden
Tappi	Technical Association of Pulp and Paper Industry, One Dunwoody Park, Atlanta, Georgia 30338

(*continued*)

Magazine	Address
Tree Planters Notes	U.S. Forest Service, Washington, D.C. 20250
Turrialba: Revista Interamericana de Ciencias Agricolas[a]	Instituto Interamericano de Ciencias Agricolas de la OEA Secretariado, Apartado 55, Coronado, San José, Costa Rica
Unasylva: International Journal of Forestry and Forest Products[a]	Food and Agriculture Organization of the United Nations, Forestry Department, Distribution and Sales Section, Via delle Terme di Caracalla, 00100 Rome, Italy *Note:* Distributed in the United States by UNIPUB, 345 Park Avenue South, New York, New York 10010
Wood Science and Technology	Springer-Verlag, 175 Fifth Avenue, New York, New York 10010

[a]Publications that are particularly suitable for tropical forestry.

Proceedings and Symposia
of Special Interest to Tree Improvement

Title	Country	Date	Organization or Publication
"Die Früdiagnose in der Züchtung and Züchtungsforschung"	Germany	1957	Der Züchter, Springer-Verlag, Berlin (collection of papers)
"IX International Botanical Congress" (Vol. II)	Canada	1959	Proc. Montreal, University of Toronto Press
"Forest Genetics Workshop"	U.S.A.	1962	Proc. Southern Forest Tree Improvement Committee, Macon, Georgia.
"The Influence of Environment and Genetics on Pulpwood Quality"	U.S.A.	1962	Annotated Bibliography, Technical Association of Pulp and Paper Industry, TAPPI Monograph Series, No. 24, Atlanta, Georgia
"Genetics Today"	The Netherlands	1963	Proc. XI, International Congress on Genetics, The Hague
"Statistical Genetics and Plant Breeding"	U.S.A.	1963	Pub. No. 982, National Academy of Science, National Research Council, Washington, D.C. 62300
"Conference on Forest-Tree Genetics, Selection and Seed Production"	Russia	1969	Synopses of Reports (translated from Russian for USDA and National Science Foundation)

(*continued*)

Title	Country	Date	Organization of Publication
"Twelve Selected Articles" (translated from Russian)	Russia	1969	Botanicheskii, Zhurnal, from U.S. Dept. Commerce and USDA and National Science Foundation
"Quantitative Genetics"	U.S.A.	1970	2d Meeting of Working Group on Quantitative Genetics, IUFRO, Raleigh, North Carolina. Published by U.S. Forest Service, New Orleans, Louisiana
"Seminar on Forest Genetics and Forest Fertilization"	Canada	1970	Proc. Pulp and Paper Research Institute of Canada, Montreal, Canada
"Effect of Growth Acceleration"	U.S.A.	1971	Symposium, University of Wisconsin, Madison, Wisconsin
"Biology of Rust Resistance in Forest Trees"	U.S.A.	1972	Proc. NATO–IUFRO, Advanced Study Institute (USDA Miscel. Pub. No. 1221)
"Working Party on Progeny Testing"	U.S.A.	1972	IUFRO Proc., Georgia Forest Research Council, Macon, Georgia
"Selection and Breeding to Improve Some Tropical Conifers" (Vols. I and II; 15th IUFRO Congress)	U.S.A.	1972	Commonwealth Forestry Institute, Oxford, England, and Department of Forestry, Queensland, Australia
"Tropical Provenance and Progeny Research and International Cooperation"	Kenya and Australia	1973	Commonwealth Forestry Institute, Oxford, England
"Population and Ecological Genetics: Breeding Theory and Progeny Testing"	Sweden	1974	Proc., Department of Forest Genetics, Royal College of Forestry, S-104 05, Stockholm, Sweden
"Advanced Generation Breeding"	France	1976	Proc. IUFRO, Bordeaux, INRA, Laboratoire d' Amélioration des Conifères, 33610 Cestas, France
"Forest Genetic Resources"	Sweden	1976	Royal College of Forestry, Stockholm, Sweden
"Management of Fusiform Rust in Southern Pines"	U.S.A.	1977	South. Forest Disease and Insect Research Council, University of Florida, Gainesville
"Vegetative Propagation of Forest Trees—Physiology and Practice	Sweden	1977	Symp., The Institute of Forestry Improvement and Department of Genetics, College of Forestry, Uppsala, Sweden
"Progress and Problems of Genetic Improvement of Tropical Forest Trees" (Vols. I and II)	Australia	1978	Commonwealth Forestry Institute, Oxford, England

(*continued*)

Title	Country	Data	Organization of Publication
"Tree Improvement Symposium"	Canada	1979	Ontario Ministry of Natural Resources and Great Lakes Forest Research Center, Toronto, Ontario (COJFRC Symp. Proc. O-P-7)
"Effects of Air Pollutants on Mediterranean and Temperate Forest Ecosystems"	U.S.A.	1980	U.S. Forest Service, Riverside, California
"Genetic Improvement and Productivity of Fast-Growing Trees"	Brazil	1980	IUFRO Symposium and Workshop, São Paulo, Brazil
"The Forest Imperative"	Canada	1980	Proc. Canadian Forestry Congress, Toronto (The Pulp and Paper Industry of Canada)
"Workshop on the Genetics of Host–Parasite Interactions in Forestry	Wageningen	1980	Proc. several international organizations

Proceedings of Meetings Held Periodically for Which Addresses Vary

Southern Forest Tree Improvement Conferences

Biology Workshops (sponsored by the Society of American Foresters)

Lake States Forest Genetics Conferences

North-Central Tree Improvement Conferences

Central States Forest Tree Improvement Conferences

Northeastern Forest Tree Improvement Conferences

Canadian Tree Improvement Association in Canada (Committee on Forest Tree Breeding in Canada)

First, Second, and Third World Consultations on Forest Tree Breeding held in Stockholm, Washington, D.C., and Canberra, Australia, respectively

LITERATURE CITED

Austin, L. 1927. A new enterprise in forest tree breeding. *Jour. For.* **25**(8):977–993.

Barner, H. 1952. Skovtraeföraed lingens muligheder (possibilities in forest tree breeding). *Dan. Skovfören. Tidsskr.* **37**:62–79.

Batzer, H. O. 1961. "Jack Pine from Lake States Seed Sources Differ in

Susceptibility to Attack by the White Pine Weevil." Technical Note No. 595, Lake States Forestry Experimental Station.

Bey, C. F. 1974. "Drought Hardiness Tests of Black Walnut Seedlings as Related to Field Performance." Proc. 9th Central States For. Tree Impr. Conf., Ames, Iowa, pp. 138–144.

Bouvarel, P. 1957. Génétique forestière et amélioration des arbres forestiers [Forest genetics and the improvement of forest trees]. *Bull. Soc. Bot. Fr.* **104**(7–8): 552–586.

Brix, H. 1959. "Some Aspects of Drought Resistance in Loblolly Pine Seedlings." Ph.D. thesis, Texas A&M College, College Station.

Buchholz, E. 1953. Neuen sowjetische Arbeiten über Forstpflanzenzüchtung and forstliche Samenkunde [Recent Soviet work in forest tree breeding and seed collection]. *Z. Forstgenet.* **2**(3):65–70.

Dietrichson, J. 1961. Breeding for frost resistance. *Sil. Gen.* **10**(6):172–179.

Duffield, J. W. 1956. Genetics and exotics. *Jour. For.* **54**(1):780.

Fischer, F. 1954. Forstliche Pflanzenzüchtung als ein Mittel zur Steigerung des Wäldertrages (Tree breeding as a means of increasing forest yields). *Schweiz. Z. Forstwes.* **105**(3–4):165–183.

Ford-Robertson, F. C. 1971. *Terminology of Forest Science, Technology Practice and Products*, The Multilingual Forestry Terminology Series No. 1. Society of American Forestry, Washington, D.C.

Goddard, R. E., Schmidt, R. A., and Vande Linde, F. 1975. "Immediate Gains in Fusiform Rust Resistance in Slash Pine from Rogued Seed Production Areas in Severely Infected Plantations." Proc. 13th South. For. Tree Impr. Conf., Raleigh, N.C., pp. 197–203.

Goddard, R. E., Zobel, B. J., and Hollis, C. A. 1976. "Response of *Pinus taeda* and *Pinus elliottii* to Varied Nutrition." Proc Conf. on *Physiol Genetics* and Tree Breeding, Edinburg Scotland, pp. 449–462.

Greeley, W. B. 1952. Blood will tell. *Am. For.* **58**(9):18–19, 28.

Haley, C. 1957. "The Present Status of Tree Breeding Work in Queensland." Seventh British Comm. For. Conf., Queensland For. Department.

Heimburger, C. 1958. Forest tree breeding in Canada. *Gen. Soc. Can.* **3**(1):41–49.

Hellinga, G. 1958. "On Forest Tree Improvement in Indonesia." Proc. 12th IUFRO Congress, Oxford, England, Vol. 1(11), pp. 395–397.

Heth, D., and Kramer, P. J. 1975. Drought tolerance of pine seedlings under various climatic conditions. *For. Sci.* **21**(1):72–82.

Hosner, J. F., and Boyce, S. G. 1962. Tolerance to water saturated soil of various bottomland hardwoods. *For. Sci.* **8**(2):180–186.

Hyun, S. K. 1958. "Forest Tree Breeding in Korea." Proc. 12th IUFRO Congress, Oxford, England, Vol. I(11), pp. 375–385.

Johnsson, H. 1949. "Experiences and Results of 10 Years Breeding Experiments at the Swedish Forest Tree Breeding Association." Proc. 3rd World For. Cong., Vol. 3, pp. 126–130.

King, R. C. 1968. *Dictionary of Genetics*. Oxford University Press, New York.

Lacaze, M. 1963. "The Resistance of *Eucalyptus* Trees to Active Limestone in the soil." Proc. World Consul. on For. Gen. and Tree Impr. 4/8, Stockholm, Sweden.

Lamontagne, Y., and Corriveau, A. G. 1973. *Glossaire des termes techniques utilisés en amélioration des arbres forestiers*. Ministere des Terres et foréts, Quebec, Canada.

Langner, W. 1954. Die Entwicklung der Forstgenetik und Forstpflanzenzüchtung in Deutschland (the development of forest genetics and forest tree breeding in Germany). *Z. Forstgenetik* 3:55–60.

Larsen, C. S. 1951. Advances in forest genetics. *Unasylva* 5(1):15–19.

Leopold, A. J. 1929. Some thoughts on forest genetics. *Jour. For.* 27:708–713.

Lindquist, B. 1948. *Forstgenetik in der Schwedischen Waldbaupraxis (Forest Genetics in Swedish Forestry Practice)*. Neumann Verlag, Radebeul/Berlin.

Matthews, J. D. 1953. Forest tree breeding in Britain. *Forstgenet* 2(3):59–65.

McCormick, L. H., and Steiner, K. C. 1978. Variation in aluminum tolerance among six genera of trees. *For. Sci.* 24(1):565–568.

Namkoong, G., Barnes, R. D., and Burley, J. 1980. "A Philosophy of Breeding Strategy for Tropical Forest Trees." Tropical Forest Paper No. 16, Comm. For. Inst., Oxford, England.

Ohba, K. 1979. Forest tree breeding in Japan. *JARQ* 13(2):138–144.

Parker, J. 1963. Cold resistance in woody plants. *Bot. Rev.* 29(2):123–201.

Perry, T. O., and Wang, C. W. 1958. The value of genetically superior seed. *Jour. For.* 56(1):843–845.

Rao, H. S. 1951. Genetics and forest tree improvement. *Indian For.* 77:635–647.

Richens, R. H. 1945. "Forest Tree Breeding and Genetics." Imperial Agr. Bureaux, Joint Pub. No. 8, Imp. Bur. Plant Breed. and Gen. Imp. For. Bur., Oxford, England.

Righter, F. I. 1946. New perspectives in forest tree breeding. *Science* 104(2688):1–3.

Roberds, J. H., Namkoong, G., and Davey, C. B. 1976. Family variation in growth response of loblolly pine to fertilizing with urea. *For. Sci.* 22(3):291–299.

Sakai, A., and Okada, S. 1971. Freezing resistance of conifers. *Sil. Gen.* 20(3):53–100.

Schönback, H. 1961. The variation of frost resistance in homegrown stands of Douglas fir. *Rec. Adv. Botany* 2(14):1604–1606.

Schreiner, E. J. 1935. Possibilities of improving pulping characteristics of pulp-woods by controlled hybridization of forest trees. *Paper Trade Jour. C,* 105–109.

Schreiner, E. J. 1950. Genetics in relation to forestry. *Jour. For.* **48**(1):33–38.

Snyder. E. B. 1972. *Glossary for Forest Tree Improvement Workers.* U.S. Forest Service, Southern Forestry Experimental Station, New Orleans, La.

Sziklai, O. 1981. "Present and Future Research Requirements." Seminar, Tree Improvement in the Interior of British Columbia, Prince George, British Columbia, Canada.

Thulin, I. J. 1957. "Application of Tree Breeding to New Zealand Forestry." Technical Paper No. 22, Forest Research Institute, New Zealand Forest Service.

Toda, R. 1970. *Abstracts of Japanese Literature in Forest Genetics and Related Fields,* Vol I-A. Noorin Syuppan Co., Ltd., Shinbashi, Tokyo, Japan. (2156 refs. before 1930) 1972; Vol. I-B, pp. 363–918 (refs. 2157 to 5385), 1931–1945.

Toda, R. 1974. "Forest Tree Breeding in the World." Bull. Govt. Forest Experiment Station, Meguro, Tokyo, Japan.

Toda, R. 1980. An outline of the history of forest genetics. In *Advances in Forest Genetics.* Ambika Publications, New Delhi, India.

Toyama, S. 1954. "Studies on Breeding for Forest Trees." Bull. Govt. Forest Experimental Station, Meguro, Tokyo, Report 24, pp. 56–269.

Wright, J. W. 1953. A survey of forest genetics research. *Jour. For.* **51**(5):330–333.

Wright, J. W. 1981. "A Quarter Century of Progress in Tree Improvement in the Northeast." 27th Northeast. For. Tree Impr. Conf., Burlington, Vt. pp. 6–15.

Zobel, B. J. 1952. The genetic approach for improving wood qualities of the southern pines. *Jour. For. Prod. Res. Soc.* **2**(2):45–47.

Zobel, B. J. 1957. Progeny testing for drought resistance and wood properties. *Der Zuchter H.,* 95–96.

Zobel, B. J. 1974. "Increasing Productivity of Forest Lands through Better Trees," S. J. Hall Lectureship, University of California, Berkeley.

Zobel, B. J. 1978. "Progress in Breeding Forest Trees—The Problem of Time." 27th Ann. Sess. Nat. Poultry Breed. Roundtable, Kansas City, Mo., pp. 18–29.

Zobel, B. J., and Zoerb, M. 1977. "Reducing Fusiform Rust in Plantations through Control of the Seed Source." Symposium on Management of Fusiform Rust in Southern Pines, South. For. Dis. and Insect Res. Coun., Gainesville, Fla., pp. 98–109.

CHAPTER 2

Variation and its Use

Without sufficient genetic variability of the correct types for traits that are of economic interest, an attempt to use genetics to improve forest trees will be unsuccessful or a failure. Therefore, the first thing to do when starting a tree improvement program is to determine the amount, cause, and nature of the variation that is present in the species of interest and to learn how to use it. Activities related to assessing variation take much of the tree improver's time, and they require continued and intensive effort. The fact that variation does exist among species, races, and individuals within species is generally not too difficult to prove, but the determination of its causes can be very time consuming and costly. A difficult but essential task is to discover what portion of the variation is genetically controlled so that a determination can be made about how best to exploit it in a tree improvement program to produce better forests with higher-quality products (Figure 2.1; see also Figure 1.4).

Foresters are lucky in this instance because tree populations are generally genetically variable. They must be so in order to survive, grow, and reproduce under the differing conditions and numerous environments that are encountered during a single generation and over generations (Antonovics, 1971; Nienstaedt, 1975). The value of this "gift" of great variability in forest trees is often underestimated. The proper kind of genetically controlled variation provides the needed conditions for a tree improvement program, giving the necessary tools for large, quick gains from the use of genetics in forestry. As compared to agricultural crops, forest tree populations have been little influenced by human activities until now. Tree breeders are working essentially with wild populations that contain the genes and gene complexes needed for breeding programs. It is a fact that most *forest tree species possess greater variability* than species of other organisms; it is reported to be almost double that of other plants (Hamrick et al., 1979). Forest tree improvers therefore possess a huge advantage by being able to draw on this variability in their breeding programs. However, it places a great responsibility on the breeder to maintain and enhance the great store of variation for future use, a subject that will later be dealt with in some detail.

SOME BASIC GENETIC CONCEPTS

General

Trees are the largest and are among the most complex organisms in a world with millions of diverse life forms. Despite the tremendous diversity that exists in nature, however, certain basic mechanisms of inheritance are common to all species, including forest trees. Although this book is not meant to serve as a genetics text, a brief review of a few basic concepts and their effects on variation will be helpful in understanding the tree improvement principles presented in subsequent chapters. The message to be gained is that genetic processes are ordered, and in many instances they are predictable. Tree improvers must appreciate this fact and use it in their programs if they are to be successful. Detailed

FIGURE 2.1

Great variability commonly occurs within a species. Shown are two loblolly pines in an area subjected to fume damage. One tree has been killed by fumes; the other appears to be growing normally. When such variation has a genetic basis, it can be used to develop strains of trees of special value and use.

explanations of inheritance mechanisms can be found in genetics textbooks, such as those by Srb et al. (1965), Gardner and Snustad (1984), Grant (1975, or Strickberger (1976).

Cells and Chromosomes

Like most organisms, trees are composed of cells. There are numerous types of cells in trees, but all living plant cells have in common a *cell wall,* a *cytoplasm,* and a *nucleus* (Figure 2.2). The nucleus is of special interest genetically because it

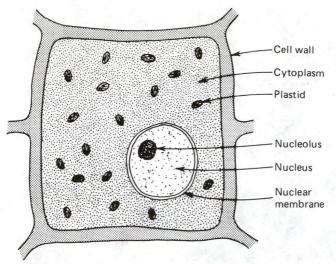

FIGURE 2.2

This schematic drawing of a cell indicates the integral parts of a
cell that consists of (1) a surrounding wall; (2) the cytoplasm; (3)
plastids, which are small bodies in the cytoplasm that aid cell
function and reproduction; (4) the nucleus, which contains most
of the material that affects inheritance; (5) a membrane around
the nucleus; and (6) a nucleolus, which apparently controls
cellular activity.

contains the *chromosomes,* which harbor most of the genetic information necessary for the growth and development of the tree. Chromosome numbers are
usually constant in number in every vegetative (somatic) cell of an organism, in all
populations of a species, and in most instances, in every individual of a species.
Chemically, chromosomes are composed of deoxyribonucleic acid (DNA), which
is the source of genetic information, and a protein sheath. Genes, which are the
functional units of inheritance, occur in a linear arrangement along the DNA
molecule in each chromosome. The location of a gene on a chromosome is said to
be a *gene locus.* Genes occurring on the same chromosomes are said to be *linked*
or to be in the same *linkage group.*

During most phases of a cell's growth and development, chromosomes exist as
long, threadlike structures that are difficult to observe except with the most
powerful microscopes. Just prior to cell division, chromosomes contract into
rodlike structures that are observable under light microscopes and are usually
countable. Species of trees differ widely in the number of chromosomes found in
their nuclei. For example, all normal members of the genus *Pinus* have 24
chromosomes, whereas redwood (*Sequoia sempervirens*) has the rare number of

66 chromosomes in its vegetative cells (Figure 2.3); Douglas fir (*Pseudotsuga menziesii*) has 26 chromosomes. Generally, the conifers have a few large chromosomes (although there are rare exceptions like redwood), whereas the angiosperms tend to have higher numbers of smaller chromosomes. As examples, American sycamore (*Platanus occidentalis*) has 42 chromosomes, whereas members of the *Eucalyptus* genus have a relatively low number of 20 to 22 somatic chromosomes. In some species, the number of chromosomes varies, depending on the population that is being sampled. In paper birch (*Betula papyrifera*), for example, chromosome numbers range from 56 in variety *cordifolia* to 84 in variety *occidentalis*. The chromosome numbers of most cultivated species are listed by Darlington and Janaki Ammal (1974); for conifers, by Khoshoo (1961); and for hardwoods, by Wright (1976). The chromosome number for some selected tree species are shown in Table 2.1.

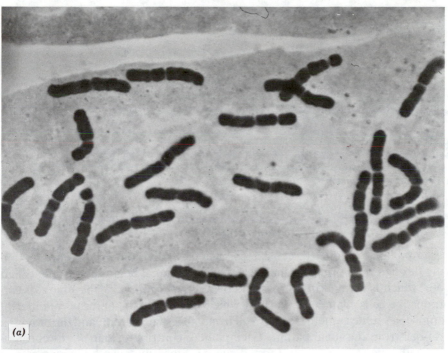

(a)

FIGURE 2.3

(a) Pines have 24 chromosomes as shown in the aceto-carmine preparations from the root tip of *P. taeda*. (b) (on next page) Occasionally, conifers are polyploid as shown by the 66 chromosomes from a cell of *Sequoia sempervirens*. (Photos courtesy of L. C. Saylor, North Carolina State University.)

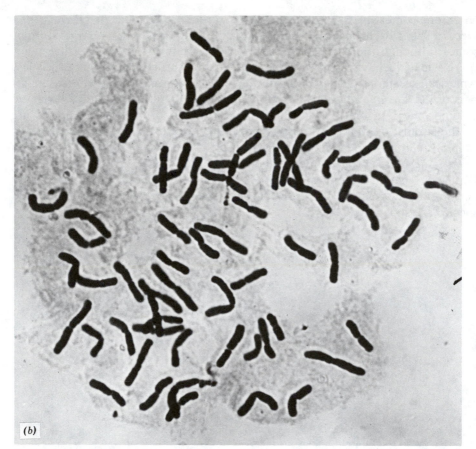

(b)

FIGURE 2.3 (*continued*)

Mitosis and Meiosis

Trees grow and reproduce by cell extension and cell division. When reproduction occurs through the normal sexual cycle, each individual begins as a single-celled zygote, which is formed in a process known as *fertilization* by the union of two reproductive cells (gametes), each of which has only half the usual number of chromosomes. One gamete is an egg cell from the female parent, and the other is a sperm cell from the male parent. Each gamete contributes one set of chromosomes and genes to the newly formed zygote. Since the zygote has two sets of chromosomes, it is said to be *diploid* (or 2n). The gametes are *haploid* (1n) because they each carry only one chromosome set, that is, one of each type of chromosome. The two chromosome sets in the zygote that carry the same genes are homologous and carry genes that affect the same function. Therefore, each gene locus is represented twice in the zygote, once on a chromosome from one parent and again in the homologous chromosome from the other parent.

TABLE 2.1
Varying Chromosome numbers of Trees.

Gymnosperms		Angiosperms	
Genus	**Somatic Chromosome No.**	**Genus**	**Somatic Chromosome No.**
Abies	24	*Acacia*	26 (52)
Araucaria	26	*Acer*	26 (and variable)
Cedrus	24	*Albizzia*	26 (and variable)
Chamaecyparis	22	*Alnus*	28 (and variable)
Cryptomeria	22	*Betula*	28 (2n)
Cunninghamia	22	*Carya*	32
Cupressus	22	*Castanea*	24
Cycas	22 (24)	*Diospyros*	30 (60)
Ginkgo	24	*Eucalyptus*	22 (24)
Juniperus	22 (44)	*Fagus*	24
Larix	24	*Ficus*	26

These are expressed either as the basic n number or the somatic number, which is usually double the n number. Some trees are polyploid and have more than $2n$ chromosomes. The gymnosperms often have small somatic numbers (22, 24, or 26,) whereas the angiosperms frequently have much higher numbers. Only a few of the forest tree genera can be shown. Chromosome values were obtained from a number of sources in the literature.

One of the unique properties of DNA is its ability to replicate itself. This is the key to the mechanism of inheritance, which allows a parent to pass along his or her genetic potential to his or her progeny. DNA also carries genetic information intact from one cell to the next as a tree grows.

Vegetative growth in a tree occurs throughout its lifetime. A mature tree contains billions of cells, all of which are direct descendants of the zygote formed at fertilization and each of which carries the same genetic information. Cells divide in all phases except the reproductive phase by means of a process called *mitosis*. Mitotic cells division begins with the replication of DNA, and therefore with a temporary doubling of chromosome numbers in the parent cell. The chromosomes then shorten into the rodlike structures described previously, and one copy of each chromosome moves to the opposite ends of the parent cell. The cell then builds a wall between them that results in two daughter cells, each with its own cell wall, cytoplasm, nucleus, and full complement of chromosomes. As trees grow and mature, numerous types of cells are formed, each with a special function and morphology, but the genetic information contained in the nucleus of each cell is identical to that in the original zygotic cell. The reason for cell and tissue differentiation is that certain sets of genes are activated in one type of cell, whereas other genes are active in other cell types.

Sexual reproduction is made possible through a process known as *meiosis*. Meiosis actually involves two cell divisions, and it results in the reduction of the chromosomes from the $2n$ number in the parent to $1n$ in the gametic (reproductive) cells (Figure 2.4). The meiotic process begins with DNA replication. Homologous chromosomes then pair. Duplication is evident at this point, and it is followed by an arrangement of the chromosomes in an orderly fashion in the center of the cell. During the time in which homologous chromosomes (*homologues*) are paired, an exchange of genetic material can occur through what is termed *crossing over*. When this happens, the chromosomes break in equivalent positions and exchange chromosome segments. This exchange of genetic material by means of crossing over is an important occurence because it serves to break up linked groups of genes, and this results in new combinations of genes and thus variation in the population. Following pairing and crossing over, one member of each homologous pair of chromosomes moves to opposite ends of the cell in a random manner so that each end contains mixtures of maternal and paternal chromosomes. A wall usually develops between the ends, and the first cell division then results in the formation of two daughter cells. Following this division, chromosomes line up again in the center of each of the two daughter cells. The two replicates of each chromosome that were formed at the beginning of meiosis then move to opposite ends of the cells, and cell division by means of a new wall occurs again. The result is that four gametic cells have formed, each with one set ($1n$) of chromosomes (see Figure 2.4). When the $1n$ gametes (usually from different parents) come together in fertilization, the resultant zygote will be diploid ($2n$), and it will contain the same number of chromosomes as it parents but with a new combination of genes, half from one parent and half from the other.

Genes and Alleles—Gene Action

Genes can be thought of as the functional units of inheritance. Each gene may be represented in the population by one, two, or more alternate forms. Each of the alternate forms for a given gene is called an *allele*. Alleles carry the genetic potential for different expressions of the same trait. For example, an allele at a gene locus that influences leaf size (allele A) might code for long leaves, whereas another allele of the same gene (allele a) might code for shorter leaves. An analogy could be made between alleles and different kinds of pickup trucks. A number of companies manufacture pickup trucks. All are pickup trucks, and they have the same function, even though each one is a bit different in appearance. Similarly, all alleles for one gene locus serve the same function, but they may cause a different expression of the same trait.

Each gene locus is represented twice in a diploid cell, once on each of the two homologous chromosomes. Therefore, if more than one allele exists for a gene locus in the population, an individual tree may have two of the same alleles or two different alleles that govern the expression of the particular trait that is influenced by that locus. If we continue with the hypothetical example of leaf size, an individual could possess two alleles for large leaves (AA), which would make it *homozygous* for that gene locus; it would be referred to as a *homozygote*. Similarly, an individual could be homozygous for the allele that codes for smaller leaves (aa). Alternatively, the tree could be *heterozygous*, which means that it

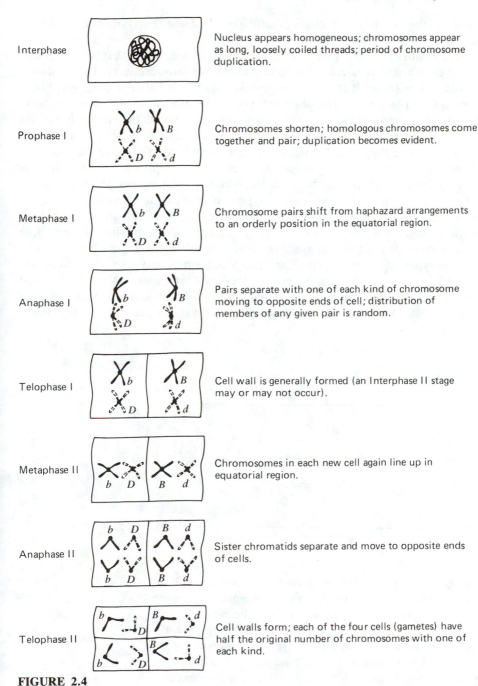

Interphase	Nucleus appears homogeneous; chromosomes appear as long, loosely coiled threads; period of chromosome duplication.
Prophase I	Chromosomes shorten; homologous chromosomes come together and pair; duplication becomes evident.
Metaphase I	Chromosome pairs shift from haphazard arrangements to an orderly position in the equatorial region.
Anaphase I	Pairs separate with one of each kind of chromosome moving to opposite ends of cell; distribution of members of any given pair is random.
Telophase I	Cell wall is generally formed (an Interphase II stage may or may not occur).
Metaphase II	Chromosomes in each new cell again line up in equatorial region.
Anaphase II	Sister chromatids separate and move to opposite ends of cells.
Telophase II	Cell walls form; each of the four cells (gametes) have half the original number of chromosomes with one of each kind.

FIGURE 2.4

Sexual reproduction takes place through the process called meiosis. It involves two cell divisions and results in a reduction of the chromosomes from the 2n number in the parent to 1n in the reproductive cell (gamete). See telophase II.

possesses the two different alleles (*A* and *a*) for leaf size. This type of individual is referred to as a *heterozygote*.

When individuals are homozygous for a trait, expression of that trait is set and straightforward. An individual with two alleles for large leaves (*AA*) will always produce large leaves if the environment is suitable. In other words, it will be a large-leaved phenotype. The phenotype of the heterozygous genotype *Aa*, however, is not so predictable. Its appearance will depend upon the interaction of the two alleles at the locus for leaf size. However, if one allele shows at least *partial dominance* over the other, the phenotype will be governed more by the dominant allele. When *complete dominance* is present, the expression of the trait is governed solely by the dominant allele. When allele *A* is completely dominant over allele *a*, then the *AA* and *Aa* genotypes will produce the same phenotypes. In the latter situation, allele *a* is said to be *recessive*, and its effects on the phenotype are observed only when the genotype is a homozygote for that allele. These various types of gene action are shown diagramatically in Figure 2.5.

Genotypes of progenies that are produced when two parents are crossed depend upon the type of gametes produced by each parent. An *AA* genotype would produce only *A* gametes, whereas an *aa* individual will produce only gametes with an *a* genetic constitution. Heterozygous genotypes (*Aa*) would

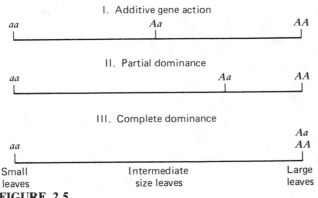

FIGURE 2.5

Effects of allelic interaction on phenotypic expression of the leaf size trait are shown when the trait is governed by a single gene locus. (I) Additive gene action. The phenotype of the heterozygote (*Aa*) is midway between the phenotypes of the two homozygotes (*AA* and *aa*). (II) Partial dominance. The phenotype of the heterozygote (*Aa*) is influenced more by one allele than the other, but both alleles have some effect on the phenotype. (III) Complete dominance. The phenotype of the heterozygote is the same as that of the homozygote for the dominant allele.

produce both *A* and *a* types of gametes. The array of genotypes that can be found in the progeny produced by a cross of two *Aa* individuals can be shown as follows:

Male Parent (♂)

Aa Genotype

Female Parent (♀)

Aa Genotype

	A	*a* Gametes
A	*AA*	*Aa*
a	*Aa*	*aa*

Gametes

Genotypes of the progeny produced by this cross are shown in the box. They occur in the following ratio:

1*AA*:2 *Aa*:1 *aa*

Other possibilities for crossing two parents exist as follows:

$$AA \times aa \rightarrow \text{all } Aa \text{ progeny}$$
$$AA \times AA \rightarrow \text{all } AA \text{ progeny}$$
$$aa \times aa \rightarrow \text{all } aa \text{ progeny}$$
$$AA \times Aa \rightarrow 50\% \ AA \text{ progeny, } 50\% \ Aa \text{ progeny}$$
$$aa \times Aa \rightarrow 50\% \ aa \text{ progeny, } 50\% \ Aa \text{ progeny}$$

The expression of any particular trait may be governed by one, two, or many gene loci. In the hypothetical leaf size example, the leaf was influenced by only one gene locus. When one or only a few genes influence a trait, those genes are said to be *major genes* or to have major effects. However, a most important concept is that *most economic traits* in forest trees are *influenced by many genes,* each of which has a small effect on the trait. Therefore, tree breeders are rarely concerned with a particular gene locus, but, rather, they work with the understanding that the characteristic represents the culmination of the effects of many genes. Under this concept, each gene has only a small effect on the phenotype, and the effects of the environment are usually large. Special techniques have been developed by geneticists to deal with traits that are influenced by many genes. These are called *quantitative characteristics.* A discussion of quantitative characteristics is presented in detail in Chapter 4.

In subsequent sections of this chapter as well as in later chapters, there will be an emphasis on observing and measuring variation within and among forest trees. Traits will be described largely as they result from the effects of many genes on a tree's appearance, and rarely will reference be made to individual gene loci; sometimes the term *gene complex* is used. It is important to remember, though, that regardless of the number of loci influencing a particular trait, inheritance at any one locus is governed by the same principles that were presented previously for the hypothetical leaf size example.

CAUSES AND KINDS OF VARIABILITY

General

Basically, all differences among trees are the result of three things: the differing environments in which the trees are growing, the genetic differences among trees, and the interactions between the tree genotypes and the environments in which they grow. Some genetic variations are predictable and useful, whereas other types are random and are more difficult for the tree breeder to use.

In forest trees, a number of categories of variation exist that can be broadly grouped into species, geographic sources (provenances), stands, sites, individual trees, and the variability within individual trees (Zobel et al., 1960b). Everyone is aware of differences among species, and this need not be discussed further. However, knowledge of the relative importance of the other categories of variation is mandatory if a tree improvement program is to be successful. For example, it has been found for characteristics related to survival and adaptability (such as cold hardiness) that geographic variation is often the most important, whereas for economic characteristics, which are not so obviously related to fitness (such as stem straightness or wood specific gravity), individual tree variability is generally the greatest.

A key point, which is often overlooked, is that a study of variation of natural stands or plantations in which parentage is unknown *tells one nothing about the genetic control* of the characteristic involved. All one sees and measures in the forest is the *phenotype* of the tree. It is not possible to assess which portion of the differences among trees, stands, or provenances are genetically or environmentally controlled without actually making genetic tests. Inferences certainly can be made that are based on the magnitude and pattern of variation, but proof of genetic control requires genetic tests in which the parentage is known. The expensive and time-consuming activity of genetic testing is the key to a determination of the kind and control of variability that exists in a species, and thus to making continued and maximum progress in tree improvement activities.

Environmental Variation

Environmental variation is understood by most foresters, and its management is the basis of most silvicultural activities. Some environmental factors that influence tree growth can be controlled and manipulated, whereas others cannot. Things such as stocking levels and tree-to-tree competition can be handled by controlling plant spacing or by thinning. Within limits, nutrient deficiencies can be adjusted by fertilization, and soil moisture can be changed by drainage. Soil texture cannot generally be altered, but site preparation can change the soil structure to a considerable extent. Operations such as subsoiling are sometimes useful to create an environment for better root development and tree growth. Site preparation and herbicides are commonly used to reduce competition that, if left unchecked, will reduce or limit tree growth. Other environmental variables, such as rainfall, temperature, wind action, soil depth, aspect, and many other segments of a tree's environment, can be little influenced by humans, but all of these forces influence the phenotype of the tree. Variation among trees caused by environmental differences cannot be used in a breeding program and is often not even predictable. However, environmental forces are the greatest cause of variability in some characteristics, especially those related to growth. Form and quality may also be strongly affected by environmental differences, but, generally, the quality characteristics in forest trees tend to be more highly inherited and less influenced by the environment than are growth characteristics.

Although foresters generally cannot easily control the environment, it is frequently possible to develop strains of trees that will grow satisfactorily under adverse environmental conditions. In fact, about the only method the forester possesses to overcome adverse temperature, rainfall, wind action, pests, or other strong environmental influences is to create strains of trees through breeding or to use those found in nature that are more tolerant to the adverse factors.

Genetic Variability

Genetic variability is complex, but if its magnitude and type are known and if it is well used, genetic variation can be manipulated to obtain good gains in some tree characteristics. Genetic variation can be generally divided into *additive* and *nonadditive* components so that genetic variance = additive variance + nonadditive variance. In simple terms, the additive variance is due to the cumulative effects of alleles at all gene loci influencing a trait. Nonadditive genetic variance can be divided into two types. *Dominance* variance is due to interaction of specific alleles at a gene locus, whereas *epistasis* variance is due to interactions among gene loci. (These concepts will be developed more fully in Chapter 4.) For now, it is sufficient to know that the additive portion is the one of value in population improvement programs. Nonadditive variation can be exploited only by use of other, more specialized production programs that involve making specific crosses or using vegetative propagation for the commercial production of planting stock. In most tree improvement programs, the nonadditive types of genetic variability have generally been given little attention, because the additive portion of genetic variance is easier to utilize.

Most characteristics of economic importance in forest trees are under some degree of additive genetic control. This is fortunate because additive variance can be successfully used in simple selection systems such as those that are most suitable to new tree improvement programs. Characteristics such as wood specific gravity, bole straightness, and other quality characteristics of trees have stronger additive variance components than do growth characteristics. Although growth traits are controlled to some degree by additive genetic effects, they also have considerable nonadditive variance associated with them. Therefore, any selection program must include testing the progeny of selected phenotypes to determine the actual genetic worth of the tree. The response to selection of characteristics with considerable nonadditive variance, such as growth, is generally less satisfactory than for the quality characteristics that are usually under strong additive genetic control (Stonecypher et al., 1973).

Although not yet thoroughly quantified, it appears that most adaptability characteristics are strongly inherited in an additive manner. Thus, excellent gains have been made in developing strains of trees that will grow suitably on marginal or submarginal sites by selecting those individuals that grow best there and then using seed from them to reforest similar areas. Pest resistance involves both additive and nonadditive variance, depending on the pest and the tree species involved, but, generally, good gains are possible through selection programs that use the additive portion of the genetic variation.

There is little that the tree improver can do in the short term to improve the amount or kind of genetic variance available for use. The initial challenge to the tree improver is to determine the magnitude and kind of variance present from natural or unimproved populations and then to use it wisely. By better control of the environment, it is possible to capture and use more of the genetic variance. This results when intensive forest management is teamed with genetic manipulation of the trees.

Mating Systems

The type of crossing system within a species has a major effect upon the variation pattern. *Outcrossing* systems, which are common in most forest tree species, usually produce highly variable (heterozygous) genetic populations. In outcrossing systems, different genotypes cross successfully with each other, and little successful crossing takes place between male and female structures of the same plant, or with closely related individuals. When pollen from a tree or a given genotype pollinates flowers on itself, the term *selfing* is applied. This also applies to pollinations resulting among ramets of the same clone. Even though the ramets (grafts, rooted cuttings, etc.) are different plants, they are genetically identical. Thus, if pollen flies from one ramet of a particular clone to another ramet of the same clone, it is the same as pollination between flowers of the same tree. This concept is of key importance in clonal seed orchards, requiring extreme care so as not to establish ramets of the same clone too close together.

Most of this book deals with outcrossing mating systems; therefore, they will not be discussed further here. It is sufficient to state that the outcrossing breeding system maintains a high degree of genetic variation; related matings reduce genetic diversity. Foresters are most fortunate in that most important tree species are outcrossers, making it easy to maintain the variation needed for simple selection and breeding programs. However, some related matings do occur in many forest trees, and they are discussed in greater detail below because of their great effect on genetic diversity (Orr-Ewing, 1976; Allen and Owens, 1972; Eldridge, 1977).

Vigor often is greatly reduced when related matings occur; it is a kind of reverse hybrid vigor. Related matings can occur within the same tree, among ramets of related clones, or among related trees. Relatedness is common in natural forest stands. It is for this reason that in most programs only one tree per stand is selected for use in operational seed orchards. Many degrees of relatedness can occur. Too little is known of the effects of sibling, cousin, or other types of related matings in forest trees, but their adverse effects are well recognized in agricultural crops. Numerous studies have been started to clarify the situation of relatedness of forest trees. Those that have been completed (Franklin, 1971; Orr-Ewing, 1976; Libby et al., 1981) all show that matings between close relatives have some adverse result and should be avoided. One of the most common effects is a reduction in seed set, although Andersson et al. (1974) found none when full- and half-sibs were mated. There is abundant knowledge about the adverse effects of selfing, the most severe form of related matings. Almost always, seed set and germination are reduced (Diekert, 1964). When viable seedlings are obtained, they often have reduced growth rates that can continue to advanced ages (Eriksson et al., 1973) (Figure 2.6).

The importance of related matings is increasing dramatically as forest tree breeding programs move into advanced generations. This will be discussed more

FIGURE 2.6

When selfing occurs, such adverse reactions as low-seed set or reduced growth rate often result. Shown are three rows of loblolly pine all having the same mother. The two outside rows resulted from pollinations by nonrelated fathers, whereas the center row was derived from pollinations using pollen from the "mother tree" itself; that is, it is a self.

fully in later chapters, but it is mentioned here in relation to the effects on variation patterns. Outstanding individual trees from good general combiners are found in every seed orchard with the result that many of the best progeny chosen for advanced generation breeding have one or both parents in common. Until the effects of related matings are known, the breeder cannot know in the advanced generation breeding program whether to use poorer, but nonrelated individuals in seed orchards or whether occasional full-sib or half-sib selections from a superoutstanding progeny can be used even though some inbreeding depression may result. The question is how much gain is sacrificed when sib relatives are allowed to mate in seed orchards.

Relatedness is common in natural populations, often occurring in neighborhood patterns (Coles and Fowler, 1976). It is particularly common in areas where abandoned fields have been populated from a very few parents that were growing along fences, near houses or roads or along areas too wet to be farmed (Figure 2.7). Most of the forests in the Piedmont area of the southern United States arose in this manner. Many have been logged two or three times, with the amount of relatedness increasing with each logging. In hardwoods that regenerate by root sprouts, such as the aspens (*Populus*) or sweetgums (*Liquidambar*), all individuals in a stand may have identical genotypes, the most extreme form of relatedness. One can assume that some degree of relatedness will exist among close neighbors within a natural stand of trees. Therefore, the only safe approach is to do as

FIGURE 2.7

Relatedness is common in natural stands of forest trees. As an example, old fields are frequently populated with trees from a very few residual parents along fences or around buildings. Among the forces that can change the variation within population is one called *genetic drift*. It can operate only in small populations. Whether it is important in forestry is debated, but the conditions required for a small population are common in stands arising from abandoned agricultural fields.

suggested in the subsequent chapter on seed orchards: Do not put more than one tree from a given stand into a single-seed orchard.

The extreme form of related matings (selfing) occurs in many species, both conifers and hardwoods (Barnes, 1964; Gabriel, 1967; Franklin, 1971; Orr-Ewing, 1976; and many others). Selfing can have a number of different results as has been indicated by the wide experience of the authors with *Pinus taeda,* and as others have found with other species like *Pinus radiata* (Bannister, 1965). Results from selfing can vary greatly by individual species and mother trees within a species. Some common results are the following:

1. No sound seed are formed.

2. Seed are formed but they will not germinate.

3. Seed germinate but the seedlings are abnormal and often will survive only a short time before they die.

4. Seedlings will survive, but they will be small, weak, often yellow in color, and grow slowly. Some of these can be recognized and removed in the nursery prior to field planting.

5. Seedlings grow more slowly than do normal trees, but they are not poor enough to be observed easily and culled in the nursery bed. This result is quite common and is most dangerous because the selfed trees will be outplanted and survive in plantations, but they will produce much less wood than would be obtained from outcrossed seedlings.

6. Seedlings grow as well, or sometimes even better, than outcrosses; selfed trees that grow as well as outcrosses are rare.

The use of inbred lines, later to be outcrossed, has been suggested as a breeding system. This method is widely used in agriculture where inbred lines are produced later to be outcrossed. The outcrossing restores the vigor lost when the inbred lines are developed and, what is most important, it results in populations that are very uniform. Because the number of selfs of forest trees that grow well enough so that they reproduce and can later be outcrossed is small, large numbers of genotypes are lost in a selfed breeding program resulting in a drastically reduced breeding population.

Inbreeding in forest trees with its resulting reduction in variability is being studied in several species. Sniezko (1982), working with nearly 100 flowering selfs of loblolly pine, made crosses between inbreds and noninbreds, inbreds × inbreds, and he also selfed inbreds. He found great difficulty in obtaining a good enough seed set to allow for further tests. Working with foliage color in jack pine (*Pinus banksiana*), Rudolph (1980) found that selfing S_1 (S refers to generations of selfing) families increased homozygosity, resulting in smaller intrafamily variation. When selfs were outcrossed ($S_1 \times S_1$), he found heterosis and full recovery from the inbreeding depression. Although some of Rudolph's results were similar to those of Sniezko (who also reported that inbreeding depression in loblolly pine increases into the second generation), they differed in that Sniezko found that outcrosses among selfs failed to equal the performances between their parents. Thus, the hoped-for restored vigor from crossing selfs was not obtained.

Some species self poorly or not at all. Examples are sycamore and sweetgum (Boyce and Kaeiser, 1961; Schmitt and Perry, 1964; Beland and Jones, 1967). Other species, like *Pinus resinosa,* seem to be unaffected by selfing (Fowler, 1964). No general statement regarding the amount or ease of selfing can be made for forest trees, because they vary widely. Obviously, selfing is no problem in dioecious species.

Some researchers, such as Franklin (1969), have concluded that an inbreeding–outcrossing breeding system is not practical in a forest tree improvement program because of the low seed set of the inbreds, the low vigor of the inbreds, and the resultant drastic reduction in the size of the breeding population.

Polyploidy

The number of sets of chromosomes a tree has is termed *ploidy,* and it can greatly affect the variability pattern within and among species. Usually, each parent contributes one set of chromosomes ($1n$) to the progeny, so the tree has two sets of chromosomes and is called *diploid* ($2n$). Sometimes species, or individuals within species, have more than the usual sets of chromosomes; these are called *polyploids.* The subject of polyploids—their causes and results—are fascinating. [This has been well covered for forest trees in a chapter in Wright's (1976) book.] Polyploid breeding has been suggested as a tree improvement tool by persons such as Gustafsson (1960). Generally, success has been limited. A discussion of the kinds and causes of polyploids is not suitable for this beginning text, but it will be summarized as follows:

1. Polyploidy is more common in hardwoods than in conifers. The poplars (Zufa, 1969), alders (*Alnus*), birches (*Betula*), ashes, (*Fraxinus*) and numerous other groups have polyploid members; it is estimated that one third of the hardwoods are polyploid derivatives (Chiba, 1968). Polyploid hardwoods, such as the triploid aspens, have attracted considerable attention as fast growers, and research programs have been developing more triploids (Sarvas, 1958; Benson and Einspahr, 1967). Generally, the polyploid hardwoods are no better than the diploids, although there are some exceptions, such as in *Alnus* (Johnsson, 1950). Usually, an intensive search will locate diploids as good as the polyploids. The best-known polyploid conifer is redwood (*Sequoia sempervirens*), which is hexaploid; that is, it has 66 chromosomes rather than $2n = 22$ (see Figure 2.3). Other polyploid conifers have been reported as having outstanding properties, like the triploid larch (*Larix*), but generally, as in the pines, the polyploid conifers usually are deformed runts, with deformed roots (Hyun, 1954; Mergen, 1958).

2. Polyploidy can complicate a tree-breeding program, depending on its type and cause. Inheritance patterns can be very complex and difficult to manipulate. Creation of polyploids by controlled crosses is quite possible in some species, but often this does not produce viable plants in others. Development of haploid plants ($1n$) has been suggested as a breeding tool, but the success to date with most species of trees has been limited (Stettler, 1966).

3. The best way to use good polyploids is by means of vegetative propagation where sexual reproduction and the resultant genetic recombination are not required. In the plant kingdom, it has been observed that many polyploids easily propagate vegetatively and that some polyploid species seem to have lost the ability to propagate by seed.

4. At the present time, polyploid breeding is of limited value in forest tree improvement but as methods to develop them are improved, and especially as methods of vegetative propagation become more refined, polyploids will very likely play an increasing role in forest tree breeding.

Genotype × Environment Interaction

A condition that is of key importance in studying variation is commonly referred to as *genotype × environment interactions*. The term is used to describe the situation *where there is a change in the performance ranking of given genotypes when grown in different environments*. Such interaction must be known if maximum progress in breeding is to be obtained. Too frequently in tree improvement, a group of families are tested in a single environment, and their performance is then extrapolated to other environments, when in fact their relative performance might have been different when grown under other conditions. The trademark and challenge of forestry is the necessity to grow trees in a variety of environments, some of which are greatly different from others.

It is essential not to confuse true genotype × environment interactions as defined with a simple response to environmental differences, where the relative ranking among the families tested remains essentially the same even though the average performances of families in the different environments vary greatly. For example, it is usual to find that the average height of Family A might be 65 ft (19.5 m) in one area and only 55 ft (16.8 m) in another whereas Family B is 70 ft (21.1 m) in the first area and 62 ft (18.8 m) in the second. The differences in height between the same families in the two differing environments are only growth responses to different environments, but there is no meaningful genotype × environment interaction because the relative performances of families A and B do not change in the two environments. A common error is to refer to performance differences from site to site as genotype × environment interaction.

Strong genotype × environment interactions are more likely to occur when environments differ widely. As forestry operations become more intensive and as the productive forestland base decreases, there is need to establish plantations on sites that formerly were considered marginal or submarginal for satisfactory tree growth. This results in offsite planting in environments that are grossly different from those to which the species is best adapted. Even in normal forestry environments, peoples' activities and attempts to improve forestland productivity through intensive management usually create a different and quite often artificial habitat that is considerably different from the one in which the original forest was growing. A major concern now becoming generally recognized is how the best species, sources, or even the best individual genetic selections that have been obtained will perform when grown in the different environments (Squillace, 1970;

Matheson, 1974; Zobel and Kellison, 1978). Expressed another way, knowledge is needed about how stable a given genotype is when grown in quite different environments (Hanson, 1970).

A detailed treatment of genotype × environment interaction will be presented in Chapter 8.

VARIATION IN NATURAL STANDS

Introduction

Foresters are exceptionally fortunate to usually work with an undisturbed pool of high natural variability that has developed over eons (Perry, 1978). Intensive studies of variation within species are necessary for tree improvement programs to be successful. Many have already been made, such as on the loblolly pine by Thor (1961), by Lamb (1973) and Barnes et al. (1977) and others on *Pinus caribaea*, and by Yeatman (1967) on jack pine. (*P. banksiana*)

A determination of the amount and kind of variability present within a species is a large job and must be done carefully. There is no one "correct" way to assess the variability patterns within natural stands, but time and experience have shown that employment of a *nested sampling procedure* is very good.

The nested sampling method consists of determining variability within a species among varying groupings from large ones through ever smaller ones to individuals and within individuals. In forest trees it generally consists of determining the variation present in the following categories:

1. Geographic (provenance) variation
2. Sites within provenances
3. Stands within sites
4. Individual trees within stands
5. Within trees (when applicable)

A study of natural variation of a given species should first determine what geographic differences are present and then the variation that might be present within the lesser categories. A knowledge of where the bulk of the variation exists can indicate much about the development of a specific characteristic and how it might best be used in a breeding program. It must be emphasized that because the genetic and environmental components of variation cannot be separated by a study of natural stands, *no definitive conclusions about degree of inheritance of any characteristics can be made* from nested sampling studies.

Geographic (or Provenance) Variation

Geographic variation will be discussed in detail in Chapter 3; therefore, it will only be mentioned here. Genetically controlled geographic differences are often large, especially for traits related to adaptability. The differences can be of key importance, and the success of any tree improvement program depends upon knowledge and use of geographic variation within the species of interest. Geographic differences within species often are not easy to define, and boundaries usually are not clear-cut, unless there is a definite environmental separation. Therefore, the determination of what constitutes a geographic source is often one of judgment and opinion.

Variability among Sites

A given provenance can sometimes contain quite large differences related to differing sites; frequently, these are not genetically fixed and only represent the effects of varied environments on the growth and development of the forest. For example, trees of a given provenance that grow like shrubs on the sand dunes facing the sea where there are constant winds may grow normally when planted inland. Whether the scrubby sand dune trees are actually genetically different from the taller inland trees can only be determined through tests made on both sites.

In general, studies of pines have shown that site differences contribute only a small amount to the total genetic variation compared to other causes of genetic variation. However, site differences within a provenance are large and common enough so that they must be considered important when natural populations are sampled, even though they usually turn out to be environmentally rather than genetically caused.

Differences among Stands within Sites

Sometimes stands of trees within a given site differ; usually the genetic differences are relatively small, but sometimes unexplained pockets of variation are found (Ledig and Fryer, 1971). This is especially true for form characteristics, which usually differ very little genetically for trees on any common site. Forces of natural selection that can cause differentiation from stand to stand are small. Sometimes stand-to-stand variation results from an accident of sampling caused by small population sizes. Usually, stand-to-stand differences within a site are of such little importance that they can be ignored, but this is not always true, especially when humans have intervened by changing populations through selective cutting, thinning, or other forest management activities. It is common, for example, to find stands of straight trees growing near stands of crooked trees, the difference having been caused by past pole and piling operations that left only crooked trees for parents in the exploited stand.

Tree Differences within a Stand

Individual trees of a species often vary a great deal from one another even when growing in the same stand. This is the major type of genetic variation the geneticist uses in a selection and breeding program. Many individual tree differences, especially quality traits such as form and adaptability, are strongly controlled genetically. It is amazing how two trees can be the same age, grow side by side with their roots intertwined, and still be so different in form, wood qualities, pest resistance, and even in growth patterns (Figure 2.8).

E 112 × E 1101
396/2 8.81

53
396/2 8.81

(a)

FIGURE 2.8

Genetic differences from tree to tree within a species can be large and are the major source of variation used by the tree breeder. Illustrated are two types of variation. (*a*) Individual differences are shown for progeny from select and wild standard pine in Finland. (*b*) The

(b)

(*continued*)

spruce variant shown for possible use as a parent tree is straight, narrow crowned, thin branched, and rapidly grown. (Photos courtesy of Lauri Kärki, Foundation for Forest Tree Breeding and Max Hagman, Forest Research Institute, Finland.)

In general, most economic characteristics of special value in forest trees have a large amount of individual tree variability that will be available to the tree breeder. This is true even for characteristics that are complex. An occasional tree species, such as *Pinus resinosa* (red pine), will show only a small amount of tree-to-tree genetic variation (Fowler and Morris, 1977), but these cases are the exception rather than the rule.

Variation within a Tree

Within a tree, variability can occur only for some characteristics. A tree is only so tall or has only one diameter at breast height (dbh); therefore, no within-tree variation for height or dbh can exist. But for other characteristics, considerable within-tree differences can occur. For example, wood specific gravity in the southern pines shows considerable differences, depending upon the height in the tree at which the wood sample is taken (Zobel et al., 1960a). Great within-tree differences occur for foliage characteristics, as, for example, in sun and shade leaves on the same tree (Figure 2.9). Within-tree variation is important where it occurs, because it influences the types of measurements and the positions where measurements must be taken to obtain statistically sound assessments of the real tree-to-tree differences.

Summary of Natural Variation

In general, provenance variation and tree-to-tree differences account for the bulk of the genetic variation found within a tree species growing in natural stands; these two variants may account for nearly 90% of all the variation observed. Wood specific gravity and cold resistance estimates for loblolly pine show the approximate distribution of the total variation (genetic plus environmental) for two very different types of characteristics.

Type of Variation	Wood Specific Gravity (%)	Cold Tolerance
Provenance (geographic)	15	70
Site	5	0
Stand	0	0
Tree to tree	70	30
Within tree	10	0

Because genetic tests have shown that for many characteristics much of the tree-to-tree variation is genetically controlled, good gains can be achieved from a selection and breeding program that concentrates on tree-to-tree differences.

It is essential to emphasize again that a study of variation in natural stands can give no proof of the intensity of genetic control of a characteristic because, in such a study, one cannot separate the effects of environment and genetics or their interactions. But the patterns of variation in natural stands can give good indications of possible genetic gains. To cite one example: Nothing was known about the inheritance of wood specific gravity of loblolly pine when intensive tree improvement programs were first started for the species. An intensive nested sampling study showed that the bulk of the variation in wood specific gravity was from tree to tree within a provenance. Despite some differences among geographic locations, nearly the same magnitude of individual tree variation was found at all geographic locations. This consistent individual tree variation in the north and south, on the coast and inland, and on sandy and on clay soils hinted

(a)

FIGURE 2.9

It is essential to recognize within-tree variation when sampling. There is no within-tree variation for characteristics like height, but there are great differences for things like leaves. Shown is the variation between sun (*a*) and shade leaves (*b*) (on next page) of *Quercus macrocarpa* from Michigan; both are taken from the same tree. (Photo courtesy of R. Braham, North Carolina State University, Raleigh, North Carolina.)

(b)

FIGURE 2.9 *(continued)*

that there was a broad and probably reasonably strong genetic control of wood specific gravity of individual trees. Based on this assumption, wood was included as a major characteristic in the operational loblolly pine tree improvement programs. It was relatively costly but, happily, later research showed that wood specific gravity was indeed strongly genetically controlled in such a manner that excellent gains could be made through selection.

DETECTING VARIATION IN PEDIGREE STANDS

Methods to assess variability in stands of trees with known and recorded pedigrees, such as half-sib or full-sib families, are essentially the same as for wild stands, but much more information is available from the analyses. In contrast to wild stands of unknown pedigree where only the total (or phenotypic) variation related to provenance, site, stand, and tree can be determined, one can assess the relative importance and contribution of the environmental and of the genetic variance in pedigree stands. From this information, it is then possible to obtain an assessment of the genetic value of the parents involved.

If the pedigree includes information on both the mother and father trees, it is then possible to divide the genetic variability into its component parts of additive and nonadditive variance and even to assess different types of nonadditive variance. Details of this breakdown become rather complex statistically, and they are handled in the special field called *quantitative genetics*. A later chapter in this book will deal with quantitative genetics in a simplified manner; more exhaustive treatments are covered by such persons as Falconer (1960) and Namkoong (1979). Calculations dealing with quantitative genetics in forest trees have been made by several authors; one of the better of these studies is by Stonecypher et al. (1973) for loblolly pine.

The question is often raised whether different provenances within a species can be treated as having a pedigree that can be assessed in a way that is similar to families in the calculation of variances. This is sometimes done, but it requires some differences in the interpretation and use of the values obtained.

In summary, phenotypic variation can be divided into its genetic and environmental components only when pedigree stands are available. The degree of separation of the genetic variation depends on the mating design used. This will be discussed in more detail in Chapter 8.

MAINTENANCE AND USE OF VARIATION

General

Although the subject will be dealt with in later chapters, it should be mentioned here that one absolute necessity in a tree improvement program is to maintain and increase genetic variability within the forest tree populations being used. Unless properly applied, an intensive selection program will reduce variability for the characteristics involved. Indeed, the objective of genetic manipulations in forestry is to produce more quickly the desired products with greater uniformity. All successful breeding programs will change gene frequencies; if this does not happen, the program will fail.

Most forest tree species contain great variability for such economically important characteristics as tree straightness or wood specific gravity, for such adaptability as tolerance to cold or drought, and for resistance to diseases or insects and for growth (Figure 2.10). As clearly stated previously, a major strength of tree improvement is that most characteristics of value to the tree breeder are complex and are essentially inherited independently, so it is possible to "tailor make" trees with the desired combination of characteristics. But continued development in

FIGURE 2.10

Shown is a mutant form of white pine; compare it with the size and characteristics of normal white pine of the same age growing adjacent to it. Mutants sometimes can have great value. The pine here would seem excellent for a Christmas tree, but it would have to be propagated vegetatively because most mutants do not produce viable seed.

later generations is not possible unless the variability in the breeding population is maintained, and this is no easy job. Maintaining and even increasing variability is a key objective of the developmental, or research, phase of a tree improvement program.

Forces That Shape Genetic Variation

All the variation in wild stands has occurred as the result of natural forces. It is available for use by the forester if it can be recognized and packaged into individual trees in the form of improved genotypes. The ultimate source of all variability is mutations, but other strong forces are at work to either increase or decrease variation within a stand. In addition to the variability found in natural stands, humans can interfere and help create either new variability or bring together genotypes to create new and useful genetic combinations.

Although variation in our forests today is primarily the result of natural forces over which the forester has little control, it is essential that these forces are understood. They determine the amount and kind of genetic variation found among and within populations. These forces form the basis for the specialized area of *speciation* and *evolution,* subjects on which many books have been written. [A few of the most lucid and easily understood are those by Stebbins (1950, 1977) and Grant (1975).]

In the most simplified terms, variability in natural stands is caused by four main forces, two that increase variation and two that decrease it. The forces in nature working to increase variation are *mutation* and *gene flow;* those that reduce it are *natural selection* and *genetic drift.* The forces at work are illustrated schematically in Figure 2.11.

Mutations   Mutations are the ultimate source of variation. A *mutation* is a heritable change in the genetic constitution of an organism, usually at the level of a gene. Since the total genetic makeup of a tree (its *genotype*) is determined by the action and interactions of thousands of genic and allelic combinations, mutations can occur somewhere in an organism with considerable frequency, but this will not happen often for any specific gene or gene complex or for a given tree characteristic. Although talking about frequency of mutations really amounts to nothing more than an academic exercise, because they vary greatly by species and loci within species, a general figure often quoted is 1 in 10,000 to 1 in 100,000 genes. When one considers that trees have tens of thousands of genes, it is not unusual for a single tree to have several mutations. Most are recessive and have little effect on the phenotype of the tree.

Mutations occur more or less randomly. Most mutations are deleterious, and many are eliminated from the population. Through time, forces of evolution have made most populations well adapted to their environments, with genes and gene complexes in the population that are the most suitable for growth and reproduction. The chance that a random mutation would improve such a well-coordinated system is very small.

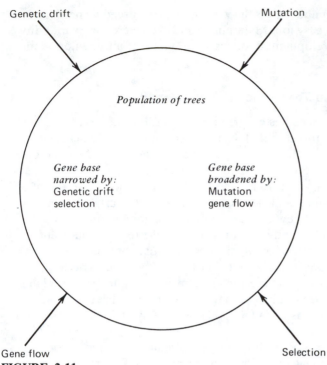

Genetic drift

Mutation

Population of trees

*Gene base
narrowed by:*
Genetic drift
selection

*Gene base
broadened by:*
Mutation
gene flow

Gene flow

Selection

FIGURE 2.11

There are a number of forces that alter the variation pattern within a population. It is increased by mutation and gene flow and reduced by natural selection and genetic drift. These are indicated by the schematic drawing.

Some mutations are retained in the population, even though they are deleterious, because they are of the recessive type and are not recognizable or detectable unless in the homozygous form. The value of this kind of mutation may not be known. It may only become significant at a much later date when different forces affect the environment and the formerly useless mutation may actually make the tree more fit to grow and/or reproduce. These recessive or neutral mutations do not normally disrupt an integrated genetic system like a dominant one would. Therefore, they can be carried along in the population for many generations. Although mutations may be rare and small, they will produce variation that can possibly make a tree more adaptable as environments change.

Gene Flow (Gene Migration)   The other action within a population that increases variation is called *gene flow;* it is the migration of alleles from one population or species into another where they may be absent or at a different frequency. Gene flow can result from several causes, but the most common is movement of pollen or seed. Occasionally, gene flow or gene transfer takes place on the species level through a process called *introgression* that sometimes occurs between two species after hybridization (Anderson, 1949). The hybridization brings together two dissimilar parental genetic complexes, thus creating a "new" genotype. This new organism may not be well adapted to compete with the parental species, but sometimes it will find an environmental "niche" that is especially suitable and that enables the new genotype to grow and reproduce. Because the new genotype is rare, or one of a kind, it will usually exchange genes with one of the parents to produce a backcross to one of the parental species. After this process occurs several times, the resulting population of trees can look very similar to the original parent species, although they will contain some genes or gene complexes that have been transferred from the one parental species to the other.

We can utilize the concept of gene flow in breeding programs (Sluder, 1969). For example, *Pinus Jeffreyi* is a well-formed species that is susceptible to the pine reproduction weevil. *Pinus Coulteri,* on the other hand, has poorer form but because of its thicker bark, it is resistant to the weevil. If we create a *P. Coulteri* × *P. Jeffreyi* hybrid and then backcross to *P. Jeffreyi* several times and select the most desired individuals, a tree that is similar to Jeffrey pine can be produced that still carries considerable weevil resistance. The gene complex for thicker bark would have been transferred from Coulter to Jeffrey pine.

Gene flow can be important in natural populations, and will cause distinct changes in patterns of variation. Gene flow in conjunction with recombination is the immediate source of increased variation patterns in many populations, even though the ultimate source of variation was mutation.

Selection   *Natural selection* is a strong force that usually reduces variability (Mason and Langenheim, 1961). Because it determines which trees will grow and reproduce, it has a directional (nonrandom) effect on the genetic makeup of trees in a population. Natural selection favors the fittest; that is, those trees with gene combinations that make them best suited to grow and reproduce in a given environment. Natural selection preserves and results in an increase in the number of those genotypes most suited to a specific environment. Although normally a process that reduces variability, natural selection can actually preserve or increase variation if selection favors the heterozygotes. Whether natural selection works to favor heterozygotes (which would maintain variability) or homozygotes (which would decrease variability) is currently a topic of considerable debate (see, for example, Lewonton, 1974), although most geneticists think that selection works to decrease variation by favoring the best alleles in a homozygous condition.

It is often difficult to assess the effects of selection because so many factors are involved in determining which tree will be best fitted to grow and reproduce. Each fitness characteristic has its own selective value, and the adaptations created by one factor may either positively or adversely affect others. In general, natural selection is considered to be a powerful force to reduce variability within a population in a given direction.

Genetic Drift   *Genetic drift* is a complex mechanism that operates through chance fluctuations (not fluctuations caused by selection pressures) in allele frequencies within a population. It is essentially a sampling phenomenon in which the gene frequencies in the progeny populations deviate by chance from those found in the parent populations. Such populations are almost always small and have a tendency toward fixation or loss of an allele that affects a characteristic. Thus, genetic drift tends toward reducing variation by fixing or losing alleles.

Genetic drift is nondirectional and tends to create "disorder." Which genes or alleles are fixed or lost is strictly a matter of chance. Although the theory of genetic drift is plausible, its operation is difficult to prove with long-lived trees, and many reasons can be cited why it cannot be a factor in the natural variation of forest trees. But despite these objections, some natural stands show variation patterns that could be the result of genetic drift if it were operating. Genetic drift is usually of consequence in small breeding populations of perhaps 25 or fewer individuals, a situation that frequently occurs in forestry due to natural catastrophies or man's influence (see Figure 2.7).

VARIATION CAUSED BY MAN

In addition to the normal variation patterns that occur in natural populations, many changes in the variation pattern in forest trees can be caused by human beings. Such things as dysgenic selection, where the best trees are removed and the poor trees are left to reproduce, or a selection method in which only the best are left, will ultimately cause a shift in gene frequencies, and thus in variation patterns. Our actions can cause a very rapid change in variability when we apply intensive selection and breeding practices.

Because the major objective of tree breeding is to change the percentage of certain characteristics in a desired direction within a population, the bulk of this book deals with variation caused by human beings or by our activities to change variation. As breeding programs progress, it will be essential for the tree improver to purposely increase variability. There are a number of options that can be followed to do this when natural variability becomes too limited for a breeding program. The first is to make sure that all variability within a species is known. Wide crosses may be made within species to bring together genotypes that could never occur under natural conditions. Interspecific hybrids and back crosses can be produced to develop new genetic combinations. Finally, programs to increase variation by means of mutations may ultimately be possible. The objective of all the above options is to assure that enough genetic variation exists so that productive breeding programs can be pursued.

The key point to remember is that our activities can cause large changes in variance relatively rapidly, either in a positive or in a negative way. Thus, it is possible through our efforts to make the large and quick genetic gains that are needed to keep tree improvement programs solid.

LITERATURE CITED

Allen, G. S., and Owens, J. N. 1972. "The Life History of Douglas-Fir." Canadian Forest Service, Ottawa Cat. No. Fo. 42-4972.

Anderson, E. 1949. *Introgressive Hybridization*. John Wiley & Sons, New York.

Andersson, E., Jansson, R., and Lindgren, D. 1974. Some results from second generation crossings involving inbreeding in Norway spruce (*P. abies*). *Sil. Gen.* **23**(1–3):34–43.

Antonovics, J. 1971. The effects of a heterogeneous environment on the genetics of natural populations. *Am. Sci.* **59**(5):593–595.

Bannister, M. H. 1965. "Variation in the Breeding System of *Pinus radiata.*" New Zealand Forest Service Report No. 145. *In The Genetics of Colonizing Species*, Asilomar Symposium, Pacific Grove, Calif., pp. 353–374.

Barnes, B. V. 1964. "Self- and Cross-Pollination of Western White Pine: A Comparison of the Height Growth of Progeny." U.S. Forest Service Research Note INT-22.

Barnes, R. D., Woodend, J. J., Schweppenhauser, M. A., and Mullen, L. J. 1977. Variation in diameter growth and wood density in six-year-old provenance trials of *Pinus caribaea* on five sites in Rhodesia. *Sil. Gen.* **26**(5–6):163–167.

Beland, J. W., and Jones, L. 1967. "Self-Incompatability in Sycamores." Proc. 9th Conf. For. Tree Impr., Knoxville, Tenn., pp. 56–58.

Benson, M. K., and Einspahr, D. W. 1967. Early growth of diploid, triploid and tetraploid hybrid aspen. *For. Sci.* **13**(2):150–155.

Boyce, S., and Kaeiser, M. 1961. "Why Yellow Poplar Seeds Have Low Viability." USDA, Central States Forestry Experiment Station Technical Paper 186.

Chiba, S. 1968. "Studies on Tree Improvement by Means of Artificial Hybridization and Polyploidy in *Alnus* and *Populas* Species." Bull. Oji. Inst. For. Tree Impr., Hokkaido, Japan.

Coles, J. F., and Fowler, D. P. 1976. Inbreeding in neighboring trees in two white spruce populations. *Sil. Gen.* **25**(1):29–34.

Darlington, C. D., and Janaki Ammal, E. K. 1955. *Chromosome Atlas of Flowering Plants*. George Allen & Unwin, Ltd., London, England.

Diekert, H. 1964. Einige untersuchungen zur selbststerilität und inzucht bei fichte und lärche [Some investigations on self sterility and inbreeding in spruce and larch]. *Sil. Gen.* **13**(3):77–86.

Eldridge, K. G. 1977. "Genetic Improvement of Eucalypts." 3rd World Cons. For. Tree Breed., Canberra, Australia.

Eriksson, G., Schelander, B., and Åkebrand, V. 1973. Inbreeding depression in an experimental plantation of *Picea abies*. *Hereditas* **73:**185–194.

Falconer, D. S. 1960. *Introduction to Quantitative Genetics.* Ronald Press, New York.

Franklin, E. C. 1969. "Inbreeding as a Means of Genetic Improvement of Loblolly Pine." Proc. 10th South. Conf. For. Tree Impr., Houston, Tex., pp. 107–115.

Franklin, E. C. 1971. Estimates of frequency of natural selfing and of inbreeding coefficients in loblolly pine. *Sil. Gen.* **20**:(5–6):141, 224.

Fowler, D. P. 1964. Effects of inbreeding in red pine, *Pinus resinosa. Sil. Gen.* **13**(6):170–177.

Fowler, D. P., and Morris, R. W. 1977. Genetic diversity in red pine; evidence for low genetic heterozygosity. *Can. Jour. For. Res.* **7**(2):343–347.

Gabriel, W. J. 1967. Reproductive behavior in sugar maple; self-compatibility, cross-compatibility, agamospermy and agamocarpy. *Sil. Gen.* **16**(5–6): 165–168.

Gardner, E. J. and Snustad, D. P. 1984. *Principles of Genetics*, 7th ed. John Wiley & Sons, N.Y.

Grant, V. 1975. *Genetics of Flowering Plants.* Columbia University Press, New York.

Gustafsson, Å. 1960. "Polyploidy and Mutagenesis in Forest Tree Breeding." Proc. 5th World For. Conf., Vol. 2, pp. 793–805.

Hamrick, J. L., Metton, J. B., and Linhart, Y. B. 1979. "Levels of Genetic Variation in Trees: Influence of Life History Characteristics." Proc. Symp. on Isozymes of N. Amer. For. Trees, Berkeley, Calif., pp. 35–41.

Hanson, W. 1970. "Genotypic Stability." 2nd Mtg., Work. Group on Quant. Gen. IUFRO, Sect. 22, Raleigh, N.C., pp. 37–48.

Hyun, S. K. 1954. Induction of polyploidy in pines by means of colchicine treatment. *Zeit. Forestgen. Forstpflan* 3(2)25–33.

Johnsson, H. 1950. On the C_o and C_1 generations in *Alnus glutinosa. Hereditas* **36**:205–219.

Khoshoo, T. N. 1961. Chromosome numbers in gymnosperms. *Sil. Gen.* **10**(1): 1–9.

Lamb, A. F. A. 1973. *Pinus caribaea,* Vol. I, *Fast Growing Timber Trees of the Lowland Tropics,* Oxford University Press, Oxford.

Ledig, F. T., and Fryer, J. H. 1971. A pocket of variability in *Pinus rigida. Evolution* **26**(2):259–266.

Lewinton, R. C. 1974. *The Genetic Basis of Evolutionary Change.* Columbia University Press, New York.

Libby, W. J., McCutchen, B. G., and Millar, C. I. 1981. Inbreeding depression in selfs of redwood *Sil. Gen.* **30**(1):15–24.

Mason, H. L., and Langenheim, J. H. 1961. Natural selection as an ecological concept. *Ecology* **42**(1):158–165.

Matheson, A. 1974. "Genotype–Environment Interactions, and Regions for Breeding." 4th Mtg., Res. Comm. of the Australian For. Council, Res. Working Group No. 1, Gambier, South Australia.

Mergen, F. 1958. Natural polyploidy in slash pine. *For. Sci.* **4**(4):283–295.

Namkoong, G. 1979. "Introduction to Quantitative Genetics in Forestry," Tech. Bull. No. 1588, U.S. Forest Service.

Nienstaedt, H. 1975. "Adaptive Variation-Manifestations in Tree Species and Uses in Forest Management and Tree Improvement." Proc. 15th Can. Tree Impr. Assoc., Part 2, pp. 11–23.

Orr-Ewing, A. L. 1976. Inbreeding Douglas fir to the S_3 generation. *Sil. Gen.* **25**(5–6):179–183.

Perry, D. A. 1978. "Variation between and within Tree Species." IUFRO, Proc. Ecology of Even-Aged Forest Plantations, pp. 71–98.

Rudolph, T. D. 1980. Autumn foliage color variation among inbred Jack pine families. *Sil. Gen.* **29**(5–6):177–183.

Sarvas, R. 1958. Kaksi Triploidista Haapaa Ja Koivia (two triploid aspens and two triploid birches). *Commun. Inst. Forest. Fenn.* **49**(7):1–25.

Schmitt, D., and Perry, T. O. 1964. Self-sterility in sweetgum. *For. Sci.* **10**(3):302–305.

Sluder, E. R. 1969. "Gene Flow Patterns in Forest Tree Species and Implications for Tree Breeding." 2nd World Consul. on For. Tree Breed., Washington, D.C., pp. 7–16.

Sniezko, R. 1982. "Inbreeding in Loblolly Pine." Ph.D. thesis, North Carolina State University, Raleigh.

Squillace, A. 1970. "Genotype–Environment Interactions in Forest Trees." 2nd Mtg., Working Group on Quant. Gen. IUFRO, Sect. 22, Raleigh, N.C., pp. 49–61.

Srb, A. M., Owen, R. D., and Edgar, R. S. 1965. *General Genetics*. W. H. Freeman, San Francisco.

Stebbins, G. L. 1950. *Variation and Evolution in Plants*. Columbia University Press, New York:

Stebbins, G. L. 1977. *Processes of Organic Evolution,* 3rd ed. Prentice-Hall, Englewood Cliffs, N.J.

Stettler, R. F. 1966. "The Potential Role of Haploid Sporophytes in Forest Genetics Research." Sexto Congreso Forestal Mundial, Madrid, Spain.

Stonecypher, R. W., Zobel, B. J., and Blair, R. L. 1973. "Inheritance Patterns of Loblolly Pine from Nonselected Natural Populations." Agricultural Experiment Station Technical Bulletin No. 220, North Carolina State University, Raleigh.

Strickberger, M. W. 1976. *Genetics.* Macmillan, New York.

Thor, E. 1961. "Variation Patterns in Natural Stands of Loblolly Pine. 6th South. Conf. For. Tree Impr., Gainesville, Fl. pp. 25–44.

Wright, J. W. 1976. *Introduction to Forest Genetics.* Academic Press, New York.

Yeatman, C W. 1967. Biogeography of jack pine. *Can. Jour. Bot.* **45**:2201–2211.

Zobel, B., Henson, F., and Webb, C. 1960a. Estimates of certain wood properties of loblolly and slash pine trees from breast-height sampling. *For. Sci.* **6**(2):155–162.

Zobel, B. J., Thorbjornsen, E., and Henson, F. 1960b. Geographic site and individual tree variation in wood properties of loblolly pine. *Sil. Gen.* **9**(6):149–158.

Zobel, B. J., and Kellison, R. C. 1978. "The Importance of Genotype × Environment Interaction in Forest Management." 8th World For. Cong., Jakarta, Indonesia.

Zufa, L. 1969. "Polyploidy Induction in Poplars." Proc. 11th Meet. Comm. For. Tree Breed. in Canada, Part 2, pp. 169–174.

CHAPTER 3

Provenance, Seed Source, and Exotics

THE IMPORTANCE OF SOURCE OF
SEED IN TREE IMPROVEMENT PROGRAMS

Success in the establishment and productivity of forest tree plantations is determined largely by the species used and the source of seed within species (Larsen, 1954; Callaham, 1964; Lacaze, 1978). The need to use the best-adapted source of seed was recognized in the early years as being important by such persons as Tozawa (1924), Wakeley (1954), and Langlet (1967). No matter how sophisticated the breeding techniques, the *largest, cheapest, and fastest gains in most forest tree improvement programs can be made by assuring the use of the proper species and seed sources within species* (Figure 3.1). This chapter will deal with the use and manipulation of seed sources within species, especially as they are used for planting exotics.

The best information available in the tree improvement field relates to seed sources. Many studies on this subject were started a number of years ago. In some species, much is now known about which are poor and which are good seed sources. Much effort, energy, cost, and thought have been expended on seed source studies. The most successful tree improvement programs are those in which the proper seed sources and provenances are used. The losses from using the wrong source can be great and even disastrous.

Many studies on provenance and seed sources have been made or are currently underway (Wright and Baldwin, 1957; Wakeley and Bercaw, 1965; Wells and Wakeley, 1970; Burley and Nikles, 1973a; Lacaze, 1977). In some countries, almost all the tree improvement effort has been expended on determining the best species and sources within species. However, even with the amount of effort already expended, ignorance about the best sources to use is still widespread for many species. When source information is available, the *reliability* and *availability* of the desired source of seed needs to be determined. As Anderson (1966) put it, "A reliable provenance would be one producing a decent forest crop with 90% probability rather than an outstanding crop 50% of the time. An available provenance is one from which seed are readily and economically available as needed."

The literature on exotics and source of seed is voluminous, and only a few references can be cited in this book. Summary publications such as those by Langlet (1938), Burley and Nikles (1972, 1973b), Dorman (1975), Nikles et al. (1978), and Persson (1980) list many study results and describe the techniques used in provenance breeding and testing. Many publications not listed include different species throughout the world.

A book could be written outlining the history and development of provenance activities in the European countries, Australia, South Africa, and numerous other areas. However, it would serve no useful purpose to discuss all of the studies in detail. Lessons have been learned about how difficult it is to define seed or planting zones in areas with complex physiographies. Some general rules on what should and should not be done have been developed. For example, the need for representative sampling to obtain seed to test sources properly is no longer

FIGURE 3.1

Attention to seed source is essential for the best success in forestry. The largest and quickest gains possible in tree improvement can be obtained from use of the proper source. The difference by source is illustrated for 2-year-old loblolly pine in Zimbabwe (Rhodesia). The row behind the man is an inland source of *P. taeda,* and the large row in front of the man is a southern coastal plain source.

questioned. Another major finding is that no single rule can be applied to all species on all areas.

One example of source studies that we know best is the southwide seed source study of the four major pine species in the southern United States. Based on European results and experience, limited seed source studies were begun in the southern United States shortly after the turn of the century. The first results on *Pinus taeda,* reported by Wakeley (1954), were so striking that much larger assessments were then initiated. Alerted to the importance of seed source, foresters quit shipping seed indiscriminately for long distances and made it a rule to obtain seed from as close to the intended planting site as possible. Several major tree improvement programs were organized around 1950, and use of the proper seed source was a major basis for their development (Figure 3.2).

One result of the larger studies showed that in the southern part of the area, some outside sources performed better in growth characteristics than did the local source (Wells and Wakeley, 1966). An important finding was that loblolly pine from part of its natural range was more resistant to fusiform rust (*Cronartium quercuum f. sp. fusiforme*) than were other species. This information is now being utilized in areas with a high rust potential up to 1000 mi (1600 km) east of the indigenous range of the resistant material (Wells, 1971). In the northern part of its range, the local sources of loblolly pine were found to be best, and tree improvement programs rely on this information. The idea that the coastal sources grew more rapidly than those from the interior Piedmont was confirmed, and now this information is utilized in large planting programs. The general rule is to use the local seed source until tests have shown the utility and suitability and advantage of nonlocal sources.

FIGURE 3.2

The large "southwide seed-source study" in the southern United States has produced some outstanding results and gives proof of the value of using the proper seed source. Shown is one of the plantings in Arkansas. The group of larger 25-year-old trees on the right are from the fast-growing *P. taeda* source from coastal South Carolina; the group of smaller trees to the left are of Oklahoma origin. There was more than 2 m difference in height that was accompanied by dramatic diameter differences. (Photo courtesy of O. O. Wells, U.S. Forest Service, Gulfport, Mississippi).

Decisions about which is the best source should not be made until extensive testing has been carried out for the greater part of the rotation age. The gains to be achieved by moving sources must be weighed against the risks involved, and this requires a period of testing that will yield reliable results. One of the most common problems related to provenance testing is to have a good initial performance that is followed by a later slowdown, lack of vigor, or even death. Once sufficient information is available, breeding zones can be identified within which tree improvement programs should operate.

In no area of forestry is the need for international cooperation stronger than in provenance testing. In fact, this has occurred on a very wide scale. Some very large international seed source tests have been made in such species as *Pinus silvestris, P. caribaea, Pseudotsuga menziesii, Picea abies, Picea sitchensis, Populus species, Tectona grandis,* the *Eucalyptus* species, and many others. Some of these studies are huge, containing hundreds of sources. Many of the studies are now old enough so that reliable results are available.

The very extensive and expensive seed source studies have been made possible by the cooperation among governments and industries. The international organizations like IUFRO (International Union of Forest Research Organizations) and FAO of the United Nations have been especially helpful. Specific groups like the CFI (Commonwealth Forestry Institute) in Oxford, England, have been instrumental in organizing numerous large-scale tests. Many governmental organizations, such as Queensland, Australia, have spearheaded international studies. Coalitions of industries and governments, like the CAMCORE Cooperative (Central America and Mexico Coniferous Resources Cooperative) have had a major influence on international seed collection and testing.

Although more work is always needed, activities in provenance testing have been most satisfying and the results most rewarding. For any tree improvement program, it is essential to obtain information about the best source of seed as soon as possible. Genetic gains from conventional tree breeding are determined or restricted by the quality of the geographic race or seed source used (Squillace, 1966). The ideal is to undertake an intensive tree improvement program only after the best geographic source is known (see Figure 3.2). In practice, this is often not possible, but good gains can still be made if proper land race development is followed, even though the original provenance used may be suboptimal.

TERMS RELATED TO SOURCE OF SEED

Terms such as *adapted, exotic, provenance, geographic source, geographic race, seed source,* and *land race* are standard for the tree improver. They are now coming to be used routinely in forestry circles, so their meanings must be clearly understood. The following definitions and illustrations are intended to simplify the very confused group of terms that have been related to exotics and seed sources.

Adapted The term *adapted* refers to how well trees are physiologically suited for high survival, good growth, and resistance to pests and adverse environments. For exotics, it refers to how well the trees will perform in their new environments. These may differ greatly from those found in the indigenous range of the species or seed source. *Adaptation* commonly refers to the tree's performance over a full rotation in the new environment.

Exotic The term *exotic* can be defined in several ways, but for the sake of simplicity, the following is useful: "An exotic forest tree is one grown outside its natural range." For example, ponderosa pine (*Pinus ponderosa*) from the western United States is an exotic when grown in the eastern United States; *Eucalyptus* and radiata pine are exotics when grown in Chile. Some persons restrict the term *exotic* to trees grown outside certain geographic or governmental boundaries, but the preferred usage is as shown here.

In the previous chapter, the major types of natural variation described for forest trees were those related to provenance, geographic source, or geographic race. These terms are similar, and they *are* used interchangeably and usually mean the same thing. A fourth term, *seed source*, appears to be similar but has an important and different meaning that must be recognized and should not be used synonymously with the other three.

1. **Provenance, geographic source, or geographic race** These denote the original geographic area from which seed or other propagules were obtained (Callaham, 1964; Jones and Burley, 1973). If, for example, seed of *Eucalyptus grandis* were obtained from Coff's Harbour, New South Wales, Australia, and grown in Zimbabwe, they would be classified as the Coff's Harbour provenance (or geographic source or geographic race).

2. **Seed source** If seed from the trees grown in Zimbabwe were harvested and planted in Brazil, they would be referred to as the Zimbabwe seed source and the Coff's Harbour provenance. The term *origin* is used by Barner (1966) in the same way as *seed source*. When the difference between provenance and seed source is not recognized, large and costly planting errors can be made. *Seed source* will be used very specifically throughout this book, whereas provenance, geographic source, and geographic race will be used interchangeably.

Divisions of source of seed that are finer than provenance and seed source are commonly employed. There is often a perplexing proliferation of terms such as *altitudinal race, climatic race, physiological race, physiographic race, edaphic race,* and many others. All of these are used in the literature to describe within-species variation and are referred to as *races*. Not only are these terms confusing in themselves, but they are often further mixed with such words as *variety, strain, ecotype,* and *cline.* What follows is an attempt to recognize and to simplify terms that are encountered.

Racial Variation

As mentioned in the discussion on provenance, geographic source and geographic race are synonymous. Barner (1966) discusses the concept of race; he feels the term *race* should be used only when natural populations are described. Races develop in response to evolutionary forces, such as natural selection, that vary in different parts of the natural range of a species. Populations thus developed will show large-to-small differences when grown together in a uniform environment. This is racial variation. Since provenance, geographic source, and geographic race have the same definition, the comprehensive descriptions of races in forest trees prepared by Wakeley[1] are most useful. Some of his descriptions are paraphrased as follows.

1. A geographic race is a subdivision of a species, differing in ways that can be demonstrated by observation and experiment from another race or races within the same species.

2. A geographic race has evolved within the species of which it is a part through the process of natural selection, and the individuals making up the race are related by descent from a common ancestor or group of related ancestors.

3. The characteristics that distinguish a race are genetically controlled; that is, they are heritable in the ordinary process of reproduction.

4. As the name implies, a geographic race occurs naturally in a fairly well-defined environment to which it normally is well adapted as a result of natural selection and has the ability to survive and reproduce in that environment.

From these four statements, Wakeley defined a *geographic race as a subdivision of a species consisting of genetically similar individuals, related by common descent, and occupying a particular territory to which it has become adapted through natural selection.*

If the specific territory or environment to which the race is adapted is a distinct altitudinal or climatic zone or soil province, it may be referred to as an *altitudinal, climatic,* or *edaphic race* rather than the more general term, *geographic race.* But differences in soil or altitude or even in climate usually do not account in total for the specific characteristics of a race. For example, a particular race may result from migration of plants from some mountain mass to a coastal site that has been recently uplifted from the sea, so that the characteristics of the race can be determined by rates of migration, or barriers to migration, as well as by extremes of heat, cold, drought, or other characteristics of the environment. In such

---

[1]Wakeley, P. C. 1959. In "Proceedings of Forest Genetics Short Course," North Carolina State University, Raleigh, June 1959 (unpublished mimeo).

instances, the general term *geographic race* is more appropriate than altitudinal, adaphic, or other specific race categories.

Geographic races occur most often in species that have a wide natural range and so encompass a large range of environments (Figure 3.3). They are determined by differences in latitude, altitude, rainfall patterns, or other environmental conditions that expose the trees to considerable variation in temperature, moisture, soil, day length, or any number of other environmental variables (Holzer, 1965). Most forest species have distinct geographic races. The species that contain the largest racial divergence afford the best opportunity for genetic gains through provenance selection, but they also are the ones with which the tree improver must use the greatest caution to assure that proper sources are identified for use. It is not unusual for differences to be less between species than between races within a given species that occurs over widely differing environments. This is seen for a number of genera, such as the pines in Mexico, where sometimes the differences between high- and low-elevation material can be greater within a species than between two species growing on similar environments at similar elevations.

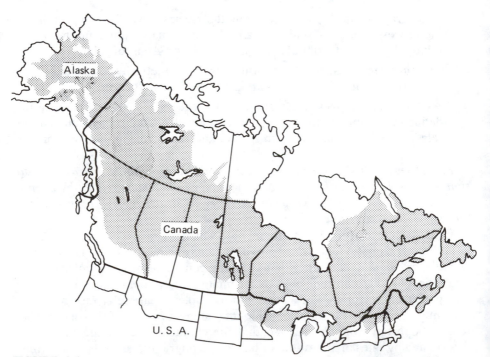

FIGURE 3.3

Geographic races almost always develop in species that have very wide geographic ranges, as is shown here for white spruce (*Picea glauca*). The diverse provenances encountered enable the tree improver to find strains of trees that are the most adapted to the area in which planting will be done.

When two species occupy the same or similar ranges, races may develop within them that are quite alike physiologically. Most high-latitude sources grow slowly, but have better tree form with straight stems and smaller limbs than those from lower latitudes. In addition, they can withstand cold weather that would kill or injure trees adapted to lower latitudes. These three characteristics of slow growth, good form, and good resistance to cold are common to most high-elevation and latitude sources, regardless of the species.

Often a determination of the boundaries of a geographic race is difficult. The gradation can be abrupt and clear where there are gaps in the species range (such as separation by deserts or mountains), or they can be gradual in cases in which the species occurs continuously from south to north or from low to high elevations. When variation is continuous, the distinction among geographic races becomes one of judgment or viewpoint (Langlet, 1959a; Farmer and Barnett, 1972); this is the most common situation. The definition of what exactly constitutes a race is not the important item; what is important is *an understanding of the patterns of variation and how this knowledge can be used* in a tree improvement program.

Characteristics that determine different races *are more frequently* physiological than morphological (Langlet, 1936). This means that racial determination based on phenotype alone without intensive testing will often be inaccurate. The need for testing is of primary importance because the physiological characteristics usually are related to survival, growth, and reproduction. Thus, they are key items in determining which geographic race of a species should be used in an operational planting program. This need is especially great when seed are moved long distances. Changing environments can cause unusual growth patterns, such as "foxtailing" of several pine species when grown in tropical environments. The foxtail produces no limbs; this situation can vary from only for a few years to the lifetime of the tree (Woods et al., 1979; Whyte et al., 1980). The ultimate proof of suitability of a source as an exotic and whether it will have such characteristics as foxtailing must be based on test results.

Although individuals within a race are somewhat similar from past heritage or selection pressures, they are by no means genetically identical. Usually, great individual differences in genotype and often in phenotype occur among trees within each race. This heterogeneity enables individual tree selection within sources to be used effectively. For example, there are differing genotypes that will produce either straight or crooked trees in the higher latitudes as well as in the lower latitudes.

Clines and Ecotypes

A whole series of categories have been proposed and used to describe patterns of genetic variation. Among these, the most important are the *ecotype* and the *cline,* which are very briefly discussed in this chapter. The terms are most widely used in the fields normally referred to as *speciation* and *evolution,* which are a whole "science" unto themselves. Many books and articles have been written in which these concepts have been discussed, for example, Turesson (1922), Stebbins (1950), Mettler and Gregg (1969), Grant (1971), and Endler (1977). Speciation and evolution are complex and often controversial subjects, and this book is no place for a detailed discussion of them. Yet, because ecotype and cline are so commonly used, an attempt will be made to define them clearly as they apply to forest trees.

An *ecotype* is a group of plants of similar genotype that occupy a specific ecological niche. In forestry, the ecotype is sometimes used synonymously with race but usually consists of a smaller discrete population. Frequently, ecotypes are not distinguishable by morphological characteristics and can only be separated by physiological differences, which are usually related to survival capabilities (Rehfeldt, 1979). The concept of ecotype was suggested by Turesson (1922) who defined it as a "genotypical response of a species to a particular habitat." The whole concept is based upon the adaptability to a specific environment. Gregor (1944) compared the environment to a sieve that sorts out the genotypes that are best able to survive.

Cline was first defined by Huxley (1938) as "a gradient in a measurable characteristic." Most people add the following to this: "which follows an environmental gradient." A cline, by definition, is based on a single characteristic that has continuous variation; it may or may not be genetically fixed (Figure 3.4). There may be a whole series of clines for different characteristics within a given population. Recognition of clines in tree improvement is very important, and such things as drought or cold resistance usually follow a clinal pattern. Since *clines* refer to a given characteristic and not to the entire genetic constitution of a population, the concept of *ecotype* would have more utility to the tree breeder. However, because ecotypes and clines both occur in nature, a successful breeder needs to understand both of them.

The *ecotype* and *cline* concepts have widespread utility in forestry. Because of the commonly observed continuous type of geographic variation, ecotypic differentiation is difficult to observe. It was to point out that ecotypes are discreet and separate that Langlet (1959a) wrote "a cline or not a cline." The key characteristic in understanding ecotypes is that they are the adaptation of *whole genotypes* or *gene complexes* to *specific environments*. They are not just expressions of single characters, and they are distinct one from another (they do not intergrade). Although they are not called *ecotypes,* we use something similar when developing land races that are adaptable to extremes of environments, such as very dry, wet, or cold habitats.

Table 3.1 was developed to aid in comparison of the two terms.

WHERE RACES ARE DEVELOPED BEST

It is of importance to recognize the conditions in which differing races are most likely to be found or developed.

1. *Species with very wide ranges over diverse environments* usually have the greatest racial development. Those species that span continents, or that grow over very wide latitudinal or altitudinal ranges contribute the most useful

FIGURE 3.4

Wood specific gravity indicates well the gradual change in a characteristic in conjunction with an environmental gradient. For example, there is a clinal change in wood specific gravity of loblolly pine from high values in the southern coastal plain to low values in the northern areas. There is also an abrupt pattern from the higher-specific-gravity coastal sources to the lower-gravity inland values, as is indicated on the map.

TABLE 3.1.
Comparison of Clines and Ecotypes

	Cline	Ecotype
Number of characteristics	One	Many (the genotype or gene complex)
Pattern	Continuous	Distinct populations
Genetics	May or may not be genetically controlled	Genetically controlled
Cause	Follows environmental gradient	Adaptation to specific environment
Use	Descriptive	Descriptive and as breeding unit, similar to race

races. The differences may be visible morphologically (Figure 3.5), or they may only be distinguished on the basis of physiological characteristics that are detected by testing. Usually, morphological and physiological differences both occur, but the physiological differences, which often are of the most importance to the tree improver because they affect survival and growth, sometimes develop without visible differences among trees. Thus, cold-, drought-, or moisture-tolerant races may develop within a species, even

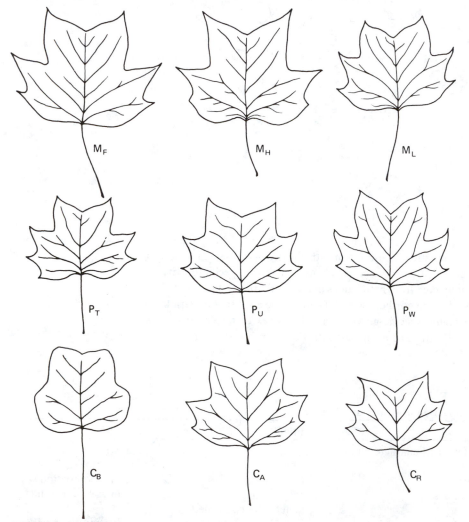

FIGURE 3.5

Many species show anatomical variation with geographic source. Shown are leaves of *Liriodendron tulipifera* from differing sources. The extreme, nonlobed form at **bottom** left is commonly associated with coastal, acid soils. (Courtesy of Bob Kellison, North Carolina State University.)

though the trees appear to be similar phenotypically. Examples of wide-ranging species with great racial development are *Pseudotsuga menziesii, Pinus banksiana, Pinus silvestris, Eucalyptus camaldulenis, Pinus caribaea* (where varieties are recognized), *Pinus oocarpa* (where both varieties and new species are debated), *Tectona grandis, Pinus taeda,* and many others (Figure 3.6). Only a few wide-ranging species do not have strong racial development. For example, it has been suggested that the very widespread aspen (*Populus tremuloides*) has limited geographic variability throughout its range when compared to most species. There are not enough data from truly in-depth provenance studies of aspen to support or refute the statement that this species has limited racial development.

2. *Species growing in a wide altitudinal range* commonly develop races, and often changes are such that one species will intergrade into another with no break that is evident between the two. This appears to occur in Australia where *Eucalyptus regnans* in the lower elevations seems to intergrade into *E. delegatensis* in the higher regions. There are many examples of such intergradation in the Mexican and Central American pines, where different species grow at different elevations with no distinct divisions between them

FIGURE 3.6

Country with severe and rapidly changing environments is ideal for the development of differing races. Shown is a view in Guatemala, where trees are growing in dry, poor sites as well as in excellent forest conditions at higher elevations. Much racial development occurs in this area.

(Caballero, 1966). Racial differences are often recognized, such as those that occur in *Pinus oocarpa* or Douglas fir where some persons feel the current races should be given subspecific or specific taxonomic ranking. Within-species differences for survival and growth that are attributable to geographic variation can be very large, such as those for yellow poplar (*Liriodendron tulipifera*) (Kellison, 1967). A well-documented example is the altitudinal races that were developed in the pines in the Sierra Nevada mountains in California (Callaham and Liddicoet, 1961).

3. *Species that grow in regions of greatly diverse* soils, soil moisture, or slope and aspect (Squillace and Silen, 1962) can develop very distinct racial differences. A good example of extremely differing soil environments that are close together are the granitic and serpentine soils in the Sierra Nevada of California; this is also true for the greatly differing environments on the western and eastern sides of the Cascade Mountain Range in Oregon. Here, in relatively short distances, natural selection caused by the differing soil and climatic environments has resulted in racial differences within species.

Where to Select

A serious problem facing the tree improver working with indigenous species is whether to use local or outside sources of seed. There is a natural attraction to prefer outside sources, and this often results in the use of exotics. However, the safest method is to use the local source until an outside source has been proven better (Krygier, 1958). Frequently, the local source turns out to be best (Long 1980), and sometimes source seems to make little difference (Talbert et al., 1980). There are also examples when an outside source proves to be superior (Namkoong, 1969). Using the local source until tests have proven that something else is better is admittedly a conservative approach, but it helps avoid large-scale losses that might occur if the nonindigenous population is poorly adapted to its new environment.

An ongoing argument is whether one should select from the center or fringe of the species range (Figure 3.7). Discussions of this kind that often involve the so-called *center of origin* of a species (Stebbins, 1950), or the *gene center theory* of Vavilov (1926), indicate that the richest genetic pool is present in these areas. There can be no completely correct answer to the question about where to select; therefore, the stock sentence "it depends" must be employed. If one is seeking a population of trees for introduction into some specific or extreme environment, then the fringe populations will often be best if the environments in which they are growing resemble those where the exotics will be planted. They may have limited total variability but may possess specially needed adaptability. If one is going to plant the introduced trees in a nonextreme environment and wishes for the maximum genetic variability, then it is best to select from the center of the range where genetic variation usually is the greatest. This problem has been addressed by Muller (1959) and van Buijtenen and Stern (1967).

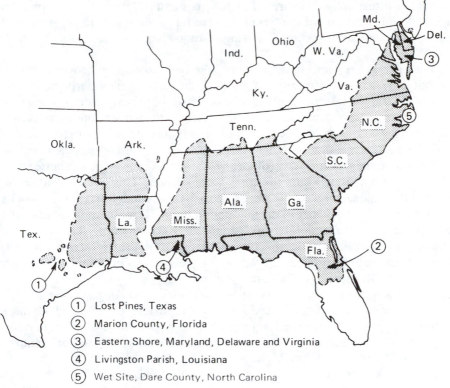

① Lost Pines, Texas
② Marion County, Florida
③ Eastern Shore, Maryland, Delaware and Virginia
④ Livingston Parish, Louisiana
⑤ Wet Site, Dare County, North Carolina

FIGURE 3.7

Shown is the natural range of loblolly pine (*P. taeda*). There are many "outlier" populations such as the lost pines in Texas and the wet-site loblolly in Dare County, North Carolina, that apparently are quite different genetically from the general species. Controversy exists about how much selection should be practiced in the outlier areas.

The Land Race Concept

The concept of a *land race* is simple and is of key importance when working with provenances planted outside their normal environments (Marsh, 1969; Pellate, 1969). A *land race* is *a population of individuals that has become adapted* to a specific *environment in which it has been planted.* The steps involved in land race development consist of planting the trees in the new environment, letting nature sort them out according to their adaptability through natural selection. This is followed by choosing the best of the naturally selected trees and then using them as a source of seed to replant the area. This can be done after a single generation, but the best land races occur after several generations of growth and selection in the new environment. The group of best-adapted individuals with desirable growth and form are collectively referred to as a *land race.* In reality, one could also call the fittest individuals from indigenous stands a land race, but common usage and understanding of the term land race has limited it to the best trees of suitable provenances of species following testing when used as exotics.

Why Land Races Are Important   When one moves a species or provenance into an exotic environment, it rarely is fully adapted to the new environment, and sometimes it is quite poorly adapted. As individuals of the exotic grow in the new environment, the most well adapted will survive and perform the best (Figure 3.8). When the best trees are selected for use as a source of propagules for planting or for the next generation, either through seed or vegetative propagation, the performance of the new forest will often be from moderately to greatly better than the original stand from which the trees were chosen. This depends upon the quality of the original trees, the selection intensity, population size, breadth of its genetic base, and the severity of the new environment. It is not unusual for a land race, even from only a moderately well-adapted source, to outperform any other provenance of the same species that is planted directly in the exotic environment. This indicates that selection can be very effective within a broadly based, large, moderately well-adapted population.

Use of land races can be the easiest and best way of making quick and large genetic gains in exotic forestry. There are numerous examples in which land races have performed well above the level expected or hoped for, and *almost always* they perform better than any newly imported sources of the species. For example, Owino (1977) found that for *P. patula* and *Cupressus lusitanica* the advanced "land race" selections were highly superior. For northwestern Europe, Edwards (1963) stated that great advances can be made when seed of exotic species can be collected from stands of plus trees growing in the exotic environment. Distinction between native and exotic species may then disappear.

Previously established plantations of the species desired are the first thing to look for when starting a large program involving exotics. When such exist, occasionally outstanding individuals may be found (see Figure 3.8), even though the plantations may be rather poor overall. Seed from these individuals can be used in operational plantings, while seed orchards are being established and further introductions and tests are being made. If applied intensively, the land race approach will lead to the development of new strains of a species with great utility in the new environments. For example, in the southern United States, a cold-hardy strain of eucalypt is being developed by planting the exotic *Eucalyptus*

FIGURE 3.8

Occasional individuals in exotic plantings grow exceptionally well compared to the average of the plantation. Shown is an outstanding 7-year-old *P. caribaea* growing on the very severe sites of the Guyano llanos in the Orinoco Basin of Venezuela. A group of such trees is used to establish a land race that is better suited to the new environment than the population used for the original planting.

viminalis, selecting the best of the trees that survive the severe cold, and bringing these together as a seed source to grow *Eucalyptus* in a region where it previously could not be grown operationally (Hunt and Zobel, 1978).

When land races are developed in one region, they often will also be useful in other, similar regions (Nikles and Burley, 1977). Too often this source of material has been overlooked for exotic plantings. A good example of a broad usage is the apparent good adaptability of the greatly improved land race *P. caribaea var. hondurensis* developed in Queensland, Australia. The land race in Queensland has performed very well in Fiji, New Caledonia, parts of Brazil, Zambia, and in other countries.[2]

In summary, land race development is most feasible when the following conditions are met:

1. The original provenance is reasonably well adapted to the environment in which it will be planted. This can be helped by a careful consideration of climatic similarities between the exotic environment and the one where the provenance was developed.

2. The populations from which the land race trees will be selected must have a broad genetic base. Usually, several hundred parent trees should contribute seed to the plantation from which the land race selections are to be made.

3. The plantations from which selections will be made should be reasonably large, generally on the order of 400 ha (1000 acres) or more. This is not always possible, but small plantations do not have enough trees to give a reasonably high-selection intensity.

4. Enough local plus trees need to be selected to form the land race. As few as 30 can be used for a production seed orchard, but 200 to 300 or more will be needed for the developmental program. Smaller populations will soon lead to trouble from related matings and will restrict the genetic base needed for advanced generation development.

5. The selection system used to choose parents for the land race must be well devised and rigorous to assure the choice of only the very most outstanding trees.

The Stress Theory

The stress theory is of great importance because trees planted as exotics are usually not well adapted to the new site; therefore, they will be growing under stress. If there is a problem with adaptation, some of the poorly adapted trees may die immediately, but the rest will grow normally until they come under a period of extreme stress caused by severe environmental fluctuations or pest attack. At that time, many trees will develop leader and branch dieback or perhaps even die, and only those that are well adapted will survive and grow normally (Figure 3.9). It is

---

[2]Personal communication, Garth Nikles, Queensland Forest Service, Brisbane, Australia.

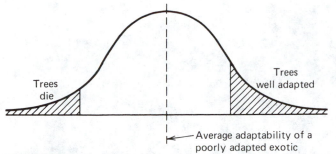

FIGURE 3.9

How well a tree can tolerate stress is the key to whether geographic races or species will grow well in an area. A *land race* is made up of those individuals that will best tolerate stress, as shown diagramatically by a population of exotics planted on a severe site.

these few best-adapted individuals that should be used to develop the land race with the assumption that most are superior genotypes and that their phenotypic superiority is not due primarily to an accidentally favorable environment.

Everywhere throughout the world where trees are planted on stress environments, such as the *llanos* (grasslands) of the Orinoco Basin in Colombia and Venezuela, dieback may occur in one form or another at some time in the life of the stand (Figure 3.10). Stress increases following drought, or as the trees become larger and come into severe competition with one another for available moisture. Dieback is a response to stress, and after the stress situation is alleviated the trees that have not died will recover to some extent. When stressed again, the trees will go through the same cycle, and the amount of deformity and death will increase with each cycle unless this situation is modified by silvicultural practices such as thinning. This reduces the stress and removes the poorly adapted individuals. *Thinning* is a major way of avoiding damaging stress in exotic plantations until a proper land race can be developed. The key to working with exotics is to develop a land race that grows better under stress than does the plantation from the original introduction.

EXOTIC FORESTRY

Exotic species are used when the local indigeous forests cannot, or do not, produce the desired quantity and quality of forest products. In countries that have very few species available, such as in northern Europe, exotics fill a special need (Edwards, 1963). In the tropics, there is often a great desire for coniferous timbers, because the indigenous forests frequently do not contain this type of wood. To fill the need, exotic plantations have been widely used. Many exotics are

FIGURE 3.10

When exotics are planted, some individuals will not be suited to the new (exotic) environment. When stress occurs, they will die or become deformed. Shown are damaged *P. caribaea* in the llanos of Venezuela following a prolonged drought. Other neighboring trees have developed and grown well despite the dry weather.

planted in the grasslands in tropical areas, such as in the llanos of the Orinoco Basin of Venezuela and Colombia and in central East Africa. Large plantations of exotics are also established in scrub forests, such as in the state of Minas Gerais, Brazil, or in large areas in Colombia and Chile. Many other regions, such as Australia, New Zealand, India, Indonesia, and the Middle East, use exotics much of the time. In many areas, especially in the Southern Hemisphere, there is no option except to use exotic conifers because no suitable indigenous species exist (Zobel, 1961, 1964, 1979).

It is important to note again that the use of a proper provenance is one key to having a successful program with exotics. Provenance differences that exist within the natural range of a species often become more evident when a species is grown as an exotic. One of many examples is for *P. taeda* in southern Brazil (Shimuzu and Higa, 1981). Exotic forestry is very widespread in tropical areas, both in naturally forested and grassland ecosystems. Tropical hardwood forests are so variable, nonuniform, and difficult to manage ecologically that, on suitable sites, foresters generally prefer to use exotic species that are more uniform, easier to handle, and whose products are known and accepted. Some of the tropical hardwoods have magnificent wood, but usually the most-desired species are slow growing and hard to manage in plantations. Currently, knowledge is generally lacking about how they should be handled. Foresters are often trained to handle the exotic species and feel comfortable with them. Because exotics grow at a rapid rate, they are commonly chosen as the heart of an economically viable forestry enterprise in tropical areas. Despite the past bias toward exotics, it is likely that certain indigenous species from the tropics will become more widely used while foresters learn to manage them, learn their biology and the value of their wood, and are able to develop suitable markets (Figure 3.11).

One serious problem related to plantation management of nonindigenous tropical species that often makes an intelligent tree improvement program impossible is a great lack of knowledge about the soils in the tropical regions. The idea that all tropical soils are fragile and that they are unsuited to forest management has been grossly overemphasized, and it has led to efforts to develop new strains of trees that really are not needed. The statements about the amount of the very intractable lateritic soils in the tropics have been especially misleading. Where these types of soils do occur, they must be absolutely avoided for exotic plantations. However as Sanchez and Buol (1975) state in "Soils of the Tropics & the

FIGURE 3.11

There has been an overemphasis on growing exotics in the tropics because foresters know how to grow them, seed are available, and their woods are known. Often such information is lacking for indigenous species. But some species in the tropical areas, such as this *Toona ciliata* growing in Brazil, have great potential. Foresters need to learn more about these species.

World Food Crisis" less than 10% of the forest soils in the tropics are laterites. They report the following values for different tropical areas.

Location	Percentage of lateritic soils
Tropical America	2
Central Brazil	5
Indian Subcontinent (tropical)	7
Tropical Africa	11
Sub-Saharan West Africa	15

Sanchez and Buol (1975) state that "on the basis of these and other estimates, we venture that the total area of the tropics in which laterite may be found close to the soil surface is in the order of 7%." They conclude with the observation that soils in the tropics are essentially similar to those in such temperate regions as the Piedmont of the southern United States. As such, tree improvement activities are suitable and are not too difficult. Based on our experience in the tropics in South America, it appears that about 50% of the tropical forest area is suitable to support plantation forestry with few soil problems or adverse ecological impacts. The areas on which plantations are to be considered should be surveyed with respect to soil types, and then actual suitable areas must be handled correctly. Ideally, tropical forestry in the future will consist of natural regeneration on the more fragile soils; on the more operable sites some plantations will be established using indigenous species, the rest will use exotics. Tree improvement allocations to these different types of management will vary, depending on soil types and wood needs. It is certain, however, that much more information is needed to plan allocations most appropriately.

Planting of exotics in tropical areas is now mainly in the grasslands or scrub forestlands, especially in South America. This is not done because those lands are better or more productive. In fact, they are often marginal for forest production, and the soils are often inferior, either structurally or chemically or in their moisture relations. Some consist of deep droughty sands, whereas others have heavy textured soils that are difficult to manage. Quite frequently, phosphorus is deficient, and sometimes a lack of other elements like boron or copper limit tree growth. All of these pose a major challenge to the tree improver to find the genetically best-adapted trees. Yet these sites are chosen for much of the exotic tropical forest plantings that are now being established because of the following factors:

1. They are usually easy to handle. Site preparation is simple and inexpensive, and competition from regrowth is limited. Converting a tropical hardwood forest to a plantation is a formidable and very expensive job, both in

removing the trees from the planting site and from the need to release the planted trees from regrowth forests.

2.   They can be directly converted to plantations without the need to use, or dispose of, the wood in the indigenous forest. Markets for tropical hardwoods are limited, and even when markets exist, only a part of the wood can be economically utilized; also, the remaining wood must be disposed of at a considerable expense. It is unconscionable to destroy the wood in tropical forests, and yet it is often not feasible to use it economically under current utilization standards. The result is that the most productive tropical hardwood forest sites are often bypassed for plantation establishment.

3.   The land is available and much of it is not suitable for agriculture.

Use of exotics in the northern forest regions, such as in northern Europe, is also widespread. The usual movement of species is from the West Coast of North America to Europe and Asia. Such magnificent species as Douglas fir (*Pseudotsuga menziesii*), Sitka spruce (*Picea sitchensis*), the pines, hemlocks, and larches have been and are now widely being used as exotics. One of the species being used on a large scale in northern Sweden in the last few years is lodgepole pine (*P. contorta*), where it is performing better than the indigenous species (Hagner, 1979) (Figure 3.12). The species has been used as an exotic in Scotland and Ireland for many years. Generally, movement of forest tree species from Europe to North America as exotics has not been successful. There are not many exotic forest tree species that grow well in the northeastern United States, although species such as Norway spruce (*Picea abies*) and larch have potential (Carter et al., 1981). In areas such as the southern and the western United States, which contain numerous valuable species, exotics are rarely of much value (Zobel et al., 1956).

A long list of species that have been successfully used as exotics could be made, but this would serve no purpose. It is important to realize how wide the choices are; for example, poplars in the Middle East, willows (*Salix*) in Argentina, teak (*Tectona grandis*) in Indonesia, and tropical pines in many countries. The most successful and widely used conifer has been *P. radiata;* in the tropics, *P. caribaea* and *P. oocarpa* are being used extensively. *Eucalyptus* species are the most widely used exotic hardwoods and, in fact, they are the most widely planted exotic forest tree, with huge areas having been established in South America, Africa, and many other regions.

Exotics are sometimes used in quite unlikely areas; an example is *P. taeda* that grows at high elevations in Colombia (Ladrach, 1980). Only a few of the many seed sources of loblolly pine that have been tried are suitable for the wet, high elevations, and cool sites in this tropical area. It would be an advantage if one could easily predict what will be a successful exotic based upon its performance in its indigenous range. Unfortunately, it is not yet always possible to do this with sufficient reliability. In most cases, it is only after tests have indicated wide adaptability, fast growth, and usable wood properties that one can be certain

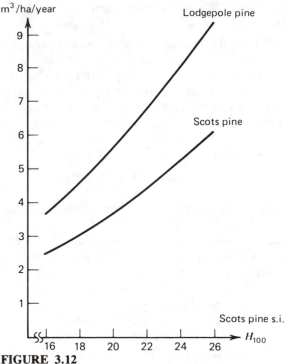

FIGURE 3.12

Sometimes an exotic species will outproduce indigenous species. This is shown for lodgepole pine (*P. contorta*) compared to Scots pine (*P. silvestris*) in northern Sweden. (Graph courtesy of Per Stahl, Swedish Forest Service, data from College of Forestry.)

about what species and sources will actually be successful exotics. The ability to predict performance in new environments is progressively improving because it is based on experience and the results of many previously established studies.

Occasionally, exotics are espoused as being of outstanding value; a recent one, which is often called the miracle tree, is *Leucaena* (Anonymous, 1978; Brewbaker and Hutton, 1979). This tree has received much publicity, but only time will prove its true value. Many trees that are proposed as exotics fail, but one seldom hears about them. In fact, only a small proportion of trees that have been tried or recommended for exotics is successful.

The advantages that exotics may present to the forester can be briefly summarized as follows:

1. More wood of greater uniformity and desirability can be produced quickly. Coniferous-type wood, which is scarce in the tropics, can be grown.

2. Rotation ages may be shortened and can be as low as 5 or 6 years for some eucalypts. This results in a large economic advantage in the cost of wood

production. Many indigenous species in the tropics are slow growing, although some, such as *Gmelina,* grow very rapidly.

3.   Exotics are suited to intensive management and plantation culture, and silvicultural methods are known. Many indigenous species cannot yet be grown successfully in plantations because seeds are not available, methodology is not known, or the biology of the species is such that pure stands will not grow satisfactorily.

4.   The wood quality produced by exotics and its utility are known. Many of the tropical hardwood species have wood that is unknown on the market or is technologically difficult to manufacture, even though it may be of high quality.

5.   Frequently, extensive study, including genetic improvement, has been done on the exotic species, and improved genotypes are available to use directly in operational planting. Such use has great potential, but must be done cautiously until the value of the improved trees has been proven in the new environment.

6.   The biology of reproduction in the exotics is known. Ignorance about the biology of tropical hardwoods is often great; sometimes, it is not even known how to collect and store seed, or what is the mode of pollination of the species.

Problems with Exotics

There will always be problems encountered when using exotic species; these can occur in many ways when poorly adapted species or unsuitable seed sources or provenances are planted on a given area. For example, pest attacks can occur immediately or be delayed for many years. It is essential that the forest manager and tree improver realize that pest problems will eventually occur, and plans must be made to cope with these. *But pest attacks will occur.* The production of an exotic in its early years should never be used to estimate future production. Problems with exotics may manifest themselves in several ways.

1.   *Immediate failure of the plantation.* Losses of this type are obvious and do not need further discussion. A surprising number of such complete failures do occur, but one rarely hears about them, and they are almost never reported in published articles.

2.   *Delayed failure* is a common problem and can occur in one of many forms that can be categorized as follows:

   a.   There is good survival and growth in the early years, but the trees never develop into a useful forest. This often occurs when high-elevation of high-latitude sources are planted at low elevations or low latitudes or when trees from a Mediterranean climate are planted in a Continental climate. For example, many of the poplar hybrids developed for the northeastern United States have been planted in the southeastern area of the United States. During the first 2 or 3 years these trees grow at an amazing rate, but after 10 years many trees are dead and often those still alive have lost vigor, the leaves are small, tops and limbs die back, and the trees literally "fall apart." Another good example of delayed failure occurs when *Eucalyptus* is planted on lands with a heavy subsoil layer a few centimenters from the surface. The trees grow well for 1 or 2 years, then they develop top dieback, and in 4 or 5 years no effective forest stand still remains. This problem of delayed failure causes severe loss, but the mistake is usually recognized soon enough to prevent continued, long-term planting of the wrong source.

b.   *There is good survival and growth rate but the wood is not suitable* (Kellison, 1981; Zobel, 1981). There are many examples of the problem of undesirable wood from exotics, especially when temperate species are planted in tropical or subtropical areas. One occurs in South Africa where in some parts of the coastal area, *Pinus caribaea* produces very low-specific-gravity wood that is not desirable either for paper or for solid wood products (Falkenhagen, 1979) (Figure 3.13). In this same

FIGURE 3.13

When trees are grown as exotics, their wood quality may vary greatly from that produced in the original environment from where the exotic was taken. Shown is a stand of *P. caribaea* from the warm low area of South Africa, which had excellent growth. However, the wood quality was quite inferior to the species in its indigenous range. Trees in this photograph had an unusually low specific gravity and an absence of summerwood, producing a wood undesirable for many products.

general forest region, *P. elliottii* produces wood with such a high density that it is also undesirable for many products. One of the most costly and common errors made when seed sources are moved or exotics are used is to plant large plantations before the quality of the wood that will be produced has been determined. The problems with wood from exotics is described more fully in Chapter 12.

c. The exotic *trees show initial good survival and growth, but there is a delayed attack by pests or adverse environmental conditions* that ultimately destroys the value of the forest. This common and very costly cause of loss is most disheartening; it is similar to item 2a, but is more long term. Many examples could be given, but perhaps the worst is the destruction of pine by the disease *Dothistroma* that is most destructive when the trees are about 5 years of age. Most of the radiata pine plantations in Brazil and central East and central South Africa have been damaged or destroyed by this disease, and damage occurs in many other areas, such as New Zealand and Chile. *P. radiata* should never be planted in a climate that has wet and warm summers because *Dothistroma* flourishes under those conditions. Another example of delayed damage is the planting of *P. caribaea* from the Guatemala source on the deep sands in Venezuela. The trees grow satisfactorily for several years, but when drought stress occurs, many trees develop a dieback of the leaders and branches, and often the whole tree dies (see Figure 3.10). Another graphic example occurs when the well-recognized Florida source of *P. taeda* is planted in northern latitudes. The trees may grow well and normally for several years, and generally they will significantly outgrow the indigenous pine sources. But when certain cold weather sequences occur, the physiology of the plant is upset so that the trees lose dominance, and the leader and branches grow in a sinuous way so that the growth of the trees is slowed and the trees become very deformed. There are innumerable examples of late decline in such hardwoods as *Eucalyptus, Populus, Gmelina, Platanus,* and other genera. One example in Minas Gerais, Brasil, is particularly worrisome. There, an unknown pest or combination of pests attacks one source of *E. grandis,* but does less harm to others. Damage does not occur until the second or third year after outplanting; then the trees develop small, yellow-white leaves, many leaves fall off, branch dieback occurs, and sometimes the trees die.

3. A major problem with exotics can be a *continued substandard performance* resulting in low production. This is a major problem with some exotics and causes the greatest losses from planting species or provenances offsite. Such losses are subtle and cannot be assessed without a comparison with the suitable sources and the best species. Often, losses can amount to 50% or more of the productive potential of the site. Unfortunately, the forester is often quite unaware of the loss because there is nothing to compare with the

planted trees. If this kind of loss is not detected, inferior sources may be used on a large scale in plantations. A good example of the subtle loss is the use of the Piedmont source of *P. taeda* in the coastal plain. Yields from the Piedmont trees will be small compared to those where the proper loblolly coastal plain source has been used. Several countries have planted large acreages with the Piedmont source of loblolly pine, not fully aware that their forests are producing 30 to 50% less than if the correct coastal plain source had been used (see Figure 3.1). *Eucalyptus* is especially sensitive to a reduced growth pattern, and there are numerous examples in Colombia, Venezuela, Brazil, Africa, and elsewhere where growth of *Eucalyptus* is much below standard because the wrong sources and species have been used in large operational programs.

4. *Growth is unsatisfactory* due to a *shortage or absence of suitable mycorrhizae.* This has been a major problem with some exotic plantings, especially in the tropics, but the importance of mycorrhizae is now generally recognized. Often the soils of the exotic environment are marginal for survival and the growth of mycorrhizal fungi.

The question needs to be asked why poor sources or wrong species are used in exotic forestry programs. The reasons are many, but some of the most important are the following:

1. **Ignorance**   Geographic differences are not well known for many species. Despite the fact that numerous studies on variation have been and are still being made, only a limited amount of information is available for some species. When a program to grow exotics has been funded and action must be taken, the only thing to do is to use one's best judgment while tests that are underway to determine the best sources are maturing. Perhaps the greatest error in determining which species to use is the lack of knowledge about provenances within the species of interest. It is common for the wrong seed source to be selected to represent a certain species. When planted, the trees do not perform well, and as a result, the whole species is tagged as being of no value for the area. Prime examples of the use of inadequate source information or poor sources resulting in rejection of the species are *P. taeda* in subtropical areas, *P. oocarpa* in the tropics, and *Eucalyptus* in many places throughout the world. Ignorance tends to build upon itself, and some of the most costly decisions made in the entire field of forestry are due to obtaining a handful of seed of a given species from an unknown source, testing it, and using the results to formulate rigid policies about which species can or cannot be used.

2. **Dishonesty or lack of concern**   It is a terrible thing, but too often seed are not of the quality or from the source from which they were supposed to be, because the supplier, and sometimes the buyer, did not care. Horror stories about buying and relabeling seed to fit an order, "salting" a good source with

poor-quality seed, and of deliberately mislabelling could be told at length. Luckily, this is becoming less common but is still a problem for some species like the tropical pines, especially where there is a seed shortage. The major cause of lack of concern is a result of the attitude that *seed is seed* and that it will not make much difference if the seed used do not exactly fit those requested. It is not unusual to find even mixtures of species in lots supposedly obtained from a given source of a desired species. The problem really is that many foresters and seed dealers simply are ignorant of or do not believe in the importance of differences in source of seed within a species. Whatever the reason, it is essential that *anyone obtaining seed from a special source* should make sure it is as labelled. Many organizations send their own personnel to the collection area to assure themselves of obtaining the desired seed. Greed also causes a part of the problem. Whole indigenous sources have been destroyed by collectors who cut the trees or hacked them to pieces with machetes because someone was offering a good price for the seed. Certification of forest tree seed, as discussed elsewhere, is a partial solution of the problem.

3. **The proper source of a species is not available**   In rapidly expanding planting programs, the best seed sources are often not available; this is one of the major stumbling blocks in the currently aggressive tropical and subtropical forestry operations. Seed is in such short supply that costs have skyrocketed, and some organizations have become so desperate that they will buy and use seed from literally any source available. Sometimes they even use the wrong species. Lack of seed from a suitable source is almost always a major problem when indigenous tropical hardwood species are used. Too often a less desirable source is chosen to fill a specific regeneration quota. Sometimes it is impossible to obtain the desired seed. In this situation, great care must be taken to obtain the next best source, and to avoid planting just any kind of trees to fill the area.

4. **Costs**   It is unbelievable, but in many instances a few cents or dollars difference per kilogram of seed become the decisive factor about whether the best source or an inferior one is used. It is known, for example, that the Piedmont source of *P. taeda* is not suitable in subtropical areas. However, seed of this source are abundant and cheap, and thousands of kilograms of it have been purchased to plant in subtropical areas simply because the Piedmont seed are less costly than the correct coastal plain source. Actually, when prorated against total plantation costs, seed costs are an infinitesimal part of the outlay to establish a plantation. Doubling or tripling seed costs will have essentially no effect on the per-hectare costs of plantation establishment.

It is a *basic fact of exotic forestry* that pest damage usually will occur; it may happen quickly after planting or may be delayed many years before it comes. Problems with pests are worse when offsite plantings are made. All exotic plantings are offsite, and some, such as those in the dry grasslands in the tropics, are extremely so.

There are three categories of pests that attack exotics. These are as follows:

1. The insect or disease is a pest on the exotic *in its natural range,* and somehow follows the exotic to its new planting area. A good example of this type of pest is *Dothistroma* on radiata pine. This disease is little more than a nuisance on radiata in its native range in California, which has cool, dry summers. However, the disease becomes a killer when it attacks radiata pine planted in areas that have warm, wet summers.

2. The pest is indigenous to the area where the exotic is planted. It may take years, but ultimately some pests will adapt to the exotic. A good example of such a pest is the cypress defoliator *Glena bisulca* in Colombia. The insect is found endemically on the indigenous hardwoods and did not seriously attack the *Cupressus lusitanica* for many years. Now, it is potentially a very serious pest. Attacks by indigenous pests seem to be the most common types that damage exotics. One example is leaf-eating beetles that commonly attack *Eucalyptus* in many regions.

3. An exotic pest, which is not from the original area of the exotic, may adapt to the exotic tree species. A reported example of this is *Sirex* on radiata pine in New Zealand and Australia, where the imported *Sirex* has adapted to the imported radiata pine.

Donor and Receptor Areas

When we deal with exotics, it is convenient for purposes of reference to speak of *donor* and *receptor* countries or areas. The donors are those areas such as Central America, Mexico, and western North America which are rich in species that have great value in the receptor countries where the material will be planted (Figure 3.14). Frequently, the donor countries are not themselves major timber-producing regions, but this does not always hold true. Examples are Mexico and western North America. The receptor countries are poor in the kinds of trees desired, or they do not have them at all; examples are Australia and Africa for pines, and South America, Africa, and many other regions for eucalypts and pines.

CHOOSING SPECIES AND PROVENANCES

How does one decide which species or sources should be obtained from a donor country to test in a receptor country? If this question could be easily answered, much labor, time, and large quantities of money could be saved. The theory for success of a transfer is quite simple; match the environment of the donor species or provenance with the one where the plantation will be established. Many helpful methods have been suggested, such as the zoning of northeastern Brazil, by

FIGURE 3.14

Species from donor countries, like those in Central America and Mexico, are widely used in exotic plantings in the tropical regions. Shown is a fine Mexican pine from which seed are being collected to test in South America.

Golfari and Caser (1977) who made a search for similar regions in countries rich in the desired species. But the decision is not that simple because rapid screening of environments is only the first step. However, some general rules can be given for choosing likely species to use in new environments. Suitability for survival, growth, and wood qualities are the key items to look for when choosing suitable material to grow in a receptor country. The ability to reproduce in the new environment is desirable but is not essential in plantation forestry if seeds can be obtained from seed orchards or seed stands that are established elsewhere. For example, *P. caribaea* will grow very well in some lowland areas near the equator, but here the species often does not produce abundant seed, therefore, it must be produced elsewhere.

In forestry, weather data are usually not available either for the donor area or for the area where new planting is to be done. Thus, matching of environments can be most difficult. When information is available, it usually is for yearly averages, although rarely, monthly or weekly averages or even daily figures have been recorded. Gross average conditions can be most misleading when making a decision about the best source to plant. Natural selection *does not operate on averages; it operates on the extremes.* It may be the 1 day in the 1 year of the plantation's life that will result in death or injury from something like cold weather. Alternatively, it may be the one dry period out of 50 years that will kill from drought. The *important aspects* in deciding about the use of *provenances* or species are things that are *unusual, extreme,* or that have *large fluctuations.* Because extremes do not occur yearly, but perhaps only once or twice during the lifetime of a given plantation, meaningful testing of the suitability of species and provenances cannot really result until the trees have grown for a full rotation. Sometimes that length of time is not even enough. For example, a "50-year freeze" may occur only once in several rotations. Even a knowledge of extremes is not sufficient. The key is the *sequence prior to or following the extremes;* for example, if cold weather occurs gradually, so the plants have a chance to harden off, results will be greatly different than if a severe cold spell immediately follows a period of warm weather.

There is always an urgent need for information about which provenance or species is suitable for use, and decisions are commonly made after only a few years of testing. The problem of too short a test period is especially critical for such exotics as *Eucalyptus* in Brazil or *P. oocarpa* in Colombia (Figure 3.15). However,

FIGURE 3.15

Tests must be of a sufficiently long duration to give the trees a chance to express their characteristics. Shown is a stand of slash pine (*P. elliottii*) grown near the equator in South America. Initial growth appeared to be normal, but later all the trees developed severe sweep; the cause was not root binding.

it can also be critical for species within their natural ranges. For example, one of the most important current questions in the southeastern United States, now that seed orchards are in full production and excess seed are available, is how far can the seed from a given orchard be moved safely for use in operational plantings? There is no immediate answer to this problem other than to apply experience with wild seed of the same origin, combined with common sense (Langlet, 1967; Rohmeder, 1959). However, for long-term answers, tests need to be made (Wells and Wakeley, 1966).

There are many rules that have been suggested as a guide to movement of provenances or species, but most fit only specific situations. A few broad generalizations that can be made are as follows:

1. Do not move provenances *from a Mediterranean to a continental climate;* it is somewhat safer to move seed from continental to maritime areas. The seasonal periods of moisture and temperature are so different that rarely will a source developed where there are cool, wet winters and warm (hot), dry summers do well where rainfall is uniform year round, or where the cool seasons are dry and the warm periods are wet (Kiellander, 1960). Failures are common when species are moved between areas with winter rains to those with winter drought and between areas with summer rains to those that have summer drought (Hillis and Brown, 1978). It sometimes is satisfactory to move trees from areas with extreme seasons to those with uniform year-around rainfall, but quite often this is not successful.

2. Do not move trees from areas of *uniform climates with small fluctuations* in rainfall and temperature to those with *severe and large fluctuations* in these factors, even though the annual averages and extremes may be similar. Many plants require preconditioning before they can tolerate extreme environments, and without this preparation period they are susceptible to damage. For example, the climate in the southeastern United States is one of the most difficult for exotic trees because of random wet and dry spells and particularly because temperatures can fluctuate from 80°F (27°C) in 1 day to15°F ($-$8°C) the same night. It is not uncommon to have several days of 80°F temperature in midwinter that are followed immediately by several days when the temperature will drop below 10°F ($-$12°C). Few plants can tolerate such large fluctuations, although with proper preconditioning they can easily withstand the absolute temperature extremes. Cold-hardy species that can normally stand temperatures as low as $-$30°F ($-$34°C), if they have been properly and gradually preconditioned, will often freeze at 10°F ($-$12°C) following a warm spell. Species, such as many of the eucalypts, are particularly sensitive to environmental extremes because they never seem to have a real dormancy. Such species start growth during warm spells, and then are killed or damaged when cold weather follows. Because of the large fluctuations in environment in the southeastern United States, most exotics have proven to be complete failures there (Zobel et al., 1956).

3. *Do not move high-elevation or high-latitude sources to low elevations or low latitudes,* or the reverse. However, high-elevation provenances from low

latitudes can often be moved successfully to lower elevations at higher latitudes and vice versa. High-latitude and high-elevation sources are usually slow growing but of good form, whereas lower-elevation and low-latitude sources have faster-growing trees with heavy limbs and crooked boles. As for all rules, this one does not always hold; for example, the inland and upland provenances of *P. caribaea v. hondurensis* tend to be larger limbed and more crooked than those from the coastal region that are straight and have better form.[3] The major question is *how far* seed can be moved without too much risk. Usually, the interest is in movement from lower elevations to higher elevations or from lower latitudes to higher in order to increase the growth rate. But with such moves, adaptability problems can occur, and a major decision must be made about whether the *gain* from the move is great enough to justify the *risk* from initially poorer adaptability. Once a large and broadly based population is established, land race development can begin.

4. Do not plant trees originating on *basic soils* on *acid soils* or vice versa. This rule also often holds for soil types, such as clay to sand or sand to clay. Some geographic races have a very high adaptability and will grow on a number of sites and soils; an example is *P. caribaea v. hondurensis* and *P. radiata*. But other races or species show very limited adaptability to differing soil environments, such as several of the best species of *Eucalyptus* and *P. oocarpa* (Figure 3.16).

No matter how carefully one may choose or match environments, the final answer regarding suitability of an exotic or provenance can only be obtained through testing. There are many subtle interactions between environment and trees that will determine the success or failure of a plantation. These often cannot be accurately predicted. Sometimes, as for some *Eucalyptus*, the species seems to have a very narrow range of adaptability in its indigenous range but grows well under many diverse environments when used as an exotic, probably because the factors that limit the natural distribution are not present in the new environment. Species like *P. radiata* come from a narrow environmental range but grow well under a wide range of environments. Most species like *P. taeda* or *Pseudotsuga menziesii* grow naturally in many different environments; therefore, the overall species has a wide range of adaptability. However, each provenance has a somewhat more narrow range of adaptability.

WHAT SHOULD BE DONE IF THE PROPER SOURCE IS NOT AVAILABLE OR IS NOT KNOWN?

The preceding question is a widespread and standard problem that must be faced by all operational organizations. Despite the ideal objective of only making small plantings while studies are underway to determine needs regarding the proper

---

[3]Personal communication, Garth Nikles, Queensland Forest Service, Australia.

FIGURE 3.16

Moving exotics without previous testing sometimes has bizarre results. Shown is a test of one source of *P. oocarpa* in Colombia that was totally unsuited to this new exotic environment. Other sources of *P. oocarpa* grow well in this same environment, and they are used operationally. Testing of sources before operational planting will prevent possible disasters.

species and provenances, the general situation, especially in the developing countries, is to go from essentially no operational forestry program to a large-scale one in just a few years. Tests to determine the proper species and sources for current planting are not available and often are not completed until the second generation of planting on a given site. Without test results as a guide, decisions are difficult to make, and some mistakes will be made. However, quick establishment of plantations is of key importance. Therefore, most forestry programs proceed with the minimal information necessary to properly guide the forester. When this situation occurs, the following procedures should be followed:

1. *Match the environment* of the potential exotic species or provenance and the new environment to be planted as closely as possible. Usually, suitable data to do this are not available, so one must make as intelligent an estimate as possible. As emphasized in the section on choosing species or provenances, the extremes and sequences of temperature, moisture, soils, or other environmental factors must be compared, not just the normal or average conditions. Often extremes are not recorded, especially in the tropical regions. Therefore, one may have to rely on the general knowledge of persons living in the area.

2. *Use common sense and experience.* The best experience is to have seen or worked with similar environmental conditions elsewhere. With knowledge of the species involved, one can make a good estimate of what will grow best on

an area being considered for use. Most of the early plantings in the tropical and subtropical areas, and in severe environments elsewhere, have been made in this way. Sometimes mistakes are made! This is inevitable, but it is absolutely amazing how close the opinions of an experienced and alert forester will be with respect to what later tests indicate are best. The *key to success* in making such recommendations is *close observation, knowing the species* being considered for use, and having a *good knowledge of the environment.*

Frequently, a situation occurs when a political decision has been made that a forestry planting program will be immediately initiated on a large scale. When this is done, the forester must do the best possible even when seed of the best species or sources are not available. There is no option but to act, and a planting program will be initiated whether or not the forester feels it is a wise decision. If one can find plantations of the desired species in the area, a land race should be immediately developed, even if the original seed source is not the best. At the same time, a series of species and provenance studies should be initiated as soon as possible (Burley and Nikles, 1972, 1973a, b; Nikles et al., 1978). These will provide the needed information and plant material necessary to develop a base for future selections. The overriding concern is time and the need for immediate action. Many programs have been scrapped because planting was delayed with the excuse that suitable seed were not available. Making operational plantings before proper information or proven plant materials are available will result in inefficiency. This is a penalty that must be absorbed to accommodate the political pressures and time needs. *Any* type of *crash program* is *inefficient,* and this is especially severe in forestry. When faced with establishing a crash program because of needs and political considerations, drastic failures can be avoided when one draws on all the experience and common sense available and applies them when making the initial recommendations.

How Far Can Seed Be Moved?

Tree improvement specialists are constantly being asked about how far it is safe to move seed. There are many answers and many formulas that have been proposed. Rules have been made for certain species, such as "it is safe to move seed 1000 ft (300 m) in elevation of 100 mi (160 km) in latitude." Wiersma (1963) uses the rule that displacement of 1° latitude is equivalent to a 100-m elevation difference; in a recent progress report of the Inland Empire Tree Improvement Program, Rehfeldt (1980) suggests that seed for reforestation of ponderosa pine in southern Idaho may be transferred about 230 m (750 ft) in elevation, 0.7° in latitude, and 1.2° in longitude. In another report, Rehfeldt (1979) states that there is no limit on elevation, latitude, or longitude that seed of white pine (*P. monticola*) can be moved in northern Idaho. In the southeastern United States, Wakeley (1963) summarized seed movement, giving cautions about the distances that are safe.

In reality, no general rule on seed movement can be made because it differs for each species and location. However, some general guidelines appear to be useful. One of these is that a wide seed transfer is safer near the center of a species range than near its edge. Thus, much more care must be taken in seed movement when species boundaries are approached (Wells and Wakeley, 1966). Also, in areas where environmental gradients are steep, such as in boreal Canada, movement of seed must be very restricted, and the well-adapted indigenous species will become more useful.

SUMMARY—STEPS SUGGESTED TO SELECT EXOTICS OR PROVENANCES

1. Make a decision about the objective of the plantings and the products desired. Then, determine the category of trees (for example, pines or hardwoods) that will best fulfill the objective.

2. Obtain all information possible, from the literature and from plantations or from tests that may be available. This informational phase should include visiting areas with similar environments and species to those that will be used for establishing plantations.

3. Survey the area for any plantations of the desired species that may be available. Immediately develop land races from these plantations for use as an immediate source of seed, unless the provenances of the plantation are obviously very unsuitable.

4. Make a systematic investigation through planting trials of potential species and provenances to determine their growth and variation patterns. Obtain seed from the best trees from these plantings to use as a good land race. Obtain improved stock through additional testing and seed orchard establishment to develop a permanent seed supply for operational planting. Choice of species or provenances to test must be made by using common sense, experience, and through matching environmental extremes and sequences.

5. Operationally, use seed from the initial land race or best potential provenance while better material is being developed through a tree improvement program.

LITERATURE CITED

Anderson, K. F. 1966. Economic Evaluation of Results from Provenance Trials. Seminar, For. Seed and Tree Impr., FAO and Danish Board of Tech. Coop. with Devl. Countries, Rome.

Anonymous. 1978. Leucaena: The miracle tree. *Africa* **86**:75.

Barner, H. 1966. "Classification of Seed Sources." Seminar, For. Seed and Tree Impr., FAO and Danish Board of Tech. Coop. with Devel. Countries, Rome.

Brewbaker, J. L., and Hutton, E. M. 1979. Leucaena—versatile tropical tree legume. In *New Agricultural Crops*, pp. 207–259. American Association for the Advancement of Science Press, Boulder, Colo.

Burley, J., and Nikles, D. G. 1972. *Selection and Breeding to Improve Some Tropical Conifers*, Vol. I. Commonwealth Forestry Institute, Oxford, England, and Department of Forestry, Queensland, Australia.

Burley, J., and Nikles, D. G. 1973a. "Tropical Provenance and Progeny Research and International Cooperation." Proc., Joint Workshop IUFRO in Nairobi, Kenya, Commonwealth Forestry Institute, Oxford, England.

Burley, J., and Nikles, D. G. 1973b. *Selection and Breeding to Improve Some Tropical Conifers*, Vol. II. Commonwealth Forestry Institute, Oxford, England, and Department of Forestry, Queensland, Australia.

Caballero, M. 1966. "Comparative Study of Two Species of Mexican Pine (*Pinus pseudostrobus* and *P. montezumae*) Based on Seed and Seedling Characteristics." M.S. thesis, North Carolina State University, Raleigh.

Callaham, R. Z., and Liddicoet, A. R. 1961. Altitudinal variation at 20 years in ponderosa and Jeffrey pines. *Jour. For.* **59**(11):814–820.

Callaham, R. Z. 1964. Provenance research: Investigation of genetic diversity associated with geography. *Unasylva* **18**(2–3):73–74, 40–50.

Carter, C. K., Canavera, D., and Caron, P. 1981. "Early Growth of Exotic Larches at Three Locations in Maine." Coop. For. Res. Unit, Res. Note No. 8, University of Maine, Orono.

Dorman, K. W. 1975. *The Genetics and Breeding of Southern Pines*. USDA Agriculture Handbook No. 471, pp. 173–175.

Edwards, M. W. 1963. "The Use of Exotic Trees in Increasing Production with Particular Reference to Northwestern Europe." World Cons. on For. Gen. and Tree Impr., Stockholm, Sweden.

Endler, J. A. 1977. *Geographic Variation, Speciation and Clines*. Princeton University Press, Princeton, N.J.

Falkenhagen, E. R. 1979. "Provenance Variation in Growth, Timber and Pulp Properties of *Pinus caribaea* in South Africa. Bull. 39, South African For. Res. Ins., Dept. of For., Pretoria.

Farmer, A. E., and Barnett, P.E. 1972. Altitudinal variation in seed characteristics of black cherry in the southern Appalachians. *For. Sci.* **18**(2):169–175.

Golfari, L., and Caser, R. L. 1977. "Zoneamento ecológico da região florestal [Ecological Zoning of the Northeastern Region for Experimental Forestry]." PRODEPEF, Technical Bulletin no. 10.

Grant, V. 1971. *Plant Speciation*. Columbia University Press, New York.

Gregor, J. W. 1944. The ecotype. *Biol. Rev.* **19**:20–30.

Hagner, S. 1979. Optimum productivity—a silviculturist's view. In *Forest Plantations—The Shape of the Future*, pp. 49–68. Weyerhaeuser Science Symposium, Tacoma, Wash.

Hillis, W. E., and Brown, A. G. 1978. *Eucalypts for Wood Production*. Griffin Press Ltd., Adelaide, Australia.

Holzer, K. 1965. "Standardization of Methods for Provenance Research and Testing." IUFRO Kongress, München, Germany, Vol. III (22), pp. 672–718.

Hunt, R., and Zobel, B. 1978. Frost-hardy *Eucalyptus* grow well in the southeast. *South. Jour. Appl. For.* **2**(1):6–10.

Huxley, J. S. 1938. Clines: an auxiliary taxonomic principle. *Nature* **143**:219.

Jones, N., and Burley, J. 1973. Seed certification, provenance nomenclature and genetic history in forestry. *Sil. Gen.* **22**:53–92.

Kellison, R. C. 1967. "A Geographic Variation Study of Yellow Poplar (*Liriodendron tulipifera*) within North Carolina." Technical Report no. 33, North Carolina State University, School of Forest Resources.

Kellison, R. C. 1981. "Characteristics Affecting Quality of Timber from Plantations, Their Determination and Scope for Modification." World Forestry Congress, Kyoto, Japan.

Kiellander, C. L. 1960. Swedish spruce and continental spruce [Svensk gran och Kontinentgran]. Föreningen Skogsträdsförädling, F.S. Information No. 3.

Krygier, J. T. 1958. "Survival and Growth of Thirteen Tree Species in Coastal Oregon." Res. Paper No. 26, Pacific Northwest Forestry and Range Experimental Station, Portland, Ore.

Lacaze, J. F. 1977. "Advances in Species and Provenance Selection." 3rd World Cons. on For. Tree Breeding, Canberra, Australia.

Lacaze, J. F. 1978. Advances in species and provenance selection. *Unasylva* **30**(119–120):17–20.

Ladrach, W. E. 1980. "Variability in the Growth of *P. taeda in* Colombia Due to Provenance." Investigation Forestal, Carton de Colombia.

Langlet, O. 1936. Studien uber die physiologische Variabilität der Kiefer und deren Zusammenhang mit dem Klima, Beiträge zur Kenntnis Ökotypen von *Pinus silvestris*. *Medd. Statens Skogsförsöksanstalt.* **29**:219–470.

Langlet, O. 1938. Proveniensförsök med olika trädslag [provenance tests with various wood species]. *Särtryck Svenska Skogsvårdsföreningens Tidskrift.* **I–II**:255–278.

Langlet, O. 1959a. A cline or not a cline—a question of Scots pine. *Sil. Gen.* **8**(1):13–22.

Langlet, O. 1959b. Polsk gran för Sverige [Polish spruce for Sweden]. *Särtryck Skogen* **5**:1–4.

Langlet, O. 1967. "Regional Intra-specific Variousness." IUFRO Congress, München, Germany, Vol. III(22), pp. 435–458.

Larsen, C. S. 1954. "Provenance Testing and Forest Tree Breeding." Proc. 11th Cong. IUFRO, Rome, pp. 467–473.

Long, E. M. 1980. Texas and Louisiana loblolly pine study confirms importance of local seed sources. *South. Jour. Appl. For.* **4**(3):127–131.

Marsh, E. K. 1969. "Selecting Adapted Races of Introduced Species." 2nd World Cons. For. Tree Breed., Washington, D.C., pp. 1249–1261.

Mettler, L. E., and Gregg, T. G. 1969. *Population Genetics and Evolution.* Prentice-Hall, Englewood Cliffs, N.J.

Muller, H. J. 1959. The prospects of genetic change. *Am. Sci.* **47**(4):551–561.

Namkoong, G. 1969. "Nonoptimality of Local Races." 10th South. Conf. on For. Tree Impr., Houston, Tex., pp. 149–153.

Nikles, D. G., and Burley, G. 1977. "International Cooperation in Breeding Tropical Pines." 3rd World Cons. For. Tree Breed., Canberra, Australia.

Nikles, D. G., Burley, J., and Barnes, R. D. 1978. "Progress and Problems of Genetic Improvement in Tropical Forest Trees." Proc. Joint Workshop in Brisbane, Australia, Vol. 1., Commonwealth Forestry Institute, Oxford University, England.

Owino, F. 1977. "Selection of Species and Provenances for Afforestation in East Africa." 3rd World Cons. on For. Tree Breed., Canberra, Australia.

Pellati, E. De Vecchi. 1969. "Evolution and Importance of Land Races in Breeding." 2nd World Cons. For. Tree Breed, Washington, D.C., pp. 1263–1278.

Persson, A. 1980. "*Pinus contorta* as an Exotic Species." Swedish Un. of Ag. Sciences, Dept. For. Gen. Research Notes, Garpenberg, Sweden.

Rehfeldt, G. E. 1979. Ecotypic differentiation in populations of *Pinus monticola* in north Idaho—Myth or reality? *Am. Natur.* **114**:627–636.

Rehfeldt, G. E. 1980. "Genetic Gains from Tree Improvement of Ponderosa Pine in Southern Idaho. U.S. Forest Service Research Paper Int-263, Ogden, Utah.

Rohmeder, E. 1959. Beispiele für die Überlegenheit fremder Provenienzen über die heimische Standorstrasse bei den Baumarten *Pinus silvestris* und *Picea abies* (examples of the superiority of foreign provenances over the native races in the species *Pinus silvestris* and *Picea abies*). *Sonderdruck Allgemeine Forstzeitschr.* **43**:1–5.

Sanchez, P. A., and Buol, S. W. 1975. Soils of the tropics and the world food crisis. *Science* **188**:598–603.

Shimuzu, J. Y., and Higa, A. R. 1981. Racial variation of *Pinus taeda* in Southern Brazil up to 6 years of age. *Bol. Pesquisa Flor.* **2**:1–26.

Squillace, A. E., and Silen, R. R. 1962. Racial variation in ponderosa pine. *For. Sci. Mon.* **2**.

Squillace, A. N. 1966. Geographic variation in slash pine. *For. Sci. Mon.* **10**.

Stebbins, G. L. 1950. *Variation and Evolution in Plants*. Columbia University Press, New York.

Talbert, J., White, G., and Webb, C. 1980. Analysis of a Virginia pine seed source trial in the interior South. *South Jour. Appl. For.* **4**(3):153–156.

Tozawa, M. 1924. Necessity of provenance test and the urgent need of a test plantation network. *Jour. Korean For. Assoc.* **22**:1–5.

Turresson, G. 1922. The species and variety as ecological units. *Hereditas* **3**:100–113.

van Buijtenen, J. P., and Stern, K. 1967. "Marginal Populations and Provenance Research." IUFRO Kongress, München, Germany, Vol. III(22), pp. 319–331.

Vavilov, N. I. 1926. Studies on the origin of cultivated plants. *Appl. Bot. Plant Breed. (Leningrad)* **16**(2):1–248.

Wakeley, P. C. 1954. The relation of geographic race to forest tree improvement. *Jour. For.* **52**(9):653.

Wakeley, P. C. 1963. "How Far Can Seed be Moved?" Proc. 7th South. Conf. on For. Tree Impr., Gulfport, Miss., Pub. No. 23, pp. 38–43.

Wakeley, P. C., and Bercaw, T. E. 1965. Loblolly pine provenance test at age 35. *Jour. For.* **63**(3):168–174.

Wells, O. O., and Wakeley, P. C. 1966. Geographic variation in survival, growth and fusiform-rust infection of planted loblolly pine. *For. Sci. Mono.* **11**.

Wells, O. O., and Wakeley, P. C. 1970. Variation in longleaf pine from several geographic sources. *For. Sci.* **16**(1):28–42.

Wells, O. O. 1971. "Provenance Research and Fusiform Rust in the Southern United States," 4th N. Amer. For. Biol. Workshop, Syracuse, N.Y., pp. 23–28.

Whyte, A. G. D., Adams, P., and McEwen, S. E. 1980. "Size and Stem Characteristics of Foxtails Compared with *P. caribaea* v. *hondurensis* of Normal Habit." IUFRO, Division 5 Conference, Oxford, England, p. 59.

Wiersma, J. H. 1963. A new method of dealing with results of provenance tests. *Sil. Gen.* **12**(6):200–205.

Woods, F. W., Vincent, L. W., Moschler, W. W., and Core, H. A. 1979. Height, diameter and specific gravity of "foxtail" trees of *Pinus caribaea*. *For. Prod. Jour.* **29**(5):43–44.

Wright, J. W., and Baldwin, H. I. 1957. The 1938 International Union Scotch pine provenance test in New Hampshire. *Sil. Gen.* **6**(1):2–14.

Zobel, B. J., Campbell, T. E., Cech, F. C., and Goddard, R. E. 1956. "Survival and Growth of Native and Exotic Pines, Including Hybrid Pines, in Western Louisiana and East Texas." Research Note 17, Texas Forest Service.

Zobel, B. J. 1961. "Pines in the Tropics and Sub-tropics." Proc. IUFRO, 13th Congress, Vienna, Austria, Vol. 1(22/10), pp. 1–9.

Zobel, B. J. 1964. Pines of southeastern U.S., Bahamas and Mexico and their use in Brazil. *Silvi. São Paulo* **3**(3):303–310.

Zobel, B.J. 1979. Florestas baseadas em exoticas (forestry based on exotics). *Bol. Tec.* **2**(3):22–30.

Zobel, B. J. 1981. Wood quality from fast grown plantations. *Tappi* **64**(1):71–74.

CHAPTER 4

Quantitative Aspects of Forest Tree Improvement

The first three chapters of this book have been largely concerned with the variation that occurs in forest tree populations and how it can be used. It has been shown that variation almost always has both a genetic and an environmental component, and that genetic tests are necessary to separate genetic and environmental influences. A major task of the forest geneticist is to obtain estimates of the genetic and environmental components and to determine how each can best be manipulated. Such estimates are crucial to tree improvement programs, and to a large degree, they will determine the efficiency of a selection and breeding program. The successful forest tree improver must have a working knowledge of various aspects of forest management and silviculture. Added to this, he or she must also have an understanding of basic genetic principles and how they are applied in tree improvement. This chapter will cover some general details of the mechanisms of inheritance in forest trees, concepts that are crucial to the development of a selection and breeding program.

GENETICAL AND STATISTICAL CONSIDERATIONS

It was emphasized in Chapter 2 that genetic variation that occurs in living organisms is inherited in a way that is common to all species. Much of the basic genetic research that elucidated the mechanisms of heredity was done with such organisms as garden peas, fruit flies, mice, and corn. Most of the early studies involved genes with major effects; that is, the expression of the trait was controlled by one or two gene loci that had a profound effect on the phenotype. Phenotypes could be classified into distinct categories, such as tall or short or brown or white, and there was rarely any overlap between them. These are defined as *qualitative* traits. The classic experiments conducted by Gregor Mendel on the garden pea (*Pisum sativum*) dealt with traits inherited in a qualitative manner, and they were the basis for the science of genetics (Strickberger, 1976). Principles derived from Mendel's experiments are included in nearly all introductory biology classes. There are many other examples of traits involving major gene effects in addition to peas. Some of the most familiar of those are eye color (i.e., blue eyed vs. brown eyed) and the A, B, and O blood groups in humans.

Very few economic traits in forest trees are inherited in a pattern that can be attributed to the effects of major genes. Most major gene effects in trees are evident when dealing with selfing, which increases levels of homozygosity for rare recessive alleles. For example, selfing studies with loblolly pine and several other conifers revealed a number of recessive alleles that produced seedlings with several unusual phenotypes (Franklin, 1968, 1970). Occasionally, an economically important trait of forest trees is influenced by genes with major effects, especially for pest resistance. One example is the inheritance of resistance of sugar pine (*P. lambertiana*) to blister rust (*Cronartium ribicola*) (Kinloch and Byler, 1981).

Almost all important traits in forest trees are influenced by several or many gene loci, each of which has a relatively small effect on the phenotype. This results in a large array of genotypes for traits influenced by many genes if there is genetic variation at the influential gene loci. When environmental effects are added to this array, a continuum of phenotypes results. An important aspect of this type of inheritance is that individuals cannot generally be placed into distinct groups. Characteristics of this sort that vary continuously are said to be *quantitative* or metric traits. These are best dealt with through *measurements* on numbers of progenies of different parents. It is important to recognize that many traits that are measured as "all-or-none" characteristics, such as presence or absence of disease symptoms, are actually influenced by multiple gene loci and are quantitative characteristics; therefore, they must be treated as such.

During the past half century, a special branch of genetics called *quantitative genetics* has been developed to deal with characteristics that are inherited quantitatively and show continuous variation. Concepts developed by quantitative geneticists form much of the basis for many plant and animal breeding programs in existence today, including forest tree improvement programs. They differ from concepts developed for qualitative traits in that they *involve large numbers of progenies* (simple ratios cannot be observed) and measurements. It is not possible to fully cover the field of quantitative genetics in one or two chapters. Indeed, that is not the purpose of this book. However, to understand tree improvement it is necessary to acquire an understanding of its concepts. In the following sections the more basic and pertinent aspects of quantitative genetics that are of primary importance to tree improvement programs will be discussed. A more complete coverage of the principles of quantitative genetics can be found in texts, such as those by Kempthorne (1957), Falconer (1960), and Becker (1975). Thorough coverage of the relationship of quantitative genetics to forest trees can be found in Namkoong's monograph (1979).

Statistical Concepts

The study and manipulation of quantitative characteristics is essentially one involving the inheritance of measurements that are analyzed using statistical techniques. For example, if tree height is the trait of interest, the tree improver begins assessment by measuring heights on large numbers of trees, either in natural stands, plantations from wild seed, or pedigreed genetic tests. Once these height–measurement data are accumulated, they are analyzed through the use of statistics.

The tree improvement specialist must have a working knowledge of statistical methods to make decisions about tree populations and how they will respond to selection and breeding. Both graduate and undergraduate training programs in forest genetics involve statistics. Although some individuals are reluctant to become involved in statistical analyses, these are not really too difficult. Also, excellent textbooks are available that provide both an introductory and advanced treatment of statistical methods (e.g., Cochran and Cox, 1957; Steel and Torrie, 1960; Snedecor and Cochran, 1967; Sokal and Rohlf, 1969; Neter and Wasserman, 1974). The statistical concepts presented in this chapter are as simplified as possible, and will already be familiar to some readers. They form the foundation for much quantitative genetics theory, and are essential for a successful selection and breeding program.

The term *population* has been used numerous times in previous chapters. In the statistical sense, a population refers to the entire group of individuals, items, or scores from which a sample is drawn. This supplements the biological definition of a population given in Chapter 1 as *a community of interbreeding individuals*. A population may be described in several ways; terms used to describe it are called *parameters*. Usually the population being studied is much too large to allow measurement of every individual in it. For example, there might be interest in the population consisting of all loblolly pine growing on the Coastal Plain of North Carolina. If specific gravity were the trait of interest, it would not be possible to measure this characteristic on every tree in the population. Therefore, to obtain an estimate of the population parameter specific gravity, samples must be obtained from the population from which analyses and inferences about the population are made. Descriptions of a population based upon samples obtained from it are estimates of the population's parameters. As successive samples are taken, they will usually be somewhat different due to sampling chance. A large number of statistical methods have been developed that will guide the researcher or practitioner in deciding if the sample obtained provides a sufficiently precise estimate of the population parameter of interest.

The most common and useful parameter used to describe a population is the *population mean,* or the average of the individuals that make up the population. Symbolically, the population mean is expressed as

$$\bar{X} = \frac{\Sigma X_i}{n}$$

where $\bar{X}$ = mean, Σ = sum of, X_i = individual observations, and n = number of observations. Simply stated, the population mean is the sum of the individual observations divided by the number of observations. A mean can be computed for any characteristic for which measurements are taken or scores are given.

Although the mean is a most useful and widely used statistic, it indicates nothing about the *distribution* of the individuals within the population. In other words, one can tell nothing about the *variation* that exists in the population just by computing the mean. The amount and pattern of variation is of vital importance in the analysis and use of information from a population. Variation patterns or distributions often are easily visualized when pictured graphically by plotting the frequency in which a measurement occurs on the vertical or Y axis against the range in values in the horizontal or X axis. This is illustrated in Figure 4.1, which shows the distribution of sampled trees having different specific gravities within a population of loblolly pine. Numerous types of distributions can occur biologically in populations, but the one most often encountered is the "normal" distribution (see Figure 4.2).

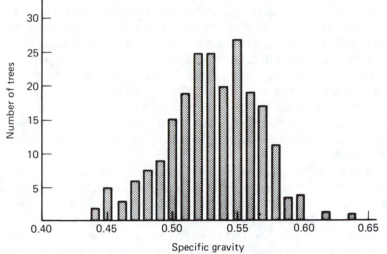

FIGURE 4.1

Variation in wood specific gravity in a loblolly pine population is illustrated. Trees sampled had the same age and were growing on land of the same general site class. The *X*, or horizontal, axis denotes specific gravity values, whereas the *Y*, or vertical, axis gives the number of trees that have a given specific gravity value. The distribution of wood specific gravity values approximates a "normal" distribution (see text).

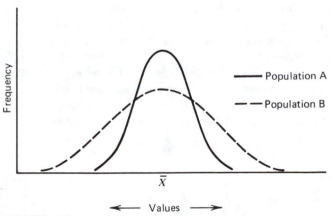

FIGURE 4.2

An example of two populations with different variances is shown. Although the mean is the same in both instances, individuals in Population B are more widely distributed than those in Population A. Therefore, Population B has a greater variance.

In a normal distribution, the measurement or score most frequently observed is an intermediate value that is equal to the population mean when the distribution is exactly "normal." Measurements or scores that differ from the mean occur with decreasing frequency the farther one proceeds from the mean. Most quantitative genetics theory assumes a normal distribution of measurements from a population. Although exactly normal distributions often do not occur, the observations usually approximate a normal distribution closely enough so that the assumption of a normal distribution is valid for analytical purposes. When other distributions occur, methods are often available to transform the measurements so that they more nearly resemble a normal distribution; then standard analyses can be used (Snedecor and Cochran, 1967).

Normal distributions are found for many characteristics in trees, especially for height and other growth factors. Sometimes, actual measurements cannot be made, but subjective scores are used to describe tree phenotypes. As an example, tree straightness may be judged on a 1-to-5 scale. A tree of average straightness would be given a score of 3, with scores of 1 or 5 given only to the straightest or most crooked individuals. Such scores are often treated as if they resemble a normal distribution.

The parameter most often used to describe the spread of individuals within a population is the *variance,* which is computed in the following manner:

$$\sigma^2 = \frac{\Sigma(X_i - \bar{X})^2}{n - 1}$$

where σ^2 = variance, X = mean, Σ = sum of, X_1 = individual observations, and n = number of observations. The *variance* is the sum of squares of the deviations of individuals from the mean divided by one less than the total number of observations. The term $(n - 1)$ in the denominator defines what is commonly referred to as the number of *degrees of freedom* for the variance estimate. A large variance occurs when individual values are widely dispersed, whereas a smaller variance results when the distribution around the mean is narrow (Figure 4.2).

Another useful population parameter is the *standard deviation,* which is simply the square root of the variance. Computationally, the standard deviation can be expressed as

$$\sigma = \sqrt{\sigma^2} = \sqrt{\frac{\Sigma(X_i - \bar{X})^2}{n - 1}}$$

The standard deviation is expressed in the same units of measurement as the mean of the population and is a very useful tool for describing the dispersion of individual values. When measurements are distributed in a normal fashion, approximately 67% of the observations will fall within one standard deviation on either side of the mean, and 95% of the observations will be within two standard

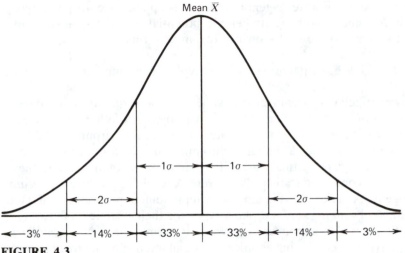

FIGURE 4.3

Illustrated is a summary of population parameters—the normal distribution, the population mean, a standard deviation, and the proportion of values within the indicated number of standard deviations from the mean.

deviations of the mean. A diagramatic illustration of a normal curve showing mean and standard deviations is given in Figure 4.3.

Populations are often described by their mean and standard deviation, or, symbolically as follows:

$$\bar{X} \pm \sigma$$

For example, if the heights of trees in a genetic test were described by the terms 14 ± 2 m, it would indicate that average tree height in the plantation was 14 m and that the standard deviation was 2 m (the variance would be 4 m). If heights in the test were normally distributed, about 67% of the trees would be between $+1$ and -1 standard deviations of the mean or between 12 and 16 m tall.

Although the mean and the variance, or standard deviations, are the statistical parameters most often used to describe a population, they do not describe modes of inheritance or the proportion of the variation that is genetic in origin. The observed or estimated variance must therefore be separated into its genetic and environmental components. Computation of these components involves partitioning the phenotypic values into genetic and environmental sources of variation.

Genetic Values

The selection phase of an applied tree improvement program has as its objective the choice of the best genotypes to use as parents for the production of improved planting stock and as a base for future breeding activities. We know that nothing certain can be said about a tree's genotype from its appearance, or phenotype, because it is influenced both by its genetic potential and the quality of the environment in which the tree is growing. In simple terms,

$$P = G + E, \text{ or phenotype} = \text{genotype} + \text{environment}.$$

The best way to tell if a parent tree is of superior genetic quality is to compare the performance of its offspring against the offspring of the other parent trees. These progeny trials are designed to separate genetic from environmental differences by giving all progenies a similar environment in which to grow. Thus, if Parent A has taller offspring than Parent B in similar environments, and these differences can be confirmed statistically, Parent A is said to produce *genetically superior* progeny. A major effort in most tree improvement programs is directed toward testing and ranking selected phenotypes for their genetic qualities.

The genetic values of parents are expressed in terms of *combining abilities*. There are two types of combining abilities of special interest to the tree breeder, and these will be explained by use of the following example.

Assume that eight trees have been selected and that it is desired to determine their genetic worth. Four of the trees are chosen as male parents, and the other four will serve as female parents. Each male is mated to each female, and the offspring are established in a progeny test. After several years in the field, the progenies are assessed. The average performance of each cross is measured in units of volume, as is presented in the following two-way table. Also indicated is the average performance of the progeny of each parent, and the overall test mean.

Female parents	Male parents				Progeny means
	1	2	3	4	
5	9	17	12	14	13
6	10	16	12	10	12
7	11	20	10	15	14
8	14	15	6	17	12
Progeny means	11	17	10	14	Test mean 13

Note that the average performance of the progeny of the specific cross 5 × 1 is 9 volume units, whereas the average performance of the offspring of all crosses with parent 5 is 13 units. The test mean, or the average performance of all trees growing in the test, is 13 units.

General combining ability (GCA) is defined as the average performance of the progeny of an individual when it is mated to a number of other individuals in the population (Falconer, 1960). Although general combining abilities may be expressed in absolute units, it is usually more convenient and meaningful to express them as deviations from the overall mean. Thus, a parent with GCA of 0 has an average general combining ability. A positive GCA indicates a parent that produces above-average progeny, whereas a parent with a negative GCA produces progeny that perform below average for the population.

The preceding table can be used to calculate general combining abilities for each of the parents. For example, the GCA of male Parent 2 would be calculated as

$$GCA_2 = \text{mean of parent 2} - \text{test mean}$$
$$= 17 - 13 = +4$$

Parent 2 therefore has a general ability (GCA_2) for volume of $+4$ units. Other general combining abilities would be calculated in the same way. For example, general combining ability for Parent 4 is $+1$ volume units, whereas GCA of Parent 3 is -3 units.

The *breeding value* of an individual is defined as *twice* its general combining ability. The difference between breeding value and general combining ability is largely conceptual in nature. Breeding values are meaningful because the parent in question contributes only half of the genes to his or her offspring, the other half coming from other members of the population. The breeding value of Parent 2 would be calculated as

$$2\ (GCA_2) = 2 \times (4) = 8$$

Specific combining ability (SCA) is a term that refers to the average performance of the progeny of a *cross* between two specific parents that are different from what would be expected on the basis of their general combining abilities alone. It can be either negative or positive. Specific combining ability *always* refers to a specific cross, and *never* to a particular parent by itself.

Specific combining ability for the cross between Parents 3 and 6 (a cross value of 12) would be calculated as follows:

1. Calculate general combining abilities for both parents.
$$GCA_3 = -3; GCA_6 = -1$$

2. The general combining abilities are added to the population mean giving the anticipated value of the cross 3×6 based upon general combining abilities.
$$\text{Anticipated value} = \text{test mean} + GCA_3 + GCA_6$$
$$= 13 + (-3) + (-1) = 9$$

3. Subtract the value calculated in (2) from the observed value of the cross. The result is the specific combining ability.
$$SCA_{6 \times 3} = \text{observed value} - \text{anticipated value}$$
$$= 12 - 9 = +3$$

This means that cross 6×3 is performing 3 volume units better than would be expected based on the GCA's of Parents 3 and 6.

It must be stressed that *nothing* can be said about the utility of a cross based solely on its specific combining ability. Because SCA is a deviation from what is expected based on general combining abilities, a cross may have a positive SCA but still not be a good performer relative to other crosses, as in the example given with Parents 6 and 3. The cross has an SCA = $+3$, but its average performance of 12 is still below the population mean of 13 because of the poor general combining abilities of the two parents involved. It should be stressed here that GCA, breeding value, and SCA are trait-specific descriptions of a parent's or cross's genetic value. For example, a parent could have an above-average GCA for volume and at the same time, a below-average GCA for wood specific gravity.

The two types of combining abilities are a reflection of different types of interactions between alleles at the gene loci. General combining ability represents an average performance of the progeny of a parent when it is crossed to many other parents. It is therefore a reflection of the parent's *additive genetic value*; that is, it reflects that portion of its genotype for a specific trait that the parent may transmit to its progeny, regardless of which other parent is involved in the cross. It represents the additive type of gene action discussed in Chapter 2. Parents who are known to have a high GCA for a trait are said to be *good general combiners*, but they are not always desirable. One can have a high GCA for disease susceptibility, for crooked tree boles, or other undesired characteristics.

Because it represents an additive effect that can be predicted, general combining ability is sometimes thought of as the "dependable" portion of a tree's genetic constitution. It is the type of combining ability that is utilized in forest tree seed orchards that are composed of many parents. The improved performance of the planting stock derived from the orchard is due to the accumulation of favorable alleles that have an additive genetic effect on the phenotypes of the trees derived from the orchard.

The specific combining ability shown by a given cross is a reflection of the interaction of the two alleles at the gene loci influencing that trait (dominance gene action) and interactions between alleles at different gene loci influencing the trait, or epistatic gene action. As discussed in Chapter 2, these two types of genetic effects are usually referred to as nonadditive genetic effects. SCA can usully be attributed in large part to the dominance type of gene action. Since specific combining ability occurs because of interactions between specific alleles, or between gene loci, its value cannot be predicted from the phenotypes of the parents before the cross is made. It cannot be utilized in a seed orchard program involving many parents because open pollination results in many different combinations of alleles across gene loci.

There are two major ways to make use of specific combining ability in a tree improvement program. One is to use vegetative propagation to produce commercial quantities of planting stock that are genetically identical to the tree from which they were taken. In vegetative propagation, the genetic makeup of the parent is passed along intact, and the specific combinations of alleles at all gene loci are preserved. The second way to utilize specific combining abilty is to make crosses to mass-produce seed from specified parental combinations; this can be done by control pollinations or by methods such as two-clone seed orchards. Both of these methods, especially vegetative propagation (Chapter 10) have been used with some species to produce improved planting stock. For most species, though, costs and technological difficulties associated with use of SCA have made general combining ability the major focus of operational tree improvement programs.

Types of Genetic Variation

Variation in tree populations can be partitioned into genetic and environmental components. The simple model described previously for individual tree values can be extended to apply to variation encountered in a *population* of individuals. If an individual phenotype is described as

$$P = G + E$$

then variation can be stated as

phenotypic variation = genetic variation + environmental variation

or

$$\sigma_P^2 = \sigma_G^2 + \sigma_E^2$$

Genetic values (σ_G^2) are influenced by both additive and nonadditive effects. Genetic variation can therefore be partitioned into additive and nonadditive components. Symbolically,

$$\sigma_G^2 = \sigma_A^2 + \sigma_{NA}^2$$

The model of phenotypic variation can therefore be extended to read

$$\sigma_P^2 = \sigma_A^2 + \sigma_{NA}^2 + \sigma_E^2$$

The additive genetic variance (σ_A^2) arises from differences among parents in general combining ability and is simply the variance of breeding values (breeding value = $2 \cdot$ GCA) in the population. Nonadditive variance (σ_{NA}^2) is the result of specific combining ability effects. The variance of specific combining abilities in a noninbred population can be shown to be equal to $\frac{1}{4}\sigma_{NA}^2$.

Most tree improvement programs are aimed at selecting parents with high general combining abilities or high breeding values. In these instances, the additive variance is the "type" of genetic variation that is utilized to produce improved propagules. Successful use of nonadditive variance depends upon vegetative propagation or using specific crosses.

Heritability

The concept of heritability is one of the most important and most used in quantitative genetics. *Heritability* values express the proportion of variation in the population that is attributable to genetic differences among individuals. It is therefore a ratio indicating the degree to which parents pass their characteristics along to their offspring. Heritability is of key importance in estimating gains that can be obtained from selection programs. The discussion here will focus on individual-tree heritability. Another type of heritability estimate, the heritability of family means, will be discussed in the chapter on genetic testing (Chapter 8).

Two types of individual-tree heritabilities are important in applied tree improvement. *Broad-sense heritability* (H^2) is defined as the ratio of total genetic variation in a population to the phenotypic variation, or

$$H^2 = \frac{\sigma_G^2}{\sigma_P^2} = \frac{\sigma_A^2 + \sigma_{NA}^2}{\sigma_A^2 + \sigma_{NA}^2 + \sigma_E^2}$$

Broad-sense heritability can range from 0 to 1. A lower limit of 0 would occur if *none* of the variation in a population was attributable to genetics. If *all* variation was due to genetics, then broad-sense heritability would be equal to 1. Broad-sense heritability has a limited application in tree improvement and is of primary use when both the additive and nonadditive variation can be transferred from parent to offspring, such as when vegetative propagation is used.

Narrow-sense heritability is the ratio of additive genetic variance to total variance. Symbolically,

$$h^2 = \frac{\sigma_A^2}{\sigma_P^2} = \frac{\sigma_A^2}{\sigma_A^2 + \sigma_{NA}^2 + \sigma_E^2}$$

The lower limit for narrow-sense heritability is also 0 (no additive variance), and the upper limit is 1 (no environmental or nonadditive variance). Narrow-sense heritability is *never* greater than broad-sense heritability; if all the genetic variance is of the additive type, narrow- and broad-sense heritabilities are equal. Most heritability estimates given in the forest genetics literature are for narrow-sense heritability, because most tree improvement programs today are aimed at improving general combining ability and thus utilize only the additive portion of the genetic variance. This will undoubtedly change as vegetative propagation methods and economical methods of producing specific crosses, such as supplemental mass pollination, become available, but as of today, narrow-sense heritabilities are of the most use to tree breeders.

An important but often overlooked aspect of heritability estimates is that they apply only to a particular population growing in a particular environment at a particular point in time. For example, estimates of heritability for a group of trees grown in a greenhouse would not be appropriate for the same trees growing in a field environment. Height in the greenhouse may not be influenced by exactly the same genes as height in the field. Even if the two traits were the same, though, estimates of h^2 obtained in the greenhouse will usually be higher than those from the field, because there is less environmental variation in the greenhouse. As can be seen in the formula for h^2, changes in the environmental variance component (σ_E^2) in the denominator will have a direct effect on the h^2 ratio. Because of the influence of the environment on the heritability ratio, the h^2 estimates for a given characteristic in a species in one geographic area probably will not be the same as those found in another region. The heritability values of a given characteristic in a population often change with age when the environment changes and when the genetic control of the characteristic changes as the trees mature. The degree of change with age has been debated, but there is now an accumulating body of evidence that suggests that heritabilities do change markedly, and perhaps in a predictable fashion, as test plantations grow and develop (Namkoong et al., 1972; Namkoong and Conkle, 1976; Franklin, 1979).

The most widely used technique in forest genetics to estimate heritability is to grow progeny from a group of parents or crosses together in the same genetic test plantation. Heritability estimates are then derived from the relative performance of the progenies within and between parent trees. Breeding schemes and experimental designs that can be used to estimate the necessary variances and heritabilities are discussed in Chapter 8. Another method of estimating heritabilities is through parent–offspring regression techniques (Falconer, 1960).

Narrow-sense heritability estimates for height, wood specific gravity, and form traits for a number of species are shown in Table 4.1. There is obviously much variation in the degree to which traits are under additive genetic control. Some characteristics, like wood specific gravity, appear to be strongly controlled genetically regardless of the species and are uniform over somewhat different environments. Other traits, like height growth, are under a lesser degree of genetic control and are strongly influenced by the environment in which the trees are grown.

Even in large experiments with many families, heritabilities are not estimated without error. All h^2 estimates should be thought of as being figures that give a general idea of the strength of inheritance. For example, a heritability of 0.15 should not be thought of as being much different than a heritability of 0.20. The imprecision of many heritability estimates can be seen in Table 4.1, where the heritability of black walnut height at age 8 was listed as $h^2 = 1.25$. Earlier discussion indicated that heritability cannot be greater than one; hence the $h^2 = 1.25$ value is an overestimate. Absolute estimates of heritability are needed for many purposes, such as in gain estimation, where h^2 is an integral part of the gain formula. However, the main value of the heritability concept to the tree breeder is to indicate the general strength of genetic control and the best approach for use in tree improvement programs. Despite its usefulness, one should always keep in mind that heritability is not an invariant value fixed to a population. Heritability itself is a variable ratio and is subject to changes. Therefore, whenever heritability is used in determining the amount of genetic gain or breeding strategies, values should be qualified with a statement that the confidence level is less than 100%

The basic and key point about heritability is that it is a ratio between genetic and phenotypic variances; thus, it is not a fixed value for a given characteristic of a given species. Estimates of heritability are not estimated without error; therefore the ratios obtained are only a relative indication of genetic control and should not be interpreted as absolute or invariant values.

TABLE 4.1
Individual Tree Narrow-Sense Heritability Estimates in Forest Trees

Trait	Heritability	Reference
Height		
Douglas fir	0.10–0.30	Campbell (1972)
Loblolly pine	.44	Matziras and Zobel (1973)
Loblolly pine	0.14–0.26	Stonecypher et al. (1973)
Slash pine	0.03–0.37	Barber (1964)
Longleaf pine Age 5	0.18	Snyder and Namkoong (1978)
Age 7	0.12	Snyder and Namkoong (1978)
Yellow poplar	0.42–0.84	Kellison (1970)
Black walnut Age 1	0.55	McKeand (1978)
Age 8	1.25	McKeand (1978)
Sweetgum Age 2	0.25	Ferguson and Cooper (1977)
Age 11	0.08	Ferguson and Cooper (1977)
Wood Specific Gravity		
Loblolly pine	0.76–0.87	Goggans (1961)
Loblolly pine	0.41	Chuntanaparb (1973)
Scots pine	0.46–0.56	Personn (1972)
Slash pine	0.50	Goddard and Cole (1966)
Eucalyptus deglupta	0.44	Davidson (1972)
Eucalyptus viminalis	0.55	Otegbye and Kellison (1980)
Form		
Loblolly pine—stem straightness	0.14–0.21	Stonecypher et al. (1973)
Loblolly pine—crown form	0.08–0.09	Stonecypher et al. (1973)
Slash pine—pruning height	0.36–0.64	Barber (1964)
Douglas fir—stockiness	0.26	Silen (1978)

Note. Heritability estimates vary with species, populations within species, age, and characteristics assessed.

QUANTITATIVE GENETICS AND SELECTION

Introduction

The primary objective of an applied tree improvement program is to change the frequency of desired alleles that influence important tree characteristics in such a way that the improved plants are superior in performance to unimproved material. The way of accomplishing this is through the process of *selection*, which can be defined as "choosing individuals with desired qualities to serve as parents for the next generation." Although selection can be a major tool for studying the way traits are inherited, in applied tree improvement programs selection is primarily used for the improvement of economically important characteristics. The following sections are developed with this in mind.

If selection is to be effective, there must be a genetic variation in the population. As shown before, for most tree improvement activities it is the additive portion of the genetic variation that is readily usable for manipulation by the tree breeder. Selection that is based on utilization of the additive variance works by increasing the frequencies of favorable alleles. The additive effects of these alleles are observed in the improved performance of progeny produced by the breeding program or in seed orchards.

The practice of selection in tree improvement is both a science and an artistic skill that must be developed by the tree improver. It will be the main subject of the next chapter in this book, in which selection will be discussed in detail as it relates to specific phases of tree improvement operations. The following paragraphs will introduce the genetic principles associated with selection activities.

Selection and Genetic Gain

Selection is based upon the principle that the average genetic value of selected individuals will be better than the average value of individuals in the population as a whole. For metric or quantitative traits, gain from selection is usually measured as a change in the population mean. The improvement that can potentially be made from selection for a characteristic is a function of the heritability of the trait, and the variation for the trait that exists in the population.

The importance of heritability in determining the response to selection was stressed earlier. A high heritability indicates that much of the variation for a given characteristic observed in the population is genetic in origin, and that the breeder has a high probability to choose parents that are good genetically by selecting those that have desirable phenotypes.

The total amount of variation for a trait is equally as important as heritability in determining gain that can be made from selection, but it is often overlooked by persons involved in tree improvement activities. The total, or phenotypic, variation is important because of its influence on the *selection differential*. Symbolized by *S,* selection differential is defined as "the average phenotypic value of the selected individuals, expressed as a deviation from the population mean." If there is much phenotypic variation for a given characteristic, then the selection differential can be large, whereas if the total variation is minimal, then the selection differential must be small. The selection differential is pictured graphically in Figure 4.4.

The hatched area in Figure 4.4 represents individuals that have been selected; that is, those that are to be used as parents to produce the next generation of progeny. The selection differential, or *S,* is the difference between the mean of the selected individuals ($\bar{X}_s$) and the population mean ($\bar{X}$). Symbolically,

$$S = \bar{X}_s - \bar{X}$$

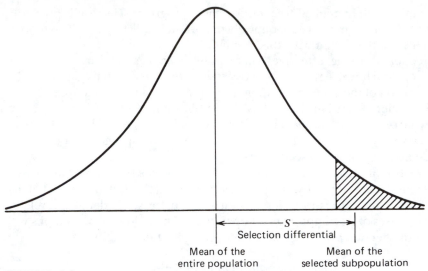

$$\overset{\longleftarrow S \longrightarrow}{\text{Selection differential}}$$

Mean of the
entire population

Mean of the
selected subpopulation

FIGURE 4.4

The selection differential is indicated as the difference between the mean of the entire population and the mean of the selected subpopulation.

When individuals are selected based only on their phenotypic values without information on relatives, response to selection can be estimated by the following formula:

genetic gain = narrow-sense heritability × selection differential

or

$$G = h^2 S$$

From the preceding formula, it is obvious that the progenies from selected parents can be no better than the mean of the selected parents and are usually much less. There are two reasons for this.

1. Usually, only a portion of the superiority of selected parents is due to genetics. The remainder is due to the environment. Superiority caused by environment cannot be passed on from parent to offspring. For example, a selected parent may have been superior to its neighbors because it was growing in a slightly better microhabitat.

2. In population improvement programs where many selected parents are mated together, only the additive genetic variance can be utilized. This is the reason why the narrow-sense heritability is utilized in the preceding equa-

tion. Even if all the variation observed was genetic in origin (no environmental variance), gain would be equal to selection differential only if all the variation was additive and none was of the nonadditive type; that is, $h^2 = 1$.

The tree breeder can influence gain from selection in essentially two ways. First, the base population can be managed so as to maximize heritability by the use of uniform sites and control of the environment. This is one of the primary factors involved in site selection and experimental design for genetic testing and will be discussed in more detail in Chapter 8. Once a population is established in a given environment, however, there is little the breeder can do to increase heritability. As a practical matter, the greatest opportunity to improve gain from selection is to increase the selection differential. This is how tree improvement programs have obtained gains by selection of trees from natural stands. Heritability values for natural stands of trees are usually low, especially for growth traits, because of extreme environmental variation, including competition, but individual trees are greatly different from one another and selection differential can be high.

The selection differential that the breeder uses is dependent upon two factors. One is the proportion of individuals in the population that are selected; that is, the intensity with which selection is done. The other factor is the phenotypic standard deviation, which, as we have seen, is a description of the variation in the population and is expressed in the same units as the population mean. Many breeders prefer to express response to selection or gain by the formula

$$G = ih^2\sigma_p$$

where

$$i = \text{intensity of selection}$$
$$h^2 = \text{heritability}$$
$$\sigma_p = \text{phenotypic standard deviation}$$

This formula indicates that both selection intensity and phenotypic variation influence gains that can be made.

A comparison of the two formulas for selection response shows that

$$G = h^2S = ih^2\sigma_p$$

Therefore, $S = i\sigma_p$, and $i = S/\sigma_p$

The selection intensity, or i, measures how many standard deviations the mean of the individuals that were selected exceeds the mean of the base population. For example, a selection differential (S) of 10 indicates that the mean of the selected population is 10 units better than the mean of the whole population. If the phenotypic standard deviation (σ_p) is equal to 5, then $i = S/\sigma_p = 2$, and the mean of the selected population is two phenotypic standard deviations better than the mean of the whole population. Because of the characteristics of the normal distribution, the equation $G = ih^2\sigma_p$ is a convenient way to calculate genetic gain. If the breeder knows the phenotypic standard deviation and the intended selection intensity, a response to selection can be predicted before selections are ever made. Alternatively, selection intensity can be varied to determine how many individuals must be chosen to obtain a certain desired gain. Selection intensity is related to the proportion of individuals in the population that are selected. If this proportion is known, it can be calculated directly from that value. Calculation of selection intensity involves a more in-depth knowledge of statistics than has been presented in this text; the procedure involves a determination of areas under the normal curve presented in Figure 4.4. When the population from which selections are to be made consists of only a few individuals, selection intensities will be lower for any given proportion saved. Selection intensities for several different levels of selection and population sizes are indicated in Table 4.2. Complete tables of selection intensities for populations of differing sizes can be found in Becker (1975).

In summary, response to selection for a given trait is determined by two factors: the heritability of the trait and the selection differential that is used. The tree improvement specialist must manage his or her population in such a way that both of these are large enough to give a useful gain from selection.

Selection Methods

There are several different selection methods available to the breeder, depending upon the types of information available. The selection systems commonly used in natural stands and unimproved plantations are discussed in detail in the next

TABLE 4.2

Approximate Selection Intensities (i) for Populations of Various Sizes and Proportions Selected

	Population Size				
Proportion Selected	**20**	**50**	**200**	**200**	**Infinite**
0.01	—	—	2.51	2.58	2.66
0.05	1.80	1.99	2.02	2.04	2.06
0.10	1.64	1.70	1.73	1.74	1.76
0.20	1.33	1.37	1.39	1.39	1.40
0.30	1.11	1.14	1.15	1.15	1.16
0.40	0.93	0.95	0.96	0.96	0.97
0.50	0.77	0.79	0.79	0.79	0.80
0.60	0.62	0.63	0.64	0.64	0.64
0.70	0.48	0.49	0.49	0.50	0.50
0.80	0.33	0.34	0.35	0.35	0.35
0.90	0.18	0.19	0.19	0.19	0.20

Note. For a given proportion selected, selection intensity increases with population size.

chapter, and methods used in advanced generations where pedigrees are known are discussed in Chapter 13. The basis for both selection procedures is introduced in the following sections.

Mass Selection   *Mass selection involves choosing individuals solely on the basis of their phenotypes, without regard to any information about performance of ancestors, siblings, offspring, or other relatives.* Mass selection works best for highly heritable traits, where the phenotype is a good reflection of the genotype. It is the only type of selection that can be used in natural stands or in plantations where tree parentage is unknown. Mass selection is rarely used when pedigrees are known, as in advanced-generation genetic tests, because more gain can be obtained using other methods. The terms *mass selection* and *individual selection* are used synonymously in this text.

Family Selection   *Family selection involves the choice of entire families on the basis of their average phenotypic values.* There is no selection of individuals within families, and individual-tree values are used only to compute family means. Family selection works best with traits of low heritability, where individual phenotypes are not a good reflection of genotypes. When family averages are based upon large numbers of individuals, environmental variance tends to be reduced, and family averages become good estimates of average genetic values. Family selection by itself is rarely used in forestry, even with traits of low heritability because more gain can be obtained from other methods that include family selection as a *part* of the method. Family selection may also lead to increased rates of inbreeding because entire families are discarded, thus reducing the genetic base of the population.

Sib Selection   *This is a form of selection in which individuals are chosen on the basis of the performance of their siblings and not on their own performance.* When family sizes are large, it is very similar to family selection. Sib selection is rarely used in forestry but may be applicable when destructive sampling must be used to make measurements, and it is not feasible to preserve genotypes by grafting or other techniques before sampling begins.

Progeny Testing   *Progeny testing involves selection of parent trees based upon the performance of their progeny.* It can be a very precise selection method, because it allows direct estimation of breeding values to use in the selection process. This is what occurs when parents from a seed orchard are progeny tested, and orchards are then rogued of parents that prove to be poor genetically. Progeny testing is not generally the initial form of selection for most breeding programs. Initial selection by progeny testing considerably lengthens the generation interval, which means a critical loss in time. As mentioned in previous chapters, the goal of tree improvement should be to achieve the maximum amount of gain per unit time. Other forms of selection are usually more efficient in accomplishing this goal.

Within-Family Selection *Here individuals are chosen on the basis of their deviation from the family mean, and family values per se are given no weight when selections are made.* Of all the selection methods, this one gives rise to the slowest rate of inbreeding, which is a major problem in most programs. In practice, family selection is rarely used in tree improvement because large increments of gain can be obtained from selection on family values. Thus, the family and within-family methods are almost always combined.

Family Plus Within-Family Selection *This two-stage method involves selection on families followed by selection of individuals within families.* It works well with low heritabilities, and is a predominant form of selection used in most advanced generation tree improvement programs. It consists of choosing the best families along with the best individuals in them. A refinement of this method is *combined selection* where *an index is computed that rates all individuals based upon their family value combined with their individual phenotypic values.* Coefficients or weights used in the index equation depend upon the heritability of the trait, with more weight given to the family average for traits with low heritabilities, and with more weight given to the individual when heritability of the trait is high.

Selection for Several Traits

Most tree improvement programs are geared toward the improvement of several traits at the same time. This requires that information developed on several characteristics be included in the selection procedure. How best to do this is one of the major areas of research in tree improvement today. Any of the methods discussed previously could be utilized to develop information on individual traits, but that information must be manipulated to develop a multitrait selection scheme. Essentially three systems have been developed that pertain to multitrait selection.

Tandem Selection When tandem selection is used, breeding is for one trait at a time until a desired level of improvement is made for that trait. After the desired improvement has been obtained in the first, and usually most important, trait (this may take more than one generation), selection and breeding efforts are then concentrated on other traits. This method of improving several traits in tandem is rarely used because of the pressure of time and the need to improve several traits simultaneously. The primary use of tandem selection is when one trait is of overriding importance, such as disease resistance, or when tropical or subtropical species are introduced into cold environments and cold hardiness must be improved before other commercially important traits can be considered.

Independent Culling Independent culling is a method of multitrait selection that involves setting minimum values for each trait of interest. Individuals must

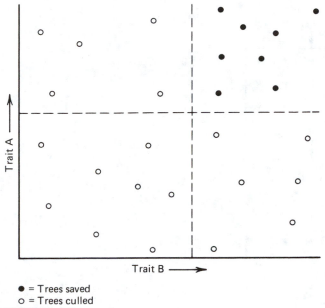

● = Trees saved
○ = Trees culled
FIGURE 4.5

The independent culling method of selection is illustrated as it would be applied for two traits. Only those individuals that meet minimum standards for both traits are saved.

meet these minimum criteria if they are to be retained. Independent culling is shown graphically in Figure 4.5. It is a very widely used form of multitrait selection in forest tree improvement.

Selection Index The selection index is a form of multitrait selection that combines information on all traits of interest into a single index. This enables the breeder to assign a total score to each individual. In addition to genetic information, it attaches economic weights to each of the characteristics under consideration. In its most complete and complex form, a selection index combines family plus individual information for all traits into one index. Index values for individuals are derived through a multiple regression equation in which the coefficients depend on the heritabilities, the correlation among traits, and the economic weights of each trait. Theoretically, the selection index method of multitrait selection can be shown to give the greatest total genetic gain for all traits combined. A major problem, however, is to have or to determine the appropriate economic weights. Derivation of the correct economic weights is still a major stumbling block to a more widespread appliction of this most useful form of multitrait selection in forest tree improvement. Use of selection indexes where economic weights are grossly incorrect can lead to very inefficient selection programs.

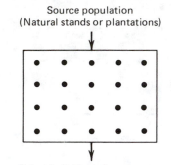

Source population
(Natural stands or plantations)

Generation 0

Select individuals from source
population and intermate to
create new progeny population
for the next cycle

Generation 1

Select individuals from new
progeny population and intermate
to create the progeny population
for the next cycle

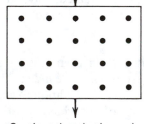

Generation 2

Continue the selection and
breeding processes as above

FIGURE 4.6

Illustrated is simple recurrent selection that
forms the basis for many tree improvement
programs. A successful recurrent selection
system will result in genetic gain for many
generations of improvement.

Recurrent Selection

Most tree improvement programs are designed to provide continued gain through each of many cycles of improvement. The selection methods discussed before have centered on improvement that can be obtained in one generation, but they are also applicable to programs that involve many generations. The selection procedure that involves many cycles of selection and breeding is known as *recurrent selection.* A number of recurrent selection schemes have been devised by plant and animal breeders in order to utilize variation in general combining ability, and, in some cases, specific combining ability. The system most often used in forest tree improvement is known as simple recurrent selection, which is shown diagramatically in Figure 4.6. With this system, improvement programs begin by selecting trees in natural stands or in unimproved plantations based upon their phenotypic values. Selected trees are then mated, and their progenies are established in such a way that they can be used as a source of selection for the second generation of improvement. In most instances, second-generation selections are made on the basis of family and individual values. These selections are then mated, creating a new progeny generation that can be used as a source of selections for the next generation. The system is one that repeats itself. Selections are made in a base population, mated in some fashion, and the resultant progenies serve as a population for the next generation of improvement.

Recurrent selection programs will be discussed more fully in Chapter 13. For now, it is important for the reader to realize that when many cycles of improvement are envisioned, extreme care must be taken at the outset of the program to ensure that genetic resources are available to make the recurrent selection program successful. The tree improver must have the following two goals in mind when beginning an applied tree improvement program:

1. Obtaining as much gain as possible as quickly as possible through selection and production of improved seed.

2. Maintaining a genetic base that is large and diverse enough to allow successful improvement programs to continue for many generations.

LITERATURE CITED

Barber, J. C. 1964. "Inherent Variation among Slash Pine Progenies at the Ida Cason Calloway Foundation." USDA, U.S. Forest Service Research Paper SE-10, Southeastern For. Expt. Sta.

Becker, W. A. 1975. *Manual of Quantitative Genetics*. Student Book Corporation, Washington State University, Pullman.

Campbell, K. 1972. Genetic variability in juvenile height-growth of Douglas-fir. *Sil. Gen.* **21**:126–129.

Chuntanaparb, L. 1973. "Inheritance of Wood and Growth Characteristics and Their Relationship in Loblolly Pine (*Pinus taeda* L.)." Ph.D. thesis, Department of Forestry, North Carolina State University, Raleigh.

Cochran, W. G., and Cox, G. M. 1957. *Experimental Design.* John Wiley & Sons, Inc., New York.

Davidson, J. 1973. "Natural Variation in *Eucalyptus deglupta* and Its Effect on Choice of Criteria for Selection in a Tree Improvement Program." Papua New Guinea, Trop. For. Res. Note no. SR-2.

Falconer, D. S. 1960. *Introduction to Quantitative Genetics.* Ronald Press, New York.

Ferguson, R. B., and Cooper, D. T. 1977. "Sweetgum Variation Changes with Time." 14th South. For. Tree Impr. Conf., Gainesville, Fla., pp. 194–200.

Franklin, E. C. 1968. "Artificial Self-pollination and Natural Inbreeding... *Pinus taeda* L. Ph.D. thesis, Department of Forestry, North Carolina State University, Raleigh.

Franklin, E. C. 1970. "Survey of Mutant Forms and Inbreeding Depression in Species of the Family *Pinaceae*." USDA, U.S. Forest Service Research Paper SE-61, Southeastern Forest Experimental Station.

Franklin, E. C. 1979. Model relating levels of genetic variance to stand development of four North American conifers. *Sil. Gen.* **28**:207–212.

Goddard, R. E., and Cole, D. E. 1966. Variation in wood production of six-year-old progenies of select slash pines. *Tappi* **43**:359–362.

Goggans, J. F. 1961. "The Interplay of Environment and Heredity as Factors Controlling Wood Properties in Conifers: With Special Emphasis on Their Effects on Specific Gravity." Technical Report No. 11, North Carolina State University, Raleigh.

Kellison, R. C. 1970. "Phenotypic and Genotype Variation of Yellow-Poplar (*Liriodendron tulipifera*)." Ph.D. thesis, Department of Forestry, North Carolina State University, Raleigh.

Kempthorne, O. 1957. *Introduction to Genetic Statistics.* Iowa State University Press, Ames.

Kinloch, B. B., and Byler, J. W. 1981. Relative effectiveness and stability of different resistance mechanisms to white pine blister rust in sugar pine. *Phytopathology* **71**:386–391.

Matziris, D. I., and Zobel, B. J. 1973. Inheritance and correlations of juvenile characteristics in loblolly pine (*Pinus taeda* L.) *Sil. Gen.* **22**:38–44.

McKeand, S. E. 1978. "Analysis of half-Sib Progeny Tests of Black Walnut." M.S. thesis, Department of Forestry and Natural Resources, Purdue University, Lafayette, Ind.

Namkoong, G. 1979. "Introduction to Quantitative Genetics in Forestry," USDA, U.S. Forest Service Technical Bulletin No. 1588.

Namkoong, G., and Conkle, M. T. 1976. Time trends in genetic control of height growth in ponderosa pine. *For. Sci.* **22**:2–12.

Namkoong, G., Usanis, R. A., and Silen, R. R. 1972. Age-related variation in genetic control of height growth in Douglas-fir. *Theor. Appl. Gen.* **42**:151–159.

Neter, J., and Wasserman, W. 1974. *Applied Linear Statistical Models.* Richard D, Irwin, Inc., Homewood, Ill.

Otegbye, G. O., and Kellison, R. C. 1980. Genetics of wood and bark characteristics of *Eucalyptus viminalis. Sil. Gen.* **29**:27–31.

Persson, A. 1972. "Studies on the Basic Density in Mother Trees and Progenies of Pine." Studia Forestalia Suecica No. 96.

Silen, R. R. 1978. "Genetics of Douglas-fir." USDA, U.S. Forest Service Research Paper WO-35.

Snedecor, G. W., and Cochran, W. G. 1967. *Statistical Methods.* Iowa State University Press, Ames.

Snyder, E. B., and Namkoong, G. 1978. "Inheritance in a Diallel Crossing Experiment with Longleaf Pine." USDA, U.S. Forestry Service Research Paper SO-140, Southern Forest Experiment Station.

Sokal, R. R., and Rohlf, F. J. 1969. *Biometry.* W. H. Freeman, San Francisco.

Steel, R. G. D., and Torrie, J. H. 1960. *Principles and Procedures of Statistics.* McGraw-Hill, New York.

Stonecypher, R. W., Zobel, B. J., and Blair, R. 1973. "Inheritance Patterns of Loblolly Pines from a Nonselected Natural Population." Technical Bulletin No. 224, North Carolina Agricultural Experiment Station.

Strickberger, M. W. 1976. *Genetics.* Macmillan, New York.

CHAPTER 5

Selection in Natural Stands and Unimproved Plantations

The objective of a selection program is to obtain significant amounts of genetic gain as quickly and inexpensively as possible, while at the same time maintaining a broad genetic base to ensure future gains. All methods of selection in an applied tree improvement program are based on the same general principle; that is, choose the most desirable individuals for use as parents in breeding and production programs. As was discussed in the previous chapter, the selection method that is used will depend on the information and plant materials that are available and on the goals of the program. The testing is done as soon as possible but is often still underway when the first use is made of the improved material.

Selection is a key part of all applied tree improvement programs. The gains can be no greater than the quality of the parents used, and the way to obtain the best parents is through intensive selection. Because this activity comes at the start of the program, many organizations become alarmed at selection costs. Although they may appear to be large, actually they generally account for a minor part of the total costs of tree improvement. Estimates from various programs range from 5% (Porterfield et al., 1975) to 11% (van Buijtenen and Saitta, 1972) to 30% (Reilly and Nikles, 1977). Selection is normally the first step in a tree improvement program and will determine how much gain will be obtained, both in the first and succeeding generations. Doing a poor job of selection to reduce initial costs certainly cannot be justified.

A number of selection methods are available to the tree improver. The one chosen for any particular program depends upon the types of genetic variations in the population, whether pedigree information exists, and the degree of urgency in establishing production seed orchards. Selection methods used for trees from stands where there is no pedigree information are almost always different than those from genetic tests where parentages are known. This chapter will concentrate on selection methods that are applicable to natural stands or unimproved plantations. Methods of selection used where pedigree information exists, as in genetic tests, will be covered only briefly in relation to beginning tree improvement programs. More thorough discussion of methods will be discussed in Chaper 13, which covers advanced-generation tree improvement.

The great variation within the important traits of most forest trees and their reasonably strong general combining ability allows a good chance for gain by selecting desired phenotypes (Figure 5.1). The best selections are then used in seed orchards, allowing favorable genic combinations to interact and produce progeny with a larger proportion of the desired characteristics. In most species, a considerable improvement in bole straightness, disease resistance, wood quality, and adaptability to adverse environments or tolerance to pests can be rapidly obtained by selecting and allowing cross-fertilization among the very best trees, as has been reported by Giertych (1967), Zsuffa (1975), Butcher (1977), and literally hundreds of others.

Of the several methods available to make gains quickly and inexpensively in a beginning tree improvement program, *individual* (*mass*) *selection* of trees is the most used and is generally the most satisfactory (Figure 5.2). It is widely applied in the initial stages of tree improvement programs and is suitable for many species.

FIGURE 5.1

Shown is an outstandingly good red maple tree (*Acer rubrum*) selected for use in a hardwood tree improvement program. Rarely does this species develop trees with small crowns and straight boles.

Occasionally, where there is little urgency for production of improved propagules for reforestation, and when time permits establishment of genetic tests, such methods as progeny test selection or family and within-family selection may be used to establish initial seed orchards. Because of its overwhelming importance in most beginning tree improvement programs, mass selection will be given the most extensive coverage in this chapter.

DEFINITIONS

To help avoid the confusion in terminology that is generally evident when selection is discussed, the following terms are defined in reference to a selection program.

1. **Candidate tree** A tree that has been selected for grading because of its desirable phenotypic qualities but that has not yet been graded or tested.
2. **Select, superior, or plus tree** A tree that has been recommended for production or breeding orchard use following grading. It has a superior phenotype for growth, form, wood quality, or other desired characteristics

FIGURE 5.2

Most tree improvement programs start with the selection of
outstanding phenotypes in wild stands or plantations. Shown is
a good loblolly pine used in a seed orchard program.

and appears to be adaptable. It has not yet been tested for its genetic worth,
although the chances of its having a good genotype are high for characteristics with a reasonable heritability.

3. **Elite tree**   A term reserved for selected trees that have proven to be
    genetically superior by means of progeny testing. An elite tree is the
    "winner" from a selection program and is the kind of tree that is most desired
    for use in mass production of seeds or vegetative propagules.

4. **Comparison or check trees**   Trees that are located in the same stand, are of
    nearly the same age, are growing on the same or better site as the select tree
    and against which the select tree is graded. Trees chosen as comparison trees

are the best in the stand, with characteristics similar to "crop" trees that would be chosen in a silvicultural operation.

5. **Advanced-generation selection**   A tree selected from genetic tests of crosses among parents from the previous generation. Some form of family and within-family selection is usually used to choose advanced-generation selections.

WHEN SHOULD INDIVIDUAL SELECTION BE USED?

Individual or mass selection works best for those characteristics that have a high narrow-sense heritability. Obviously, it is the only method that can be used to select trees in stands where pedigrees are unknown. To be most successful, mass selection should be used in stands that have a large proportion of good trees and that have not been subjected to logging operations in which the best trees have been removed (Figure 5.3). Examples of characteristics showing relatively high heritabilities are wood specific gravity, resin yields in pine, and most adaptability

FIGURE 5.3

Mass- or individual-tree selection should be made in good stands that have not been high graded, such as the loblolly pine stand shown.

characteristics. Straightness of tree bole and disease resistance are intermediate, whereas for most characteristics related to growth, individual selection is less effective because of low heritability (Shelbourne et al., 1972). For some characters with very low narrow-sense heritability, such as cellulose yield in loblolly pine, individual selection may not be a suitable method to make gains (Zobel et al., 1966; Jett et al., 1977).

Gain from an individual tree selection program can be indicated as G = heritability × selection differential, or $G = h^2S$. Heritability is generally quite constant for a given characteristic at a given age in a given environment, and the tree breeder can not do much to improve it other than to create an environment that is more suitable for the tree to express its genetic potential. However, the selection differential can be manipulated (within limits) by the tree breeder by varying the intensity with which selection is applied. As explained in Chapter 4, a major objective is to increase the selection differential that, in turn, increases genetic gain (see Figure 5.4).

As selection intensity is increased, a point of diminishing returns is reached when gains become less per unit increase in selection intensity (Shelbourne, 1970). However, the intensity of selection used in operational programs is usually less than optimal. For example, in the North Carolina State University–Industry Tree Improvement Cooperative, which used a relatively intensive selection system for pine in the first generation, the selection intensity could have been increased by three times, yielding greater genetic gains but with still the desired return on the investment (Porterfield et al., 1975).

It is essential to reemphasize that an individual selection program is based solely on the phenotype of the tree. For most characteristics, individual selection should be followed by progeny testing to determine if the selected tree is in fact genetically superior; this is especially true for traits with low heritabilities. Van Buijtenen and Saitta (1971) found that individual selection was very effective to the extent that the greatest value of the genetic tests was to serve as a source from which to select for advanced generations. For characteristics that have a high heritability and large amounts of variations, one can be sure that gains from a careful selection program will be reasonable. This was true for loblolly pine (*P. taeda*) where progeny from parents intensively selected for straight boles in the first generation were straight enough so that this characteristic was not emphasized in the following generations (see Figure 5.4). Although the trees could be straightened more by additional selection, the increased economic value from this small improvement would be minimal; this enabled placing more emphasis on other characteristics.

SELECTING SUPERIOR TREES

The techniques used in tree improvement to find and select superior trees depend on the types of stands in which selections are to be made. Numerous references exist on selection techniques for various species and stand types (Langner, 1960;

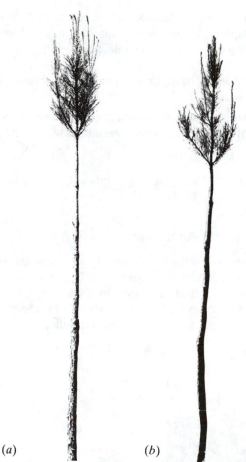

(a)   *(b)*

FIGURE 5.4

Characteristics with a high heritability respond well to selection. Shown is a progeny from a cross of two straight loblolly pine parents *(a)* and from two crooked parents *(b)*. Response to straightness selection was so good in this species in the first generation that straightness was given less emphasis in advanced generations.

Wright, 1960; Vidakavic, 1965; Morgenstern et al., 1975). There are many references related to selecting conifers, such as those by Cook (1957), Walters et al. (1960), and Andersson (1963). Numerous suggestions have also been made for selecting hardwoods. These trees differ considerably for each species or condition (Eldridge, 1966; Clausen and Godman, 1967; Beineke and Low, 1969; Pederick, 1970; Bey et al., 1971; Schreiner, 1972).

A determination of the best selection techniques depends on several factors, including species characteristics, past history, the present condition of the forest, variability and inheritance pattern of important characteristics, and objectives of the particular tree improvement program. There are two major kinds of forest stands, each of which require differing first-generation selection systems.

1. Even-aged wild stands or plantations from unimproved seed where the parentage of the trees is unknown.

2. Uneven-aged, scattered, or sprouting species where the parentage is not known. These include stands with species growing intermixed where check trees are not available.

Selection techniques for each of these kinds of forest stands will be discussed now.

Selecting from Even-Aged Stands

Individual selection works best when good even-aged stands of the proper age are available (see Figure 5.3). This allows efficient comparisons to be made among selected trees and checks. Individual tree selection is best in even-aged natural stands composed primarily of one species or in plantations. This is by far the most common method of first-generation selection and has been applied worldwide (Figure 5.5).

FIGURE 5.5

Selection of outstanding trees from plantations is very successful and has been widely used throughout the world. Shown is an outstanding *P. radiata* selected for use in a New Zealand seed orchard.

There are several advantages to selecting in even-aged stands rather than in uneven-aged or mixed stands when practicing individual tree selection. First, the breeder can be sure that age will not differ greatly among trees, and that relative expressions of growth, form, disease tolerance, and adaptability will not be confounded with age effects. Second, trees are growing under competitive situations similar to those that will be encountered when improved trees are established in commercial plantations. Also, it is in these types of stands where the "comparison tree" system of selection can be used that trees considered for selection are graded against the best trees in the stand. All of these factors work to increase selection differential, and this results in greater gain.

Generally, plantations are preferable to natural stands in selection efforts if plantations of suitable seed source are available (Figure 5.6). In plantations, all trees are exactly the same age. In natural stands, even slight differences in age cause differential competition that can result in large differences in volume and form within the stand. It is known, for instance, that in densely stocked pine stands, a difference of one or two years in age among neighboring trees will usually result in the younger trees's never becoming dominant or codominant trees in the stand. An additional advantage of selection in plantations is that spacing among trees is more uniform. Competition is in essence an environmental force that affects the tree's phenotype. When competition is equalized, heritabilities are raised.

General guides for locating select individuals in even-aged natural stands and plantations follow. These guides have been very useful in choosing superior trees in first-generation tree improvement programs.

1. The search should be concentrated on stands and plantations that are average or better in growth, pruning, straightness, branch angle, and other characteristics of interest. An occasionally acceptable tree may be found in a poor stand but this is rare, and search efforts are more efficient when they are carried out in good stands. Outstanding stands of trees are sometimes referred to as *plus stands*.

2. Stands and plantations in which candidate trees are sought should be located on the same variety of sites where plantations from improved seed will ultimately be established. This is true unless there is evidence that sites have no effect on the performance of the genotype. If the majority of an organization's landholding are on average sites, then the majority of selections should come from such sites. *There should never be a concentration of selections from the very highest site quality lands*, if the plantations are to be established on average or poor sites.

3. When selections are made from plantations, information about the suitability of the seed source used in the planting should be obtained. Selections should not be made from stands planted with seeds from areas known to be poorly adapted to the area where planting will be done.

4. In older stands, the search effort should be confined to trees that have an age range of no more than 10 to 15 years younger or older than the projected

FIGURE 5.6

Selecting trees from plantations, as shown for slash pine, is relatively easy, and good trees can be obtained compared to wild stands because all trees are of the same age and were established at about equal spacings. Care must be taken that the trees used are from the correct seed source.

rotation age of the plantations that are to be established. For species that are harvested at an early age, the trees must be old enough to have shown their potential. For tropical pines, the stands need to be a minimum of 10 to 12 years old before they exhibit development that enables efficient selection, whereas in the same areas some eucalypts as young as 3 years of age can be easily selected if very short rotations are used.

5. Selections should be made from stands that are as pure in species composition as possible. Differential growth rates among species can severely

complicate selection through differential competition if the stand has a sizable component of two or more species.

6. Stands must be avoided that have been logged for poles or piling or that have been otherwise high graded or thinned from above. If the stand has been thinned from below, or if it has suffered fire damage, allow crown competition to be reestablished before selections are made. Stands that are mechanically thinned or thinned in a truly silvicultural manner are suitable for an individual selection program.

7. The minimum size of a stand or plantation in which a candidate can be located is immaterial. If the stand is large enough to locate a good candidate tree and to allow choosing comparison trees, then it is large enough to search for select trees.

8. Preferably only one select tree should be accepted from any one small natural stand to reduce the possibility of obtaining candidate trees that are close relatives. This restriction does not apply to selecting in plantations.

9. Although it is highly desirable for candidate trees to exhibit a heavy flower or cone crop, these characteristics are generally not given much emphasis. This is particularly true in young and dense stands where many trees show no sign of flowering because of insufficient light on their crowns to stimulate flower production. Usually, these will flower heavily in the seed orchard environment.

10. Once the decision has been made to look over an area for candidate trees, a thorough, systematic search should be made. Experience has shown that excellent trees are often missed when a stand is searched haphazardly. Experience has also shown that select trees are generally found only by people who are specifically looking for them. Although an occasionally acceptable tree is located during routine woods work, this is the exception. The only efficient way to locate candidate trees is to be specifically on a selection mission.

11. A comparison or check tree selection system should be used when feasible. This helps to account for environmental differences within stands and permits more efficient and objective selection of superior trees. A method of evaluating candidate trees without the use of check trees was suggested by Robinson and van Buijtenen (1971), but usually the use of check trees is desirable.

Selection in Uneven-Aged, Mixed Species, or Stands of Sprout Origin

Forest stands are quite frequently not of types allowing use of the individual selection program that was just described for even-aged stands. There are several reasons for this: (1) Stands may be truly uneven aged; (2) the desired species may be so scattered that comparison (check) trees are not available; (3) the species may be a vigorous sprouter and trees growing near the candidate tree can be on a common root system and have the same genotype; and (4) the stand is composed of mixed species.

The comparison tree system does not work when trees are growing in all-aged stands. Since growth curves within a species vary with age, it is not suitable to use ratios such as height or diameter growth per unit time for comparison purposes. In addition, the form of the tree often changes drastically with age. Therefore, quality characteristics cannot be compared among trees of different ages from uneven-aged stands. Rarely are truly all-aged stands of either conifers or hardwoods common under natural conditions outside of the tropical forests. Many foresters make the mistake of assuming that stands containing trees of varied sizes are all-aged. The major exception to this generalization that forests are usually even-aged is when stands have been manipulated by humans into an all-aged condition by selective cutting. Even within the true uneven-aged stands, there is a tendency for a storied age class to be present.

Trees sometimes grow in *mixed stands* with relatively few individuals of a given species found in a specific area. This condition is most common for hardwoods. A comparison tree selection system will not work in this case because the scattered individuals of a species are growing under different environments. This is by far the most common situation that requires a grading system other than the standard comparison tree method (Pitcher and Dorn, 1966).

The importance and frequency of relatedness among trees from stands of *sprout origin* is often not understood. Usually, sprouts from a single tree are limited to those individuals that are adjacent to the tree, but, sometimes, sprouts from a common root system can be quite extensive. There are records of aspen stands (*Populus tremuloides*) as large as 40 acres (16 ha) having a common root system (Baker, 1925). When stands of sprout origin are large enough and of sufficient genetic diversity to enable use of a comparison tree system, the check trees must be carefully chosen so they will not be related to the candidate tree. This is sometimes difficult to do accurately.

Stands composed partially of sprouts and partially of seedlings also pose the problem of growth differential between trees from sprout and seedling origin. Initially, sprouts usually grow much faster than seedlings because of the established root system and stored food. However, sprouts often culminate growth at a younger age than do seedlings. After a few years, it commonly becomes impossible to distinguish sprouts from seedlings, but selection results will not be good if the two types of trees are mixed.

Although they are not sprout stands, occasionally trees form root grafts that can also make tree selection difficult (Yli-vakkuri, 1953; Bormann, 1966; Schultz and Woods, 1967; Eis, 1972). The result of root grafting usually is that the large tree benefits at the expense of the small tree by taking nutrients from the small tree with which it shares a common root system. This serves to inflate the superiority of the larger tree and is an environmental effect that cannot be captured through selection.

The Regression Selection System

The most useful method of tree grading for the uneven-aged or mixed-species-type stands described previously is the *regression system*. This requires the development of tables relating the characteristic of interest to tree age. The regression method is of particular value for growth characteristics because quality characteristics can often be determined on the basis of the phenotype of the candidate tree alone without need for comparison trees.

A regression selection system is built by sampling a number of trees for a desired characteristic, such as volume growth on a given site, and then plotting them against age (Figure 5.7). It is of key importance that different regressions are developed for different sites. A reliable regression curve for height or volume can be made with about 50 trees, if there is a reasonable age-class distribution. Once the curve has been developed, the regression is used as follows:

1.   A candidate tree is chosen, based on the judgment of the selector and measured for the characteristics desired, such as height or volume.
2.   The trait is plotted on the regression graph using the proper age and site. If the candidate tree falls at some defined distance above the regression line, it is acceptable and the higher above, the more desirable it becomes (Figure 5.7). When the value of the characteristic falls below the acceptable level, the tree is rejected.

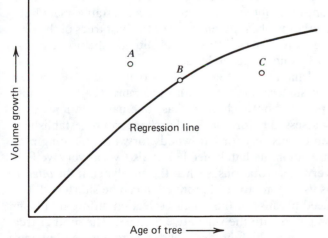

FIGURE 5.7

The regression system of selection is particularly suitable for all aged- or mixed-species stands. It consists of developing a curve of production (growth is illustrated) for different ages of trees on a given site. Candidate Tree A falls above the curve, therefore, it has the desired growth for its age. Tree B is average, therefore, its use depends on other characteristics, whereas Tree C has inferior growth for its age and should not be used. The regression line should be based on at least 50 trees if the age spread is considerable.

Although the regression system works well, it requires considerable preliminary work. To complicate the situation, some species, especially the wet-site diffuse-porous hardwoods, do not have discernible growth rings, thus making age difficult to assess. When this occurs, and if the history or age of stand establishment is not known (which is the usual situation), a reasonable estimate of age can often be made if ring-porous trees are growing intermixed in the stand. These can be used to determine the age of the stand. As an example, tupelo gum (*Nyssa aquatica*) usually has indistinguishable growth rings. Often growing adjacent to it is green ash (*Fraxinus pennsylvanica*) that produces easily determinable rings. The age of the ash is then applied to the candidate tupelo gum tree with reasonable accuracy because it is known that such stands are usually even aged. A major exception is some tropical forests that have developed in a truly all-aged condition.

The regression system is more difficult to use than the comparison tree method, but there is no doubt that it will become more commonly employed as hardwood tree improvement becomes more widely practiced.

The Mother Tree System

When there is no immediate urgency to obtain large amounts of improved seed, the mother tree system of selection may be best. It consists of locating "good" trees that are usually not as good as select trees in the comparison tree or regression systems. Then, one must obtain seed from these and establish seedlings in genetic tests. After this, either the best parent trees or the best trees of the best families can be used in a vegetative orchard. If suitably laid out, the progeny test may be thinned to create a seedling seed orchard.

The main disadvantage of the mother tree system is that time is lost before commercial quantities of seed are available for planting programs. The testing must be carried on for a long period, approaching at least one-half rotation age if growth is to be reasonably assessed before the seed orchard can be established. This method has been used extensively for hardwoods for which planting programs are small and seed are not immediately needed. It also works very well for species that are grown on very short rotations, such as the eucalypts. If the *interim seedling orchard* concept is used, time to seed production can be shortened, but potential gain will be reduced because of the smaller selection differential from the seedling orchard as compared with the vegetative orchard. The mother tree system may also be best in seriously high-graded stands where few good phenotypes are available. It may also be best for characteristics such as disease resistance that often can only be determined through testing.

The Subjective Grading System

Some persons who are familiar with a species feel that an acceptable job of selection can be done based only on the judgment of the grader about what constitutes a good tree. This is certainly possible, but the grader must *know the species* intimately and must be as unbiased as possible. This system is used successfully but has also failed. The tendency, when the subjective system is used, is to spend less time seeking the candidate trees, thus choosing less outstanding trees with smaller selection differentials. This results in less gain. The subjective grading system is frequently used for hardwoods but is successful only if the grader is exerienced and dedicated to finding the best trees possible.

TRAITS DESIRED FOR SELECTION: DEVELOPING A GRADING SYSTEM

Two factors are of paramount importance when developing a grading scheme for the selection of superior trees. First, the trait under consideration should be under at least *moderately strong genetic control*. Second, the trait *must have considerable economic value*. Regardless of the magnitude of either of these factors, a characteristic is of little use in a selection program when either the genetic control or economic value is too low. For example, if a characteristic such as needle length or leaf size were under strong genetic control, it would have little value in a commercial tree improvement program. This would be the case if trees with long needles or large leaves yielded no greater amount of the desired products than did those with small leaves or needles. The numerical value allocated to each characteristic in a tree-grading process is determined by weighting the heritability of the characteristic against its economic worth. For example, bole straightness usually is considered more important than limb size because of its stronger genetic control as well as its greater economic worth.

Growth rate is nearly always the *key characteristic* in a selection program (Bouvarel, 1966; van Buijtenen, 1969), but other characteristics are usually also important. In areas, such as the southeastern United States where the environment is favorable for tree growth, the major initial objective for first-generation orchards often is to improve quality characteristics with only a modest attempt to increase growth rate. This priority of quality characteristics over volume should be used when selections are from wild, unmanaged stands where heritability for growth is low. In advanced-generation tree improvement or in plantations where selections come from more uniform stands of the same age and equal spacing, the heritability for growth rate will be larger, and a greater concentration on improvement of volume growth should be made.

Most first-generation selection programs have a *"threshold" value* for each characteristic, below which no tree is accepted for an operational orchard no matter how excellent the rest of the tree's characteristics might be. Trees that have one marginally acceptable characeristic are often held in breeding clone banks for possible later use.

It is not appropriate to produce a list of characteristics for which selection should be done because these will vary with species, the product desired, and the objectives of each program. However, to illustrate the kinds of characteristics sometimes used and their weighting, a modification of the selection method used for the first generation in the North Carolina State–Industry Tree Improvement Cooperative is shown in Table 5.1. Note the differential weights given for different characteristics. Not shown are the "all-or-none" characteristics, such as

TABLE 5.1
Select Tree Grading Sheet for Conifers[a]

Organization _____	Tree Number _____
Location _____	Species _____
Grader _____	Plantation _____ Natural _____
Date _____	Age _____

Five Best Crop (Comparison) Trees[b]   *Selected Tree Score*

	Ht.	DBH	Vol.	Age
1				
2				
3				
4				
5				
Total				
Average				

1. Height _____
2. Volume _____
3. Crown _____
4. Straightness _____
5. Pruning ability _____
6. Branch diameter _____
7. Branch angle _____

Total Score _____

Select Tree (s) Average Crop Tree (c)   *Remarks*

Vol. _____

Ht. _____

DBH _____

Wood _____

Specific Gravity _____

1. *Height Superiority*

Less than 10%	0 points
10–11%	1 point
12–13%	2 points
14–15%	3 points
16–17%	4 points
18–19%	5 points
20%	6 points
Over 20%	7 points

2. *Volume*—formula: Vs/Vc; s = select; c = check—Average of 5. Select tree is given one point for each 10% excess in volume over the checks.

3. *Crown*—judged subjectively from the standpoint of the individual select tree as compared to the five checks and scored as follows:

   a. Based on conformation, density of foliage, dominance, and crown radius 0 to 5 points depending on superiority.

4. *Straightness*—judged subjectively for the individual select tree and not compared to the checks. No tree accepted with excess spiral, any crook in two planes, or a crook in

TABLE 5.1 (*continued*)

one plane that will not allow a line from merchantable top to stump to stay within the confines of the bole. Scored subjectively allowing 0 to 5 points, the number of points dependent upon the relative straightness of the individual.

5. *Pruning ability*—ability of select tree to shed its lower limbs (dead and alive) as compared to checks: If similar to the checks, it receives 0 points; above the checks, 1 to 3 points, depending on superiority. Pruning ability is judged subjectively by comparing the select tree with each of the checks.

6. *Branch diameter* is judged subjectively—select versus checks. If branch diameter is average, 0 points are given; 1 or 2 points for branches smaller than checks.

7. *Branch angle* is also judged subjectively. When branch angle is average, 0 points are given; if angle is flat, 1 or 2 points are awarded.

If any of the preceding seven categories is poorer than the checks, points are deducted by the same scale as they are added when the select tree is superior to the checks. A tree with a minus score in one characteristic is usually not acceptable except under certain special conditions.

Specific gravity is handled separately. No points as such are given for specific gravity, because the desires of each organization differ. The value of a tree for specific gravity is judged by two criteria:
1. Comparison of the select tree with its five check trees. This gives an indication of the tree's wood quality when compared to trees growing under the same environmental conditions.
2. Comparison of the select tree with the regional average for that particular species and organization.

[a]This is a revision of Form S-2 used by the N. C. State–Industry Cooperative Tree Improvement Program.

[b]Ht., height; DBH, diameter breast height; vol., volume; age is the age of each individual tree.

presence or absence of disease or insect damage. Also not shown is a grade for wood specific gravity. In the first-generation system used by the North Carolina State Cooperative the weight placed on specific gravity was determined by each organization, depending on its own product. This was done because some desired high gravities and others low gravities, whereas still others preferred trees with intermediate gravities. Adaptability is also not included in the grade sheets due to the fact that it is automatically taken into account because trees are selected and tested in the same areas in which they will be planted. The assumption was made that healthy, vigorous trees in natural stands are well adapted to the environment in which they are growing. A somewhat different grading system is needed for hardwoods. A method of reporting selections in hardwoods has been described by Pitcher and Dorn (1966).

The general tendency is to *include too many characteristics* in a selection system. The more things that must be graded, the more difficult it becomes to find suitable trees. The objective of tree selection and grading is to emphasize the few most important characteristics, such as volume production, bole quality, adaptability, and pest resistance. The lesser characteristics that are assessed should be kept above an acceptable level, but selection intensities need not be as great. The objective is to give the greatest weight to characteristics that have the best combination of economic importance and heritability and to give less weight to other characteristics. No tree is used that has a characteristic that falls below the level of acceptability.

The main objective of a grading system is to force the grader to look critically at the tree. The grade given does not represent the final decision about whether a tree will be used forever in the improved seed orchard. This decision is finally arrived at after progeny testing. All efforts to obtain the best phenotypes should be taken, but some mistakes will be made. The method described in Table 5.1 was used at North Carolina State University to select phenotypes for immediate use in first-generation seed orchards. This was followed by progeny testing and roguing of the undesirable genotypes to upgrade the genetic quality of the first-generation orchards.

The grading system shown in Table 5.1 is relatively simple to use and is one version of the comparison tree system discussed earlier. It works as follows:

1. Once the decision has been made to grade a tree, the five best crop trees in the stand in which the candidate tree is located are chosen as comparison or check trees. They, like the candidate tree, must have a dominant or codominant crown position and must be growing under conditions of competition that are similar to the candidate tree. The comparison trees are chosen for desirable characteristics in much the same way as the candidate tree. It is helful to consider the checks as crop trees that would be retained if the grader could leave the five best trees in the stand and not including the candidate tree. Comparison trees can occur at varying distances from the candidate tree, but they are selected on a site and under an environment similar to that of the candidate tree. If the candidate tree is located on a relatively uniform site, an attempt is made to locate the comparison trees in a circle around it. If suitable trees are not available in a circle, one or all of the comparison trees can be chosen in any sector of the circle. When the candidate tree is located on sloping terrain, the comparison trees should be selected on approximately the same contour as the candidate. In cases where this is not possible, the comparison trees should be located on the downhill or better side of the candidate to ensure that it is never compared to trees growing on a poorer site.

   The candidate tree is awarded points for each characteristic shown on the grading form, based on the importance of the character and relationship of the candidate tree to the five comparison, or check, trees. Height and diameter are actually measured for the candidate and the comparison trees.

Crown conformation, pruning ability, branch angle, and branch diameter are subjectively scored by visually comparing the characteristics of the candidate and the comparison trees. Straightness and disease or insect infection are subjectively scored on the candidate tree *only* and are not judged in relation to the comparison trees. Thus, a tree must meet a given level of straightness if it is to be graded, no matter how crooked or straight the check trees may be. In even-aged natural stands, the candidate tree is automatically rejected if it is more than three years older than the average of the five comparison trees; conversely, it is awarded points if it is more than 2 years younger than the average of the comparison trees.

2. No tree is accepted if it is infected by serious diseases or insects. In the southern pines, rejection of some of the very best trees results when they are infected by fusiform rust. The disease is genetically controlled strongly enough to make good gains by selecting only nondiseased trees for use in seed orchards, if infection levels are high (Goddard et al., 1975).

3. An ll-mm bark-to-bark increment core is extracted from the candidate tree at the time of grading to be used for wood analysis. A large core, approximately 11 mm in diameter, is necessary for the determination of tracheid lengths in conifers, but smaller cores will be satisfactory for fiber length in hardwoods.

The question arises about the feasibility of using the same scoring or grading system for different species. Although the same system is used quite frequently, there is a differing emphasis for different characteristics. For example, some pines will be graded more rigorously for prunability than will others, and some will be graded more severely for straightness of bole. Different grading sheets with different weightings usually need to be developed for hardwoods, especially to include the characteristics of epicormic branching and leader dominance.

It should be stressed that the grading system discussed here was developed only for use with pine in the North Carolina State Tree Improvement Cooperative and is shown only as an example of a comparison-tree grading system. Other tree improvement programs, including some that deal with the same species as the North Carolina State Cooperatives, employ somewhat different systems, which the organizations feel are best for their particular circumstances. The bottom line is that each organization must develop systems that fit into its own genetic, environmental, and economic constraints.

INDIRECT SELECTION

For some characteristics, it has been found that it is easier to use indirect selection rather than selecting directly for a specific character. This approach is especially valuable for forest trees because of their long life span and large size. Development of techniques to select at very young ages for performance at rotation age would result in a much shorter generation interval and greater genetic gain per unit of time and would speed up the tree improvement efforts greatly.

In forest trees, selection for most growth characteristics at older ages based on the performance of very young trees has not proven to be feasible generally. The problem relates primarily to the difficulty of obtaining good juvenile–mature correlations as described in Chapter 1. There it was stressed that assessment for growth characteristics should not be made before about one-half rotation age in natural stands or in plantations. Assessments of growth may be feasible at 6 to 10 years of age in genetic tests, when the normal rotation age is 25 years or more (Lambeth, 1980). The determination of seedling characteristics that would be used to make growth predictions for older trees would be very useful, as has been stressed by Kozlowski (1961). However, most tests to predict future growth based on gross or net photosynthesis have generally given poor results for both seed sources (Gordon and Gatherum, 1967) and for families (Burkhalter et al., 1967; Ledig and Perry, 1969; Ledig, 1974). Better methods of predicting growth at young ages will undoubtedly be developed, and much research is now being expended in this effort (Shimizu et al., 1976). A determination of reasonably accurate methods to assess the growth of mature individuals from seedling characteristics is one of the most urgent needs in research in tree improvement.

A second problem related to indirect selection is the relative genetic independence among most characteristics of forest trees. For indirect selection to be effective, the two characteristics being compared must be closely correlated. Such correlation seems to be relatively low for many traits of forest trees.

Some indirect selection methods have been tried for pest resistance with only indifferent success (Lewis, 1973; Rockwood, 1973; von Weissenberg, 1973; Wilkinson, 1980). All researchers reported that both the pests and physiological characteristics used for indirect selection were reasonably and strongly controlled genetically but were not sufficiently correlated to be effective in an indirect selection method.

As the scientific information on forest biology expands, indirect selection may well become feasible for certain traits. But, as of now, this is largely in the area of research and is not usable on a large scale.

LITERATURE CITED

Andersson, E. 1963. "Directions for Selection of Plus Trees and Phenotype Control; *Pinus silvestris* and *Picea abies*." World Cons. on For. Gen. and Tree Impr., Section 9, Stockholm.

Baker, F. S. 1925. "Aspen in the Central Rocky Mountain Region." USDA Bulletin 1291, Washington, D.C.

Beineke, W. F., and Low, W. J. 1969. "A Selection System for Superior Black Walnut Trees and Other Hardwoods. Proc. 10th South. Conf. of For. Tree Impr., Houston, pp. 27–33.

Bey, C. F., Hawker, N. L., and Roth, P. L. 1971. "Selecting Trees for Growth and Form in Young Black Walnut Plantations." 11th Conf. on South. For. Tree Impr., Atlanta, Ga.

Bormann, F. H. 1966. The structure, function and ecological significance of root grafts in *Pinus strobus. Ecol. Mono.* **36**(1):1–26.

Bouvarel, P. 1966. "Economic Factors in the Choice of a Method of Forest Tree Breeding." Sexto Congreso Forestal Mundial, Madrid.

Burkhalter, A. P., Robertson, C. F., and Reiner, M. 1967. "Variation in Photosynthesis and Respiration in Southern Pines." Ga. For. Res. Paper 46, Macon, Ga.

Butcher, T. B. 1977. "Gains from *Pinus pinaster* Improvement Program in Western Australia. 3rd World Consul. on For. Tree Breed., Canberra, Australia.

Clausen, K. E., and Godman, R. M. 1967. "Selecting Superior Yellow Birch Trees." North Central Forest Experimental Station Research Paper NC-20.

Cook, D. B. 1957. "Criteria for Judging "Plus" Larch Trees." Proc. 7th Northwestern For. Tree Impr. Conf., Burlington, Vt., pp. 40–42.

Eis, S. 1972. Root grafts and their silvicultural implications. *Can. Jour. For. Res.* **2**:111–120.

Eldridge, K. G. 1966. Genetic improvement of *Eucalyptus regnans* by selection of parent trees. *Appita* **19**(6):133–138.

Giertych, M. 1967. Genetic gain and methods of forest tree seed production. *Sylvan* **110**(11):59–64.

Goddard, R. E., Schmidt, R. A., and Vande Linde, F. 1975. Effect of differential selection pressure on fusiform rust resistance in phenotypic selections of slash pine. *Phytopathology.* **65**(3):336–338.

Gordon, J. C., and Gatherum, G. E. 1967. "Photosynthesis and Growth of Selected Scotch Pine Seed Sources." 8th Lake States For. Tree Impr. Conf., pp. 20–23.

Jett, J. B., Weir, R. J., and Barker, J. A. 1977. "The Inheritance of Cellulose in Loblolly Pine." TAPPI For. Biol. Comm. Meet., Madison, Wisc.

Kozlowski, T. T. 1961. "Challenges in Forest Production—Physiological Implications." 50th Anniversary of the State University College of Forestry, Syracuse, N.Y.

Lambeth, C. C. 1980. Juvenile-mature correlations in Pinaceae, and their implications for early selection. *For. Sci.* **26**:571–580.

Langner, W. 1960. "Improvement through Individual Tree Selection and Testing Seed Stand, and Clonal Seed Orchards." 5th World For. Cong., Seattle, Wash.

Ledig, F. T., and Perry, T. O. 1969. Net assimilation rate and growth in loblolly pine seedlings. *For. Sci.* **15**(4):431–438.

Ledig, F. T. 1974. "Photosynthetic Capacity: Developing a Criterion for the Early Selection of Rapidly Growing Trees." Bull. No. 1985, Champion International Corp. Lectureships, pp. 19–39.

Lewis, R. 1973. "Quantitative Assessment and Possible Biochemical Indicators of Variation in Resistance to Fusiform Rust in Loblolly Pine." Ph.D. thesis, North Carolina State University, Raleigh.

Morgenstern, E. K., Holst, M. J., Teich, A. H., and Yeatman, C. W. 1975. "Plus-Tree Selection—Review and Outlook." Department of Environment, Canadian Forest Service, Pub. No. 1347, Ottawa, Canada.

Pederick, L. A. 1970. "Selection Criteria." Proc. 2nd Mtg. Beerwah, Queensland, Australia.

Pitcher, J. A., and Dorn, D. E. 1966. "A New Form for Reporting Hardwood Superior Tree Candidates." Proc. 5th Central States For. Tree Impr., Wooster, Ohio, pp. 7–12.

Porterfield, R. L., Zobel, B. J., and Ledig, F. T. 1975. Evaluating the efficiency of tree improvement programs. *Sil. Gen.* **24**(2–3):33–34.

Reilly, J. J., and Nikles, D. G. 1977. "Analysing Benefits and Costs of Tree Improvement: *Pinus caribaea.*" 3rd World Cons. For. Tree Breed., Canberra, Australia.

Robinson, J. F., and van Buijtenen, J. P. 1971. "Tree Grading without the Use of Check Trees." Proc. 11th Conf. on South. For. Tree Impr., Atlanta, Ga., pp. 207–211.

Rockwood, D. 1973. Monoterpene-fusiform rust relationships in loblolly pine. *Phytopathology* **63**(5):551–553.

Schreiner, E. J. 1972. "Procedures for Selection of Hybrid Poplar Clones for Commercial Trials in the Northeastern Region." Proc. 19th Northeastern For. Tree Impr. Conf., Orono, Me., pp. 108–116.

Schultz, R. P., and Woods, F. W. 1967. The frequency and implications of intraspecific root-grafting in loblolly pine. *For. Sci.* **13**(3):226–239.

Shelbourne, C. J. A. 1970. "Breeding Strategy." Research Work Group No. 1, Res. Comm. of the Aust. For. Council, Proc. 2nd Mtg., Beerwah, Australia.

Shelbourne, C. J. A., Thulin, I. J., and Scott, R. H. M. 1972. "Variation, Inheritance and Correlation amongst Growth, Morphological and Wood Characters in Radiata Pine." Forestry Research Institute, Genetics and Tree Improvement Report No. 61, New Zealand Forest Service, Rotorua, New Zealand.

Shimuzu, J. Y., Pitcher, J. A., and Fishwick, R. W. 1976. Early selection of superior phenotypes in *Pinus elliottii. PRODEPEF,* Brazil.

van Buijtenen, J. P. 1969. "Progress and Problems in Forest Tree Selection." Proc. 10th South. Conf. on For. Tree Impr., Houston, Tex., pp. 17–26.

van Buijtenen, J. P., and Saitta, W. W. 1972. Linear programming applied to the economic analysis of forest tree improvement. *Jour. For.* **70**:164–167.

Vidakovic, M. 1965. "Selection of Plus Trees." Sumarski List, Internacionalni simpozij, IUFRO, Zagreb, pp. 7–20.

von Weissenberg, K. 1973. Indirect selection for resistance to fusiform rust in loblolly pine. *Acta For. Fenn.* **134** 1–46.

Walters, J., Soos, J., and Haddock, P. G. 1960. The Selection of Plus Trees on the University of British Columbia Research Forest, Haney, British Columbia." Research Paper No. 33, University of British Columbia, Vancouver, Canada.

Wilkinson, R. C. 1980. Relationship between cortical monoterpenes and susceptibility of Eastern white pine to white-pine weevil attack. *For. Sci.* **26**(4):581–589.

Wright, J. W. 1960. "Individual Tree Selection in Forest Genetics." Proc. 4th Lakes States For. Tree Imp. Conf. Stat. Paper No. 81, Lake States Forestry Experimental Station, pp. 25–44.

Yli-vakkuri, P. 1953. Tutkimuksia puient valisista elimillisista juuriyteyksista mannikoissa [Studies of organic root-grafts between trees in *Pinus sylvestris* stands]. *Acta For. Fenn.* **60**(3):1–117.

Zobel, B. J., Stonecypher, R., Brown C., and Kellison, R. C. 1966. Variation and inheritance of cellulose in the southern pines. *Tappi* **49**(9):383–387.

Zsuffa, L. 1975. Broad sense heritability values and possible genetic gains in clonal selection of *P. griffithii x P. strobus. Sil. Gen.* **25**(4):85–88.

CHAPTER 6

Seed Production and Seed Orchards

The applied aspect of tree improvement consists of the development of improved trees followed by mass production of the improved stock. No program will be successful until both have been achieved. Too often, improvement of forest trees is obtained without sufficient concern about how the improved material is to be reproduced and used on an operational scale. The better trees can be multiplied through seed regeneration or by vegetative propagation. (Vegetative propagation is covered in Chapter 10; this chapter will concentrate on the production of improved seed for operational planting, although there will be reference to seed production for breeding or developmental activities.)

All tree improvement programs must have seed production at some stage of their development if continued gains are to be achieved. This is true even for programs using vegetative propagules for large-scale operational planting; seed is needed for the development of outstanding trees from which vegetative propagules can be obtained.

Organizations with extensive planting programs need large quantities of improved seed immediately. The approach followed in the circumstances of immediate need will be somewhat different from that in which the need for seed for operational planting is still sometime in the future. Even when results from the breeding program may not be used until some years in the future, it is essential to set aside or establish areas for seed production early.

The first and most difficult problem related to seed production for an operational program is to determine the amount of seed needed. In young programs, this estimate must often by made without a good knowledge of the seed-producing capacity of the species or how to manage or treat trees for greater seed production or even how to handle the seed in storage and in later germination. This problem of seed production and handling is especially critical for many tropical and subtropical species for which very little biological information is currently available. Information on the characteristics of many types of seeds for genera and species in the more temperate areas of the world is available in the European literature. Also, much is known about seeds of a number of Asian species. Especially well known are the characteristics of seed of species used in exotic forestry programs in South America, Australia, Africa, and elsewhere. In the United States, the U.S. Forest Service publication "Seeds of Woody Plants in the United States," which was edited by Schopmeyer in 1974, is comprehensive. We know of no similar publication that covers seed of tropical tree species.

Plans for seed production should exceed currently known requirements. Often overlooked in assessing needs for a given program is the continuing trend toward shorter rotations and continued expansion of forest area under management that results in the use of greater quantities of seed than were initially required. Of key importance is the fact that a buffer must be calculated for loss or failure of production of seed. A normal, effective, and conservative method is to assume that good seed crops will not be obtained every year. In addition, all organizations should keep at least a 3-year supply of seed in storage (when storage is possible). The prudent approach is to plan to produce an amount of seed that at least is 30% greater than is currently needed.

There is no method of seed production that is suitable for all species and all conditions. A number of good summaries of differing systems have been developed and published. Two of these, which are for temperate species, are those of Thielges (1975) and Faulkner (1975).

MEETING IMMEDIATE SEED NEEDS

There are several methods that can be used to obtain genetically improved seed for immediate planting. These are usually interim in nature in that they are used only until the more permanent seed orchard becomes available. Often seed from the interim procedures will not yield large volume gains, although they sometimes greatly improve tree quality and pest resistance, and they can ensure good adaptability. Too often, the short-term methods of obtaining improved seed are ignored, and nothing is done to take advantage of the potential genetic gain until the longer-term seed orchards have been developed.

Seed from Individually Good Phenotypes

If seed is needed immediately for an operational program, one viable approach is to choose outstanding phenotypes from natural stands or plantations, mark the trees, and collect seed from them (Figure 6.1). Collection from the good trees is usually done during a logging operation or ahead of the logging crews. In the latter case, the selected trees may be felled and seeds collected when they ripen. The cut trees are then salvaged when the logging operations come to them. In some instances, the marked trees are climbed for seed collection in stands where logging is not planned.

Seed from good phenotypes in which the male is not known and is unselected will usually yield only limited volume improvement because of the low heritability for growth in most species. The very best individuals from natural stands are well adapted to the areas where they are growing. If trees are selected from plantations of native or exotic species, a land race will result that will be more adapted to the plantation site than to the original plantation. Improved adaptability alone usually makes collection from individual trees worthwhile. In most species, certain quality characteristics like stem straightness, and to some extent limb quality, will also be improved. When a change in wood specific gravity is desired, considerable improvement will occur by choice of the correct parents. If disease or insects are serious problems, a good degree of tolerance often can be achieved when seed are collected from healthy trees that are selected from stands that are heavily infected.

Many persons feel that the method of collecting seed from individually good phenotypes is not worth the effort. Although only one parent has the known desired characteristics, if selection is modestly rigid, the cumulative gains, especially for adaptability, will usually repay the effort. When planned ahead and if collections are made only when there is a reasonable seed crop on the desired

FIGURE 6.1

The first step in making genetic improvement is to select good trees from good stands, such as this *P. radiata* plantation in Chile, and to collect seed from them for operational planting. The better pheno-types are marked and cut just before or during a logging operation when the seed are ripe; therefore, seed can be taken from the best trees that had been marked. Gains will be mostly in improved form and better adaptability.

phenotypes, the cost of seed from individually felled trees is not much higher than standard seed collection. Costs will be somewhat more when trees must be climbed. The number of trees selected for seed collection per unit area will vary with species, quality of stands, availability of stands, and selection intensity. Usually, however, not many more than 5 to 10 trees per acre (12–25 trees per ha) will be of suitable quality to use for seed collection.

Seed from Good Stands

Although the gains are not well documented and the method is little discussed, a number of organizations use the practice of making mass seed collections only from the very best stands; these are sometimes referred to as *plus stands* (Faulkner, 1962). The types of gains are generally like those outlined previously for individual tree selections, although the magnitude of gains may be less because of less intensive selection. The method of collecting from plus stands is a worthwhile practice to follow when the individual tree method cannot be used. Seed will be cheaper to obtain with this method, but collecting from one or a few good stands will yield bulked seedlots with a higher degree of relatedness than seedlots collected from individual trees growing in many different stands.

Seed Production Areas

Seed production areas, also called *seed stands,* are quite widely used in young programs, especially for exotic species. Seed production areas have only limited application for those organizations with advanced tree improvement programs. In seed production areas, the poor phenotypes are rogued from the stand and the good trees are left to intermate. Eventually, seed are collected from them (Andersson, 1963; Dyer, 1964). Seed production areas are rarely progeny tested; therefore, both of the parents are selected only on their phenotypic qualities. Based on a number of tests, only limited genetic improvement for volume growth has been found for pine seed production areas in the southern United States (LaFarge and Kraus, 1981). However, growth improvement has been reasonably good from a number of seed production areas from plantations of exotic species, including pines and *Eucalyptus.* In this instance a land race results when good phenotypes are used as parents.

Because of the considerable value of seed production areas to new programs using exotic species, they will be treated in some detail in this chapter. It must be emphasized that seed production areas are generally used as interim sources of seed in forest tree improvement and that they are phased out as better genetic seed becomes available from seed orchards. Seed production areas have great utility in a number of tropical and subtropical countries.

Seed production areas have three attributes that are vitally important. These are as follows:

1. Seed collected will have better genetic qualities than seed from commercial collections, especially in adaptability, bole and crown characteristics, and pest resistance.

2. When seed production areas are established in natural stands (and in some plantations), the geographic origins of the parent trees are known, thus yielding seed from a suitable source. This is not true for many exotic plantations, but selection of the best individuals in an exotic plantation will result in the development of a land race.

3. Seed production areas are reliable sources of well-adapted seed at modest cost. Assurance of seed supply is becoming of ever-increasing importance, especially in rapidly expanding plantation programs that are using species suitable to tropical or subtropical conditions.

Specifications for a Seed Production Area   The best natural stands or plantations that are near full stocking are used for the development of seed production areas. There are no specific age limitations, other than the fact that the stand must be old enough to produce seed, and the individual trees must have sufficient crown surface areas so that they can produce large seed crops. For the southern pines, stands between 20 and 40 years of age are acceptable for use as seed production areas, whereas 10 to 20 years is a good age for *P. caribaea* and *P. oocarpa* (it can be 3 to 4 years for some eucalypts). Usually, seed production areas should contain a minimum of 10 acres (4 ha) in size because managing small stands is inefficient, and the danger from contamination from outside pollen is great. Exceptions to the 10-acre minimum are when only limited amounts of seed from a species are required or for species with very heavy seed production, such as sycamore (*Platanus occidentalis*). Other exceptions are some eucalypts where a couple of acres will supply all the seed needed.

For collection efficiency and to assure adequate cross-pollination, it is usually best if 50 seed trees of acceptable phenotype per acre (125/ha) can be retained, although the optimum number of "leave" trees will depend on tree size and selection intensity. It is sometimes impossible to obtain 50 good trees per acre. However, an area with 20 to 30 seed trees per acre (50 to 75/ha) after thinning is considered to be stocked well enough for operational use. Economic and biological considerations generally dictate that if less than 10 to 12 trees per acre (25/ha) cannot be retained after roguing has been completed, the potential seed production area should not be used. A very real danger for some species on certain sites is "windthrow" that is caused by opening up the stand too much.

Stands that look good for a seed production area based on a casual inspection are often found to be quite unsuitable when they are closely checked. Usually, there are very few stands of the correct age and location with sufficient good trees to ensure the minimum number of trees per unit area. When the stand chosen is judged to be suitable for seed production, it must also be situated so that it can be cleaned and managed for optimum seed production.

Selection of Trees for a Seed Production Area   Desired attributes of the trees left in a seed production area are similar to, but less rigorous than, the qualifications required for a select tree to be used in an intensive tree improvement program. Both because of their growth and seed-producing potential, only trees in the dominant and codominant crown classes are considered for retention. The seed production area tree must have a high level of vigor, be straight, have desirable limb form and pruning, and be free from insects and diseases. No tree below the desired standard should be left, regardless of spacing. Trees exhibiting the potential for good cone production are given preference, although the evidence of past seed production is not essential if the trees have been growing in a tightly closed stand. Excellent seed crops are often produced after heavy thinning by trees that exhibited little seed production prior to thinning (Figure 6.2). However, it is necessary that stands be thinned at a young enough age so that the trees retained will have crowns that are large enough and healthy enough to produce large seed crops.

It is essential that the crowns of the crop trees be released to full sunlight on at least three sides if good seed production is to be realized. When several good phenotypes occur in a group, enough of them must be removed so that the remaining ones will receive enough light to respond to the release. In spots where

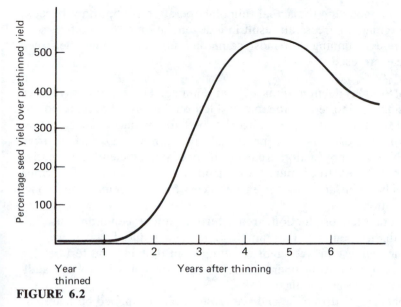

FIGURE 6.2

Full-seed production in a seed production area is delayed for 3 to 4 years after thinning. The amount of seed available following release is shown above for a loblolly pine seed production area. After the sixth year, the seed yield drops off because the stimulation from release disappears.

the only trees available are inferior phenotypes, *all trees must be removed,* even if this results in a fairly large opening in the stand. Also, even if seeds were not collected from the inferior trees that might be left to fill space, the value of the seed crop will be reduced because of contamination with the pollen from the inferior tree.

For most species, an isolation zone or pollen dilution zone should completely surround the seed production area. It is virtually impossible to eliminate completely all contaminating pollen; the purpose of the dilution zone is to reduce it to negligible amounts. Because of wind dynamics and the "dumping" effect of wind currents when they encounter an open area, it is usually best that pollen dilution zones be left unplanted or maintained in low-growing annual or perennial species. If trees are to be grown in the dilution zone, they must be of a species that does not generally hybridize with those in the seed production area.

Thinning the Seed Production Area Timing and caution are of the utmost importance when removing undesirable phenotypes from seed production areas. Timing is important because it determines the year when the seed production area will produce the first commercial crop, and also because of its possible effect on pests that might appear following thinning. Usually, heavy cone crops of pine are not obtained until the fourth or fifth year after thinning the stand, because of the time required for development of a large, vigorous crown (see Figure 6.2).

Extreme care in conducting the actual thinning operation is important because damage to the remaining trees can result in a degeneration of trees left as seed producers. Careless thinning is a most common cause of trouble when seed production areas are used.

Management of Seed Production Areas After thinning, it is necessary to remove the logging residue and to reduce the material left on the forest floor. Removing the residue allows for easier access into the area for management activities, reduces the potential dangers from pests and gives a measure of safety from "wildfires." If the seed production area is to be operated efficiently, vegetative material under the seed trees must be controlled. The forest floor will then become relatively open, allowing for easier access for management and cone-harvesting operations.

For permanent seed production areas, fertilization is commonly used in conjunction with the opening of the stand by thinning to induce heavy flower and cone crops. The increase in tree vigor resulting from thinning and fertilization enables the development of heavier and denser crowns that will produce additional male and female primordia.

Pesticide sprays to control cone and seed insects can be applied both aerially and from the ground. Efforts to control pests are sometimes not totally successful because of the difficulty of getting good spray coverage on large trees and because of the difficulty of timing the sprays to coincide with insect infestations. Spraying is expensive, and, coupled with the uncertain success, it is often not considered to be economical, although some insecticides have very effectively controlled certain cone and seed insects in pine, especially when applied aerially.

Harvesting Seed from a Seed Production Area Seed production areas can be classified into *temporary* and *semipermanent* types; the type used will be dependent upon the need for seed and the extent of good stands. A *temporary* area is established and managed so that when a heavy cone crop is obtained, the seed trees are felled to harvest the seed, cones, or fruit. This method can be used where stands acceptable as seed production areas are numerous, so that new seed production areas can be established to replace the older ones when they are harvested.

The *semipermanent* seed production area is operated on the principle that several crops will be harvested following the response to initial thinning before the seed production area trees are felled. It is necessary, from an economical perspective, to determine the sound seed yield per tree. Each crop tree should be roughly inventoried from the ground to determine if a sufficient quantity of seed is available to warrant collection. The best procedure is to scan a specified sector of the crown through field glasses, and then to extrapolate the total number of cones or fruit on the tree from the number observed in the sector. Usually less than half of the cones or fruit present within the tree's crown can be observed from the ground, but specific rules need to be developed for each species.

Collecting cones and fruits from semipermanent seed production areas is more expensive than collecting from temporary ones. Most organizations prefer to hire professional climbers because of their proficiency, although for some species, mechanical tree shakers are very effective. Some damage to the crown of the tree is inevitable during harvesting operations, but heavy damage will reduce seed crops materially in future years. Specifications regarding the method of getting into the crown, the amount of damage to the crown that can be tolerated, the availability of climbing crews when the cones or fruits are ready for harvest are essentials that should be agreed upon and included in the seed collection contract when it is written.

Seed from Proven Sources

One of the most common methods of obtaining large quantities of seed quickly is to go back to the original source or provenance that has been tested earlier and that has proven to be suitable. This can be a great success or an equally great failure. If the source information is correct and is indeed the same as reported for the initial successful planting, the method works well. Much depends on the integrity of the seed supplier and how accurately the sources are matched. Many "horror stories" can be told of second collections that were purportedly from the same area that produced a successful introduction but which was totally different in reality and produced quite worthless stands of trees. The only completely safe rule to follow, if operational quantities of seed are to be obtained based upon earlier successful introductions, is for an organization to send its own personnel to assure that the collections really are made from the desired quality of trees from the correct area.

LONG-TERM SEED NEEDS—SEED ORCHARDS

The standard method of producing genetically improved seed in operational quantities is to use the seed orchard approach (Andersson, 1960). Of the many definitions for seed orchards, two are given here. A *seed orchard* is an area where "seed are mass produced to obtain the greatest genetic gain as quickly and inexpensively as possible" (Zobel et al., 1958; Zobel and McElwee, 1964). Feilberg and Soegaard (1975) use the following definition: "A seed orchard is a plantation of selected clones or progenies which is isolated or managed to avoid or reduce pollination from outside sources, and managed to produce frequent, abundant, and easily harvested crops of seed" (Figure 6.3). Seed orchards are not always solely for genetic improvement of specific characteristics but can be used to produce quantities of seed that are adapted to a specific planting location (Gerdes, 1959; Nanson, 1972). The seed orchard definitions given here apply specifically to situations in which seed are needed immediately for large operational planting programs (Figure 6.4). The objectives and methodology of seed orchard estab-

FIGURE 6.3

The objective of a seed orchard is to produce maximum amounts of seed with as much genetic improvement as quickly and as efficiently as possible. To do this large crops are needed. The orchard is managed to produce the maximum number of seed. Shown on top are female strobili on a pine tree; on the bottom are mature pine cones.

FIGURE 6.4

Shown are two seed orchards. (*a*) shows a young *P. silvestris* orchard in Sweden, (*b*) shows a producing loblolly pine orchard in the southern United States.

lishment may be modified when seed are not needed for immediate use but where there is a perceived future need for seed.

When there is an urgent need for large quantities of seed because large planting programs are involved, shortcuts must be employed to obtain seed as soon as possible, even though some genetic gain may initially be sacrificed. The method showing how this is done was detailed in the section "The Importance of Time" in Chapter 1. The method consists of initially establishing orchards based only on the phenotype of the parent trees and then later roguing out the poor genotypes from the orchard based on progeny test results rather than waiting to establish an orchard only with parents that have already been tested for their genetic worth.

Seed orchards are usually established with a number of assumptions that are not fully correct. Chief among these is that flowering and pollen exchange among genotypes in the orchard will be uniform and equal. In reality, this is rarely the case. Some genotypes or clones produce many more flowers or pollen than others, and time of flowering is such that certain genotypes rarely ever mate because they flower out of synchronization (Barnes and Mullin, 1974; Beers, 1974).

The utility of seed orchards has been widely documented for many kinds of benefits. Orchards have produced meaningful gains in disease resistance, growth, wood qualities, adaptability, and in tree form.

Types of Production Seed Orchards

There are numerous kinds of seed orchards, but they generally fit into one of two broad categories. *Vegetative orchards* are those established through use of such vegetative propagules as grafts, cuttings, tissue culture plantlets, or other methods. Vegetative orchards are the most common type used operationally. The other general type is called a *seedling seed orchard*. These are established by planting seedlings followed by a later roguing that will remove the poorest trees, generally leaving the best trees of the best families for seed production.

There has been much discussion about the best types of seed orchards to use. Sides have been taken, often inflexibly. Discussion was so animated during the early 1960s that a special issue of *Silvae Genetica* (Toda, 1964) contained articles on the pros and cons of the two types of orchards. Since then, many articles have been published that champion one of the two types of orchards. The simple fact is that there are situations in which either type of orchard may be particularly well suited. It is most important, however, to make the decision about which kind or orchard to use that is based on the pros and cons that will be listed later; some of these are given in some detail to help the reader make better judgments about which type of seed orchard is best for a particular situation (Barber and Dorman, 1964).

Under certain conditions, a genetic test plantation can be converted into a seedling seed orchard, thus fulfilling both the testing and seed production functions at one time. When this is possible, the advantage and efficiency of the seedling orchard is evident. This advantage was initially overemphasized, and many exceptions are now recognized in which both activities cannot be accomplished on the same site. One consideration is that genetic tests should be planted on lands that are typical of those that are to be reforested with improved trees; these lands may not be conducive to good seed production or efficient orchard management (Kellison, 1971). As an example, if the progeny are tested in a swampy area or at a high elevation, the seed orchard function also cannot be followed.

Management criteria will differ for genetic tests and seedling seed orchards. If a tree is to produce heavy cone crops, it must have its crown fully developed and be in full sunlight. On the other hand, a suitable progeny assessment of growth can be made only after the trees have grown long enough to assess how they will grow when competition is strong. Therefore, the final choice of the best individuals from the best families cannot be made until that time. However, if one waits for competition before thinning for seed production, the crowns of the residual trees will be so reduced that seed production will be seriously impaired and costs of collection will be greatly increased. If thinning is done at a suitably young age so that full crown development, and thus heavy seed production, is achieved, the efficiency of the selection phase can be increased greatly. When the trees are planted at initially wide spacings to produce well-developed crowns, the progeny test function is impaired.

A partial solution to this dilemma is to establish several tests of the same families simultaneously in different areas. One can be managed for eventual seed production (i.e., wide spacing, intensive culture); the other tests are used strictly to assess family performance. The seedling seed orchard is then rogued of poor families, based upon their performances in tests designed for that purpose. A disadvantage of this alternative is that only family selection can be efficiently practiced in the orchard. Individual trees cannot be accurately assessed because they have been managed for the purpose of seed production, not for the assessment of their timber-producing characteristics. For crops where early decisions about superiority are possible before competition and for very short rotations, such as for Christmas trees, the seedling seed orchard approach is quite satisfactory. Many different designs and schemes have been worked out to take advantage of the *seedling orchard* concept. Among these are Klein (1974), Riemenschneider (1977), and Cameron and Kube (1980).

Most vegetative orchards have been established by grafting. Quite frequently, incompatibility between scion and stock develops, with resultant sick or dead grafts (Figure 6.5). For some species, such as Douglas fir, incompatibility has been a serious problem, and very heavy losses have been encountered. Such species as loblolly pine have about 20% graft incompatibility, which creates problems but can be tolerated. Most tree improvers have simply accepted this amount of incompatibility and have worked around it.

Methods have been developed to assess graft incompatibility early and to try to avoid it (Slee and Spidy, 1970; Copes, 1973a, 1981; McKinley, 1975). Currently, some vegetative orchards are being established using rooted cuttings in several

FIGURE 6.5

Shown is a typical incompatible pine graft. The graft union is abnormal, the needles are short, the tree lacks vigor, but flowers heavily. It will soon die. Incompatibility can occur almost immediately after grafting, or it can be delayed, as shown, and seriously affect trees that are 20 years or more in age.

conifer and hardwood species, such as spruce and the eucalypts. In the future it is also possible that tissue culture plantlets may be used. These types of vegetative propagation avoid graft incompatibility, but root deformation and imbalance and flowering problems sometimes develop. If severe problems are encountered with the health of vegetatively propagated orchards, then *seedling orchards* have a *clear superiority*.

The magnitude of genetic gains obtainable from the two different types of seed orchards has been debated. There are many qualifications that can be made, such as whether open-pollinated or control-pollinated seed are used for the genetic tests and seedling orchards, and which generations of improvement are being compared. If seed used for the seedling orchard are from selected parents, there will be a generation of difference between clonal and seedling orchards. If a seedling seed orchard is established as a joint test and orchard, both family and within-family selection are possible. If the progeny test is separate from the seedling orchard, then the roguing of the seedling orchard is based almost entirely on family performance.

A seedling orchard has a larger number of parents involved than the 30 to 50 that are commonly used in a vegetative orchard, thus giving it a broader genetic base. However, the selection differential is less than that for the more intensively selected parents used in the vegetative orchard. Perhaps what is the most important fact of all, but which is often not considered, is that an outstanding genotype will appear only once in a seedling seed orchard, whereas it can appear many times in many orchards when reproduced vegetatively.

Testing of each parent used in a vegetative orchard is mandatory for later roguing of poor parents. Roguing can significantly increase the genetic gain from a given orchard by removing the genotypes that have poor characteristics, such as disease susceptibility or poor growth. Testing in seedling seed orchards is not done by individuals but on a family basis, and one never can pinpoint the occasionally outstanding good general combiner parent.

The age at which seed production on seedlings and grafts occurs is highly important. Seedling orchards are very suitable for species like most *Eucalyptus,* black spruce (*Picea mariana*), the early flowering pines, and many hardwoods that produce seed at a young age. For species that have flowering delayed for 10 to 20 years, the vegetative orchard is usually best, because grafts tend to retain the physiological age of the parents from which the scions were taken, and early flowering results when grafts are made from scions from mature individuals. When young material is needed to obtain satisfactory rooting, flowering is sometimes delayed several years longer from cuttings than from grafts from mature parents. The advantage of early flower production on grafts from species that have delayed flowering is a major advantage to the vegetative approach when selections have been made from reproductively mature trees.

Until other information is available, related matings must always be avoided in all types of orchards. The danger of selfing is considerably enhanced in a vegetative orchard if the ramets of a given clone are not well separated (see Figure 2.6). Using only a few clones in the orchard makes this problem more severe, because it is not possible usually to keep a suitable separation in a vegetative orchard with less than 10 clones (van Buijtenen, 1981). In seedling orchards, selfing can occur only on the individual tree, which is what happens on the individual graft in a vegetative orchard. Great care must be taken to have related families properly spaced to avoid mating among relatives; this is a frequent oversight in seedling seed orchards.

One advantage often claimed for vegetative orchards is that they can be placed in a location that is most suitable for efficient operation, or in a climate proven to be suitable for early, heavy, and dependable seed production. This capability is now increasingly being used, even to the extent of establishing seed orchards from Northern Hemisphere species in the Southern Hemisphere to produce seeds for use in the northern areas. Seedling orchards can also be established to take advantage of such an improved flower productivity, but the dual testing and seed production function is then lost, and only family information can be exploited to rogue the seedling orchard.

Seed Orchard "Generations"

Seed orchards are commonly categorized by generation; that is, first-, second-, or more advanced-generation orchards depending upon how many cycles of improvement they represent. No matter what kind of orchard is established, pedigree records must be kept to minimize deleterious related matings and to help ensure use of only the best genetic types. The *first-generation orchard* usually results from selection from natural stands or unimproved plantations, most often by using individual tree selection methods. The pedigrees of the parent trees are usually not known. First-generation orchards are improved by roguing; that is, by removing the less desirable genotypes as determined by progeny testing. The removal of trees for spacing or health reasons in seed orchards is simply a thinning operation, not a genetic roguing. Because a first-generation orchard is started with parents whose genetic worth is unknown, the orchards are usually established at a close spacing to allow roguing of the poor genotypes and still leave a fully productive seed orchard area. Often 50% or more of the initially established clones will be removed. If the orchard is established at the desired ultimate spacing and then rogued, there will be large gaps resulting in inefficiency in seed production and often in poor seed yield and quality because of reduced cross-pollination.

There is a type of orchard that is now commonly, but erroneously, referred to as a *1.5-generation orchard*. It consists of taking the very best genotypes (best general combiners) from a number of orchards of similar geographic backgrounds and bringing them together into a new, greatly improved first-generation orchard. This common method for combining the very best, which results in excellent genetic gains, has been described by the senior author as a *1.5 generation* to emphasize that the orchard will produce greatly improved seed because it is composed of only the very best genotypes. The orchards are *not 1.5 generation:* They are only greatly improved *first*-generation orchards.

Seed Orchard Location, Establishment, Size, and Management

Many things must be considered when establishing a seed orchard. Location, size, type of orchard, and management are all of vital importance; these will be discussed later. If each has not been properly considered, an orchard may fail completely or be inefficient. A basic rule in orchard establishment is to plan ahead; 2 to 3 years' preplanning for initial establishment is necessary, and a planning period of many years is required for the total orchard complex. There have been many publications relating to the various phases of the development and management of seed orchards. Some of these are Miyake and Okibe (1967), Chapman (1968), and Sprague et al. (1978). Most information that is available is for conifer orchards, although a few publications (e.g., Taft, 1968; Churchwell, 1972; Dorn and Auchmoody, 1974) deal with hardwoods.

Location   A crucial early decision relates to the location of the seed orchard. Accessibility, potential labor supply, soil texture and fertility, air drainage, water supply, geographic location, isolation, insects, disease, and destructive animal problems must all be considered. Of key importance (but unfortunately determined only after a long experience) is whether the environment of an area favors the production of flowers. A major consideration when an orchard site is chosen is whether it is an area where seed production will be reliable from year to year. Another important consideration is how soon the orchard will come into commercial seed production. Many sites will ultimately produce seed, but the environment of some sites may be particularly good for early and heavy flowering. Some of the great differences in flowering and seed production potential can be predicted, whereas other differences occur for unknown or rather obscure reasons. The decision about where an orchard will be established must also be influenced by the risk involved from adverse environmental factors and by the financial considerations of the organization that is involved.

Often overlooked is the public relations' value of a seed orchard. Therefore, within the biological and economic considerations that are related to orchard locations, attention should be given to the accessibility and value of the orchard with respect to public relations. One of the most certain and most effective methods of obtaining support for a tree improvement program is to have a well-established, well-kept, and well-planned seed orchard that is easily accessible to those who are involved in the allocation of financial resources.

Another major consideration in location of seed orchards relates to the possible loss or alternate usage of the seed orchard land for roads, airports, dams, supermarkets, pipelines, and so forth. This may sound trivial but it is not; a whole list of stories could be cited in which programs were stopped because the seed orchard area was needed and was taken over for other purposes.

The most efficient orchard is one that is close to headquarters, with equipment and manpower that are centralized and readily available. With proper planning and with sufficient personnel, seed orchards associated with nurseries work well. A concentration of a group of orchards in a single location is efficient but increases the risk of a heavy or complete loss due to natural catastrophes such as tornados, hurricanes, hail, or ice storms (Figure 6.6). Scattered, small orchards reduce the risk of total loss but involve more expensive establishment and operational costs (Vande Linde, 1969). The decision whether to increase security by having several scattered orchards or to have larger, concentrated, and more efficient ones with a higher risk of a major loss must be made before a program starts. Most organizations have chosen the concentrated approach. When this is done, they establish their orchard clones in separate clone banks so that the desired genotypes will not be lost in case of a catastrophe to the operational orchard.

Soils, Topography, and Geography   The appropriate soil type for a seed orchard will vary with species. In general, the orchard site should be average in fertility. It is recognized that poor sites are unsuitable for seed orchards, but reasons for the poor seed production often observed on highly fertile sites are not as well understood. They may be the result of freeze damage to the reproductive structures due to the fact that the trees do not harden off or because the resting bud may not form soon enough for proper reproductive development to be initiated before cold weather (Greenwood, 1981). Finally, they may begin growth too early in the season. Good sites often need an extended time for the orchard to become suitable for commercial seed production because of heavy vegetative and poor reproductive growth. Orchard productivity can be partially manipulated by fertilization and sometimes by irrigation, but very fertile, moist sites do not leave the orchard manager the management options that are available on a less fertile site.

Frequently, orchard success or failure can be directly related to the physical properties of the soil as they affect operating conditions. Abandoned agricultural fields have generally proven to be good seed orchard sites. When seed orchards cannot be established on flat terrain, gently sloping land that is suitable for operating equipment should be sought. In cold regions, the orchard aspect should usually face toward the equator and to the west to obtain maximum light and warmth, especially during the critical flowering season. There are, however, always exceptions with different areas and species that require cool, humid conditions for the best flowering. In several parts of the world, especially those in cooler climates, the method of temperature sums is used to assess the suitability of an area to produce reproductive structures.

An orchard must be located where there is good air drainage, especially in areas where frosts occur. The orchard should also be protected from high winds in the more exposed areas (Dyson and Freeman, 1968) (see Figure 6.6). It is desirable that the orchard be close to a water source because of a possible need for sprays, irrigation, and fire protection. An orchard should not be located in the coldest part of the species range, and it *should never* be established outside the natural range of the species without prior testing. Flowering can be erratic and losses from extreme environmental conditions are common. This does not mean that orchard locations outside the range cannot be excellent. Orchards can be put south (or north in the Southern Hemisphere) of the species range (Gansel, 1973; Schmidtling, 1978; Gallegos, 1981). Excellent seed from the pines in the south-

FIGURE 6.6

Orchards must be located to be as free as possible from damage by environmental factors. Shown (*a*) is a seed orchard that had been badly damaged by an ice storm. (*b*) shows everything that was left of a mature, fully producing seed orchard following destruction by a tornado.

eastern United States are being produced in South Africa and Zimbabwe. Moving orchards can be profitable but *should be done only after tests have shown that normal fruit and seed are produced* in the new environment. In the southeastern United States, experience has shown that the geographic area around Savannah, Georgia, produces consistently good loblolly pine cone crops with the result that a number of operational seed orchards have been concentrated there.

As a rule of thumb, the orchard should be located first in the main portion of the geographic range of the species for which the orchard was established. In some locations with extreme environments, such as cold areas, seed crops may be obtained periodically, but they will be less frequent, and a high proportion of immature or nonviable seed are sometimes produced. For hardwoods as well as pines, freeze damage to the flowers is common. In no circumstances should an orchard be established in a recognized ice belt or in regions that have extreme fluctuations in temperature or rainfall. Precautions should be taken to avoid the location of an orchard in areas where the incidence of pests, such as seed and cone insects, or diseases are known to be serious or where the population densities of deer, voles, rabbits, or other destructive animals are high. This includes human activities. Orchards should be located where the danger of loss to housing, roads, power lines, or other developmental activities of humans are minimal.

Pollen Dilution Zones   An orchard must be protected from contamination by outside pollen. Although most tree improvement workers set the width of a dilution zone, an absolute minimum for pine at 400 ft (122 m) and preferably 500 ft (152 m), it is recognized that these distances are insufficient for complete isolation (Wang et al., 1960; Squillace, 1967; McElwee, 1970). Foreign pollen will still be found in an orchard with a dilution zone of this size, but studies have shown that the bulk of the pollen from outside sources will be dissipated within the dilution zone. The considerations for the maintenance of dilution zones outlined for seed production areas also apply here. Economic losses from pollen contamination can be considerable, and despite large costs, proper maintenance of dilution zones is almost always worthwhile (Sniezko, 1981). Dilution zones are most critical for advanced-generation orchards because of the greater potential loss of genetic gain, and hence profitability from pollen contamination.

Effective pollen dispersion distances of hardwoods are not yet well documented; therefore, the wise approach is to keep hardwood seed orchards as distantly separated from contaminating pollen sources as possible. The distance pollen can be transported by insects from entomophylous species is not known, but it undoubtedly varies greatly by insect and species.

Dilution zones should always be maintained between orchards from different physiographic regions and between advanced-generation and early-generation orchards. Many species, even some in the same genus, do not need to be isolated from each other because they flower at different times, or they will not cross. This is common in the pines. Such information about crossbility is badly needed for all important species, especially for those pollinated by insects. An orchard should be blocked as much as possible to permit maximum cross-pollination among members of the orchard and to reduce edge effects. It is best to orient the long axis of the orchard with the direction of the prevailing wind at the time of pollen dispersal if a rectangular configuration is used.

Seed Orchard Size   The appropriate orchard size is determined by the number of seed needed; a good coverage of methods to do this for the southern pines was the colloquium on seed orchards edited by Kraus (1974). To be certain that enough seed will be available, production capacity in excess of anticipated needs is usually developed. The actual size of the orchard needed will vary by species, location of the orchard, availability of seed from other sources, and seed needs and costs. As an example, most organizations do not consider an orchard of southern pines operational unless it is at least 5 acres (2 ha) in size. In terms of regeneration needs, if planting is to be the method used, it is not usually considered to be feasible to establish a seed orchard complex for southern pine (with a 25-year rotation age) for a forest area smaller than about 150,000 acres (60,000 ha), or an annual planting of 6000 acres (2400 ha). Small landholders usually can obtain improved seed more economically by purchase from the government or from other private organizations. Because many seed orchard costs are not directly related to orchard size, the costs per unit of seed produced will decrease when the size of the orchard is increased.

Contrary to the preceding rule, smaller orchards for specialty purposes are often established as part of a larger orchard complex. Under most circumstances, with species where long rotations are used, a minimum planting program of at least 1000 acres (400 ha) per year must be underway to justify a full-fledged seed orchard program. The rules related to size for conifers often are not suitable for hardwoods. For example, genera such as sycamore (*Platanus*) or *Eucalyptus* produce large numbers of viable seed per parent seed tree, and all the seed needed can be produced in a very small area. If enough clones are to be included in order to minimize related matings by having a suitably large genetic base, at least 1 acre (0.4 ha) of orchard is usually needed.

The most difficult stage in planning a seed orchard is making a good estimate of the number of seedlings that can be produced from a mature seed orchard of a given size. For species with unknown flowering habits, the orchard planner must make an "intelligent guess." The number of cones or fruits, or even the number of seeds produced, are not a good guide. The only real criterion is the *number of plantable seedlings that will be obtained per unit area* of seed orchard. For example, Bramlett (1974) devised a system to determine seed potential and seed efficiency; that is, how many actually useful seeds are obtained relative to the potential per cone. When this and the number of cones are known, the seed yields can be estimated quite accurately. Methods of analyzing cones and seeds have been outlined by Bramlett et al. (1977). Therefore, to estimate the needed size of a seed orchard realistically, it is necessary to know the number of cones or fruits, the number of viable seed per unit area of the orchard as well as the seed-to-seedling ratio that will be obtained in the nursery.

The nursery practices that are used will strongly influence how many plantable seedlings will be obtained from a given area of seed orchard. When calculating seedling yields per unit weight of seed, it is essential to consider that seeds from a fertilized seed orchard are often larger, more vigorous, have better germination, and often have more seed per seed-bearing structure than do those from wild stands that develop under a minimal nutrient status and without insect control.

Determining proper orchard size is difficult and the needed area may vary by two- or threefold, depending on the management given to the orchard. In *Pinus taeda,* for example, seed yields were more than doubled when control of seed-destroying insects was achieved (DeBarr, 1971, 1978). In fact, the economics of seed orchards can be greatly affected by the level of seed production and the additional value from an added amount of seed per unit area of the seed orchard (Beers, 1974; Weir, 1975).

Clonal Dispersal in Seed Orchards   For species that are wind pollinated, orchards should be designed in such a way that a minimum of relatedness will result from crossing among the parents, and so that parent trees will have an opportunity to mate freely with each other. These ideal objectives are easy to define but they are hard to achieve, especially to assure a minimum of related matings and selfing (Lindgren, 1974). There are reports of high levels of selfing in orchards, but with proper spacing of the ramets of the clones, it can be kept to a modest and acceptable level. As one example, Moran et al. (1980) found 90% outcrossed seed, based on 3 years' assessment of a radiata pine seed orchard.

For vegetative orchards, the original number of clones that are used should be enough to ensure a suitably broad genetic base after roguing has been completed. Most first-generation vegetative seed orchards have been established with 25 to 40 clones. If selected correctly, this number should be sufficient to provide a suitable genetic base for operational planting. After testing and roguing, this number may be reduced to 20 or fewer clones. For planning purposes, the conservative approach is to assume that about half the clones will be rogued from the seed orchard following testing (Figure 6.7). It is essential to recognize that the operational seed production function cannot be efficiently combined with the clone bank or breeding functions. Many errors have been made trying to combine these functions, and orchards have been established with 300 to 400 clones, with the assertion that these many clones are needed in the breeding program. Large numbers of clones are needed in the breeding population, but gains from production orchards with too many clones will be seriously restricted because of a low selection differential.

The necessity for separating the clones in a seed orchard to prevent problems caused by related matings and the rate of inbreeding must be understood (Stern, 1959). Although it varies greatly by species, many forest trees, such as *Liriodendron tulipifera,* are rather highly self-compatible. Only a few, such as *Liquidambar styraciflua* and *Platanus occidentalis,* are known to be reasonably self-incompatible (Schmitt and Perry, 1964; Taft, 1966; Beland and Jones, 1967). Some genera, like *Eucalyptus,* contain some species that self easily and some that do not. In many forest tree species self-compatibility is the rule; when selfing occurs there is a greatly reduced seed germination and often a dramatic loss in growth (Barnes, 1964; Diekert, 1964; Sorenson and Miles, 1974; Franklin, 1969.

FIGURE 6.7

After progeny testing has indicated the parents with the most desired genotypes, seed orchards are rogued of the undesirable genetic material. Shown is a young seed orchard with its first roguing.

Consequently, ramet dispersal in orchards of self-compatible species is an extremely critical factor in orchard establishment. Clonal dispersion can be particularly difficult for some hardwoods (this will be discussed later in this chapter).

A myriad of designs for vegetative seed orchards have been worked out, which range from the trial-and-error type to sophisticated computer designs. Systematic seed orchard layouts have the advantage of simplicity in establishment, and they allow easy movement from one ramet of a clone to another. They can cause problems when genetic roguing is done, because good trees do not occur systematically in the seed orchard; that is, there is no assurance that good- and poor-performing parents will not be established in groups in the orchard by chance. When this happens, wide openings will occur in the orchard following roguing, and some good genotypes will need to be removed, strictly for proper spacing. The key to determining the proper number of desired individuals is an orchard is to have enough to allow for roguing the poorer genotypes, to have the desired spacing, to maximize seed production by having enough good trees to have adequate pollination, and to ensure for a minimum of relatedness.

A good seed orchard design must have flexibility for the improvement of the genetic quality of the orchard by roguing as well as to minimize the potentials for inbreeding. Enough trees must be established in the orchard to permit several roguings and still leave enough trees for seed production. Each clone should be represented by approximately equal frequencies per unit area. Also of impor-tance is to avoid "repetitive neighborhoods" in which the same clonal pattern is repeated several times. When this is not done, clusters of good or poor clones can be located together, thus making it most difficult to do a good job of roguing. Also, when repetitive neighborhoods are used, certain clones tend to pollinate their neighbors, and the proper within-orchard mixture is not achieved.

The preceding criteria were used to develop a design for second-generation loblolly pine seed orchards in the North Carolina State University–Industry Tree Improvement Cooperative. It was recommended to start with 145 trees per acre (338 trees per ha), remove 35% of the remaining trees at each of four roguings, and ultimately to rogue 60 to 70% of the clones and to maintain a distance of 90 ft (28 m) between ramets of the same clone or related individuals. To achieve the foregoing, not less than 30 nor more than 40 clones are recommended for use. Roguing, as recommended, results in 10 to 12 clones left with a stocking of 36 to 60 trees per acre (80 to 150 trees per ha). In actuality, about 25 trees per acre (63 trees per ha) would probably make up the final orchard. Although the preceding recommendations are for one species, loblolly pine, they do give an indication of the methods used in designing a seed orchard. It is impossible to list and describe all the varied designs that have been suggested for establishing seed orchards, but a few differing types are listed as follows:

1. Goddard (1964) described some ideas about the genetic distribution in a seedling orchard following selection both within and among families.

2. Giertych (1965) early developed a systematic layout for orchards.

3. Burrows (1966) developed a theoretical model of clonal dispersion to give the greatest possible gains.

4. Van Buijtenen (1971) described the theory and practice of orchard designs for the southern pines.

5. Klein (1974) developed a special design for a jack pine seedling seed orchard.

6. Gerdes (1959) covered possible methods to use for Douglas fir, a species that has several difficult problems, especially with graft incompatibility.

7. Hatcher and Weir (1981) discussed the advanced-generation design used by the North Carolina State–Industry Tree Improvement Cooperative.

No matter what design is chosen, it must assure that related matings are minimized, and it should enhance random pollination. An orchard that is estab-lished incorrectly or carelessly becomes a horror to rogue. Instead of upgrading the genetic quality of the trees in the seed orchard, all the roguing accomplishes is to remove related individuals that are too closely spaced. *Much genetic gain is lost by poor orchard designs or sloppy planting in which a reasonably good design is not followed.*

Seed Orchard Management   Much of the advantage of a tree improvement program is lost if the seed orchards do not produce seed to their maximum potential. It certainly is beneficial to use genetically superior genotypes that are phenologically synchronized to assure cross-pollination and to use those that are inherently heavy seed producers to obtain maximum seed production. It is of equal importance, however, to understand the environmental factors and management practices that enhance seed production. An orchard that is suffering from a soil-nutrient deficiency, soil compaction, or overcrowding will not produce seed to its potential, regardless of the inherent superiority of the stock contained therein.

Seed orchard management is extremely complex. Proper procedures will vary according to the species, location of the orchard, and conditions encountered from one year to another within the same orchard. All that will be covered in this short section is a general discussion of a few of the most important items. Special articles have been published relating to the management or orchards of hardwoods (Churchwell, 1972) and for pine (Fielding, 1964; Swofford, 1968; van Buijtenen, 1968).

Soil Management   Soil *texture* (i.e., the proportion of sand, silt, and clay) is essentially unchangeable through manipulation, and as a result, the quality of the soil is one of the most important considerations in the establishment of the orchard. Soil texture has an influence on the moisture- and nutrient-holding capacity, compactibility, erodibility, and other soil characteristics. The soil texture most desired will vary with species, although most species are similar to loblolly pine, where there is evidence that a sandy loam overlying a friable subsoil, such as sandy clay, is conducive to flowering (Gallegos, 1978). Regardless of the soil texture, normal operating traffic in the seed orchards may change soil *structure* (i.e., how the sand, silt, and clay are aggregated), usually in an undesirable direction. Such activities cause formation of compaction layers ("hardpans") (compaction is worse in the clays). These pans are frequently responsible for a general decline in the vigor and seed production of the orchard and, if left uncorrected, they can result in outright death of the trees from root-penetration problems, drainage problems, and from excess concentration of salts on the pan where the roots are concentrated. The most obvious symptoms of decline in the health of orchard trees that are caused by a pan are a rounding and thinning of the crowns, shortening and poor coloration of foliage, a flattening or "fluting" of the bole of the rootstock portion of the grafts, and the emergence of the tops of large roots on the soil surface (Figure 6.8). Some of these symptoms are also indicators of graft incompatibility and may be misinterpreted as such. Incompatibility is strongly clonal, whereas general decline, because of hardpan problems, usually affects all clones to some extent. However, both can be related, and symptoms of incompatibility will appear much more rapidly under the stress conditions that result from soil compaction.

Subsoiling in established orchards helps alleviate conditions of soil compaction and associated pans (Gregory, 1975; Gregory and Davey, 1977) (Figure 6.9). It severs surface roots, resulting in greater root proliferation to greater depths, and it reduces surface water runoff, thereby improving soil moisture in the seed

FIGURE 6.8

When the soil is compacted or hardpans are present, grafts in the seed orchard often grow poorly, break down or die, (*a*). When subsoiled, the large surface roots are severed, and then reproduce a whole matrix of small feeder roots as shown (*b*) where subsoiling had been done 2 years previously.

orchard. Recent studies show that in some cases subsoiling improves flower production as well as plant development and vigor (Gregory and Davey, 1977). It may also help prevent damage by diseases that spread by root grafts through severance of the roots that would be used by the disease for movement.

The normal procedure is to subsoil on two sides of the tree in 1 year. About 2 years later the process is repeated at right angles to the original direction. It is absolutely essential that a coulter (rolling cutter) precede the subsoiler (see Figure 6.9) so that the surface roots are severed and not torn loose at the root collar of the tree. Usually, the subsoiling operation is done just prior to floral initiation. Subsoiling prior to orchard establishment is *strongly* recommended for all new orchards and is considered essential on land that was formerly cultivated or pastured, which usually will have problems of soil compaction and plow-pan formation. Initial subsoiling should be done in a grid pattern corresponding with the intended planting location of the graft. It is important to understand,

(b)

FIGURE 6.8 (*continued*)

however, that subsoiling in an established seed orchard should be done upon prescription to alleviate detrimental soil conditions and should not be applied as a "blanket" treatment in all orchards.

Surface of the Orchard   Much attention must be given to the surface of the seed orchard. Efficiency and speed of operation are increased with well-prepared smooth surfaces. In some instances, the requirement for a good orchard surface will necessitate filling and leveling and the establishment of a sod cover prior to orchard establishment. However, it is essential that good soil characteristics be maintained or restored in areas that have been altered.

The orchard floor should be protected from wind and water erosion, and the soil's organic matter should be maintained at adequate levels for proper nutrient

FIGURE 6.9

Seed orchards must be intensively managed. One most common
treatment is subsoiling; a subsoiler is shown. Seed orchards must be
managed for what they are; intensively operated orchards.

and water relations. These objectives can be met through the establishment and
maintenance of a good sod cover that will reduce soil compaction resulting from
traffic in the orchard. It will also greatly enhance trafficability during inclement
weather. Sometimes a sod cover is disadvantageous when mice or voles are
prevalent; when this is the case, sod should be kept away from the immediate
vicinity of the tree (about 1 m). Although native or naturalized grasses will
eventually colonize a new seed orchard, it usually is advisable to sow a sod cover to
protect the soil from erosion and compaction (Bengston and Goddard, 1966;
Schultz et al., 1975).

Frequent mowing of the sod and weeds in seed orchards favors sod maintenance and allows for the best use of fertilizers by recycling the nutrients to the
trees. Grasses and herbaceous material rapidly utilize the fertilizers during the
early part of the growing season; if not mowed while in a succulent stage, recycling
of the nutrients may be delayed considerably. Mowing during the latter part of the
growing season is done to control vegetation, to alleviate fire hazards, and to
facilitate cone or seed collection. Fire is fatal to thin-barked grafts, and several
seed orchards have been destroyed by only a light burn. Even burning under
controlled conditions has resulted in serious damage to the grafts. Grazing the
grass in the seed orchard has been suggested, but this should never be done
because of the resultant compaction and injury to the orchard trees.

Orchard Fertilization   Based upon soil analyses or in some instances foliar analyses, soil amendments are applied when needed to maintain plant vigor and to promote flowering. Fertilization, particularly applications of nitrogen and phosphorus, has promoted flowering for nearly every species for which trials have been established, especially for various hardwoods (Steinbrenner et al., 1960; Webster, 1971; Jett and Finger, 1973; Greenwood, 1977; Hattemer et al., 1977).

Soil acidity (pH) is of key importance and has a direct effect on many reactions in the soil and on the behavior of soil organisms and plant roots. The optimum pH will vary with species; for most conifers the desired range is 5.5 to 6.5. When the acidity drops below 5.2 or rises above 6.5 for the pines, corrective action should be taken. For some hardwoods, the desired pH is higher and for a few conifers much lower values can be tolerated. In seed orchards, the objective is to have the pH at the desired level, not at just what can be tolerated.

Lime is commonly used to raise a low pH and an acid-forming fertilizer such as ammonium sulfate or ammonium nitrate is commonly used to reduce pH. Elemental sulfur will also lower the soil pH when it is oxidized to sulfate. The orchard manager must adequately and representatively sample the orchard to assure the correct fertilizer prescription. It cannot be overemphasized that *the information obtained through testing is no better than the samples upon which it is based.* Large uniform areas of the orchard up to 5 acres (2 ha) in size may be represented by one composite sample. However, separate samples should be obtained from every area that is distinguishably different from other areas as a result of topography, soil structure, moisture, or native vegetation. The 0- to 6-in. (15-cm) analyses are usually taken from the upper zone, but in addition, the 15- to 21-in. (38.1- to 53.2-cm) zone is occasionally investigated for physical properties in newly established orchards. The condition of the subsoil where anchor roots of the tree are commonly found is important relative to the drainage, fertility, and workability of an area. During the early years of the orchard, it is recommended that nutrient samples be taken annually. Once apparently satisfactory nutrient levels have been stabilized as a result of repeated applications of lime and fertilizer, biennial sampling will often suffice.

The amounts of lime and fertilizer suggested for use in a seed orchard are based on the age of trees, species, geographical location, and soil type, and no single application is suitable to all. However, as a general guide, the following minimum fertility standards have been outlined by Davey (1981) for loblolly pine:

Calcium	400	lb/acre	(400 kg/ha)
Magnesium	50	lb/acre	(50 kg/ha)
Potassium	80	lb/acre	(80 kg/ha)
Phosphorus	40	lb/acre	(40 kg/ha)
pH	5.5		—

The timing of fertilization application is critical if satisfactory results are to be obtained (Schmidtling, 1972). It should be applied just before the initiation of floral buds if immediate, increased flowering is to result. Fertilizers also help keep the trees healthy and to grow to a large size, resulting in more reproductive bud locations. Usually, seed orchards are fertilized for maximum growth and vigor when young; the application is changed to favor flowering at a later date.

Irrigation   The installation of an irrigation system is expensive, and the question often posed is, "Is it worth it?" On the basis of increased seed production alone, the answer seemed at first to be doubtful; however, new tests make irrigation look increasingly attractive (see Figure 6.9). When increased success of establishing orchard trees, fire protection, more rapid development of trees, better sod cover, and increased seed production are combined, irrigation now appears to be a good investment for some species (Grigsby, 1966). In one loblolly pine orchard, irrigation as well as fertilization was responsible for an approximate 30% increase in seed production over the fertilizer-only area and for a 100% increase over the area that did not receive irrigation or fertilizer (Harcharik, 1981). In droughty years, irrigation may mean the difference between a poor or a good seed crop (Dewers and Moehring, 1970; Long et al., 1974; Gregory et al., 1982).

It has been found from experience that irrigation sometimes delays flower and fruit as well as cone maturity and increases pollen production; it has been used effectively to prevent freezing of flowers during critical periods. It has also been used in Douglas fir orchards to delay maximum strobilus receptivity until after maximum flight of the pollen in surrounding stands (Fashler and Devitt, 1980).

Like fertilization, irrigation is used at young ages in seed orchards to maintain optimal growth and vigor. To accomplish this, irrigation is used at any time during the year when the soil is dry enough to warrant it. As the trees come into flowering, the timing of irrigation appears to become very critical. There is some indication that southern pine trees should be under moisture stress during the period prior to the initiation of reproductive structures (Harcharik, 1981). The actual timing of stress and irrigation is still not clear for most species.

If the decision is made to install an irrigation system, it is important to have proper equipment to determine when and how much to irrigate. Tensiometers of some type are usually used for this purpose; they are simple to install and easy to calibrate and read. They allow the orchard manager to irrigate without guesswork and thus represent an economic savings.

Pest Problems   Any time seed orchards are established, pests in one or another form will become evident. Pests are different for each species and in each geographic area, so there is no point in listing and discussing them in more than a general way. Pests in seed orchards can be divided into those that attack the flowers, fruits, cones, and seeds, those that attack the tree foliage, bark, and limbs, and those that attack the roots. They vary from insects to diseases to animals, birds, and even to human beings.

Since pests often attack after an orchard has been established for some years, their most important effect is often overlooked. As Bergman (1968) and others have emphasized so strongly, the economic returns from tree improvement are closely tied to the amount of seed produced per unit area of seed orchard. In southern pine orchards, for example, DeBarr (1971) and others have shown a dramatic increase in the value of seed orchards resulting when seed losses were reduced through use of effective systemic insecticides (Table 6.1) (Figure 6.10).

The value of pest control was graphically illustrated by Weir (1975) who showed that the sizes of the potential losses from pests destroying seed-orchard seed were greater than the economic value of the timber destroyed by the very destructive southern pine beetle. A *major factor determining whether a seed orchard is economically feasible or not depends upon the success of control of orchard pests.*

Many methods have been developed to control pests in seed orchards, varying from manual removal, shooting or trapping, spraying, and using chemicals. By far the greatest damage to seed orchards is done by insects; their control is not easy. They vary from small insects and diseases that attack flowers, fruits, and cones (DeBarr and Williams, 1971; Miller and Bramlett, 1978; Hedlin et al., 1980; Cameron, 1981) through those that kill and destroy the whole tree. Insects can often be controlled by systemics; that is, chemicals that are taken into the plant that repels or kills the insect when it eats the foliage, cambium, or reproductive structures (Koerber, 1978; Neel et al., 1978). The methodology of systemic usage is being developed rapidly and may also hold promise for control of some fungal infections. In some cases, spraying can be the most effective method of insect control. The use of spraying is being developed, using both fixed-wing aircraft and helicopters.

The most insidious seed orchard pests are the seed, fruit, and cone destroyers. Some of them have been known for many years, but nothing was done to control them because they were considered to be of little importance and of nuisance value only. They can become of major importance in seed orchards, and their control can spell the difference between success and failure.

This short section on orchard pests in no way indicates their great importance.

TABLE 6.1

Available Pesticides for Preventing Losses to Genetically Improved Seed

Treatment	Percentage Sound Seed	Percentage Germination	Dead Cones by Coneworm	Conelets Surviving May–August	Sound Seed per Cone
Furadan	86	97	12	95	72
Guthion	79	99	17	89	78
Control	74	100	20	92	59

Note Shown are the results of a study designed to assess the utility of the pesticides Furadan and Guthion for controlling coneworm (*Dioryctria* sp.) losses in a loblolly pine seed orchard. Results are only for protection in the second year of development (data courtesy Union Camp Corporation).

(a)

FIGURE 6.10

Great damage is done by cone- and seed-destroying insects. (*a*) shows a seed bug *Leptoglosus corculus* on a loblolly pine cone. It punctures the developing seeds and destroys them (*b*) shows the dramatic effect on seed per cone from control of insects. (Photos courtesy Gary DeBarr, U.S. Forest Service, Athens, Ga.)

To the seed orchard manager and for the efficiency of a tree improvement program, they are a key problem. When losses from adverse or unusual environments are added to losses from pests, keeping the orchards healthy and fully productive is indeed a challenge.

Other Management Methods to Increase Flowering   There are numerous actions that can be taken to speed up or improve seed production (Heitmuller and Melchior, 1960); two are mentioned here, as follows.

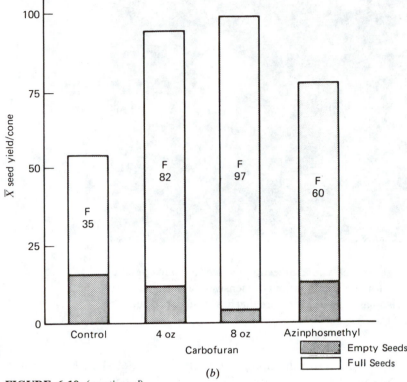

FIGURE 6.10 (continued)

Partial girdling of the stem often will result in increased flower and cone crops (Bower and Smith, 1961; Melchior, 1961; Hansbrough and Merrifield, 1963). Although partial girdling works, either as a girdle or a band, it is a severe method that is not usually recommended for orchards that are to be kept for long time periods. It is very useful to obtain heavy seed crops from isolated trees that will be used only temporarily.

Top pruning to keep tree height down and to make seed collection easier has been widely discussed and tried, but with indifferent success (Figure 6.11). It is also used in the hope of keeping the trees low for ease of control pollination. Generally, seed crops from top-pruned trees, compared to normal trees, vary from little or no reduction (Gansel, 1977) to as much as one half on pines (van Buijtenen and Brown, 1962; Copes, 1973b) and to much more for genera like *Abies* that bear most cones at the top of the tree. Although still being tried, top pruning is not a generally accepted seed orchard management method for most species. A problem is that the limbs below the severed portion tend to curve up to make a new top or a multiple top. This can be partially prevented by intensive

FIGURE 6.11

Topping trees to enable easy cone collection and seed orchard operation has been tried in many areas but only extensively used in southern Africa. Shown is a slash pine seed orchard that has been topped and kept low in South Africa.

measures such as tying down the upturning branch (Toda et al., 1963; van der Sijde, 1969). Shaping to increase flower buds on resulting new shoots has been tried with some success (Shibata, 1968).

Seed Orchard Records

The importance of maintaining good records on the seed orchard cannot be overemphasized. The records provide a history of the orchard upon which present and future recommendations are based. They identify the genetic material contained therein, and they reduce the possibility of errors. Of great importance is the fact that they provide a record of what environments and management practices have influenced the orchard, how these were handled, and what the results were.

Basically, two types of records are needed: (1) those related to the orchard as a unit and (2) those related to individual trees or clones within the orchard. A minimal set of information for the orchard as a unit would include the following:

A. Fertilization and liming
   1. Type—formulation
   2. Rates
   3. Date of application
   4. Method of application—broadcast or single tree, ground or aerial

B. Irrigation (if applicable)
   1. Dates required, tensiometer readings
   2. Amount and timing

C. Subsoiling—date, depth, direction

D. Insect and disease control
   1. Material used for control
   2. Rates used
   3. Method of application
   4. Date of application
   5. Effectiveness

E. Pruning
   1. Date
   2. Type

F. Roguing or thinning
   1. Date
   2. Clones and trees removed—trees remaining
   3. Type of cut (silvicultural and/or genetic)

G. Conditions—unusual environmental and biological phenomena. Dates and actions taken should be kept for the following:
   1. Ice storms
   2. Drought
   3. Late frosts
   4. Floods or extra heavy rain
   5. Unusual winds

Complete weather records greatly facilitate seed orchard management. A weather station should be maintained at each orchard to obtain precipitation, relative humidity, temperature, wind velocity, and wind direction. A maximum–minimum thermometer is of great benefit. Many organizations prefer to have additional hygrothermographs, anemometers, and other equipment.

The preceding records provide a general framework of the orchard that can be referred to in order to see what has happened, what was done to rectify any damage done, and how successful was the action taken. Of immediate use, such records can help to explain why there are small seed crops or why seed germination is below par and can even be used to help predict seed crops. They provide the necessary general framework of knowledge that will ensure that the management of the orchard is done correctly in order to obtain maximum returns from the tree improvement program. Management practices such as fertilization, subsoiling, and irrigation must be done at a specific time and in a specific way to obtain the greatest benefits.

Clonal or individual tree records are more detailed and provide a history of every tree in the orchard; such detailed records are often not kept by new orchard managers because their utility and need is often not appreciated until operations for advanced-generation seed orchards are undertaken. As improved first-generation orchards are established, detailed clonal information becomes essential for the selection of trees and clones to use. It is of no benefit to include a genetically superior clone in an orchard if it does not produce flowers or fertile seeds. Of even more immediate importance in vegetative seed orchards, the clonal records, combined with progeny test results, provide the basis for the roguing and consequent upgrading of first- and advanced-generation orchards. Detailed clonal records on flowering and cone or fruit maturity dates also greatly facilitate control-pollination and cone-collection operations. Minimal records for each clone in a vegetative orchard should include the following:

A. Method and date of propagation
B. Degree of incompatibility
C. Flowering
    1. Age when started (males, females)
    2. Heaviness of males and females
    3. Dates of pollen shed and female receptivity
D. Cone, fruit, and seed production
    1. Seed production (light, medium, heavy)
    2. Date of cone or fruit maturity
    3. Average number of seeds per cone or fruit
    4. Soundness of seed
    5. Germinability
E. Particular susceptibility of seed, cone, or fruit to insects and diseases
F. Specialty handling of individual ramets within a clone
    1. Problems such as abnormal growth, abnormal cone, or fruit development, abortion, and so forth
    2. Special fertilizing or subsoiling for fluted bark

The preceding suggested record system is minimal, but it will provide the essential information for an operational vegetative seed orchard. Along with a good record system, it is necessary that the ramets and trees in the orchard are positively identified and labeled. An accurate and convenient labeling system will greatly facilitate orchard operations such as making control pollinations and cone collecting. Individual trees should be marked with permanent and easily readable tags. A good map of the orchard showing the exact location and identification of each tree in the orchard is essential for operations and to reidentify plants from which the identification may have been lost.

Species with Special Orchard Problems

The bulk of the discussion on seed production has dealt with *wind-pollinated species* in which the pollen is transported from one individual to another in the air.

These were mainly *monoecious* species, which means that both sexes are on the same tree. The emphasis has been on this group of trees because they are the most common in economically important forest trees and because we have the best information about how to handle and control seed production in wind-pollinated species.

There are two major types of breeding and pollination systems that cause special problems in orchard design. These are for *dioecious species* where the two sexes are on different trees and for species in which the pollen is transported by insects (*entomophilous species*) rather than by the wind.

Designs for orchards of dioecious species basically consist of pollinators (males) who are surrounded by a group of females. The problem relates to how many females should be arranged around the males. There is, of course, no danger from selfing, although there is danger from other types of related matings. The best design must be determined for each species. Problems in a tree improvement program using dioecious species usually are about tree selection, because the sex of the tree usually cannot be determined until it flowers. This sometimes results in an excess of males relative to females, as was found for green ash (*Fraxinus pennsylvanica*) (Talbert and Heeren, 1979). This problem is magnified if the males are better formed or have better growth than the females, as has been hypothesized for several dioecious species. Nevertheless, obtaining a good genetic balance for a seed orchard is difficult in dioecious species such as ash (Talbert and Heeren, 1979). Dioecious species are common in hardwoods, especially in the tropical species. They also exist in gymnosperms, for example, in some *Cupressaceae* and in genera like *Ginkgo*.

The most difficult groups of species to work with are the insect-pollinated species (Figure 6.12). Some are pollinated both by wind and insects, such as some eucalypts, whereas others, including *Liriodendron,* are strictly insect pollinated. The latter group tends to have large, sticky pollen that is difficult to collect, store, and work with. Although it is very incompletely understood, it is evident that many of the tropical hardwoods are insect pollinated, some with very specialized pollination systems.

Designs suitable for seed orchards of entomophilous species are a puzzle, and it is not possible to recommend any one system. Everything depends on the insect and its habits. For example, Taft (1961) found that bees did not randomly range throughout a yellow poplar seed orchard, but they tended to concentrate on one tree at a time. Most bees from a hive would feed on one tree; then they would go to another. This results in flying from flower to flower on a single tree, and the main outcome is selfing. Schemes have been tried to overcome this problem with yellow poplar by grafting several clones on one rootstock so that the insects would go from clone to clone on a single tree and thus effect cross-pollination. But this has not worked well, because of the differential growth rate of the different clones that are grafted onto the common rootstock. Also tried was the agricultural practice of having traps in front of the beehive containing the desired pollen or mix of pollens through which the bees walked as they left the hive and thus caused

(a)

FIGURE 6.12

Seed orchard designs for some hardwoods are complex. For example, a really suitable design has not been found for insect-pollinated species, such as the *Liriodendron tulipifera* (*a*). Dioecious species also require special designs for efficient seed production. Such problems are particularly acute in the tropical hardwoods. For hardwood species, such as *Liquidambar styraciflua* (*b*), which are wind pollinated, the normal conifer seed orchard designs are suitable.

control pollination. This has worked well for certain horticultural crops, but often in tree species the pollen is so sticky, so hard to collect, and has such a short period of viability that the trap method is most difficult to apply.

The most difficult of all would be dioecious, entomophilous species, but there are no known seed orchards composed of such species. At the present time, it can be said for the insect-pollinated species that it is not yet known how to handle them efficiently in a forest tree seed orchard situation.

RESEARCH SEED ORCHARDS—CLONE BANKS

Most orchard descriptions are for operational seed orchards whose objective is to produce large quantities of seed for production plantings. The clone bank, breeding orchard or research seed orchard, is also of great importance and is essential for long-term programs (Figure 6.13). These are used to preserve and test large numbers of genotypes, not to produce massive quantities of seed for operational planting. The objectives and operations of the research orchards are

FIGURE 6.12 (*continued*)

described in Chapter 1. These types of orchards are mentioned here only to point out that they may be considered to be seed orchards but must be handled somewhat differently than the production orchards that have been described in detail before.

The question is always asked, "How many trees within a species are necessary as a basis for a breeding orchard?" The stock answer is "the more the better." Realistically, however, some persons feel that 200 nonrelated individuals will form a suitable base; however, most calculations have shown that 300 to 400 would be more appropriate. Many programs have 400 or more different genotypes as an objective.

A few items in which the breeding orchards differ from production orchards are summarized as follows:

1. Related individuals can be placed together, if efficiency of establishment and ease in movement from one ramet of a clone to another is desired for controlled pollinations. Planting in this manner means that testing by open-pollinated seed collected from the clone bank is impossible because there will be related matings and selfs.

2.   Pollen dilution zones are not required if open-pollinated seed are not to be used.

3.   No less than three, nor generally more than six, ramets of the same clone or closely related individuals need to be maintained.

4.   The breeding orchard can be placed wherever the trees will grow best and produce seed. It can often be established close to research headquarters for ease of operation. Usually, a breeding orchard is put in a location that is different from the production orchard as an added insurance against destruction or loss of genotypes by weather extremes or other catastrophes. Many organizations exchange plant materials for breeding orchards to assure greater safety. Representatives of each genotype of interest are put into the breeding orchard.

5.   The research orchard must be managed well enough to keep it healthy so it will produce seed. This implies intensive management similar to that used in production seed orchards. *Records and labeling* must be kept up to date to prevent loss or mixing of identities or pedigrees.

FIGURE 6.13

Breeding orchards have the function of the conservation of genetic material for use in the breeding program and for advanced generations. Shown is a hardwood clone bank that contains several clones of several different species that are being held for use in future breeding programs.

Breeding orchards or clone banks need a broad genetic base to avoid inbreeding in future generations and to preserve genes and genotypes that might be useful as the tree improvement program develops. They serve as a base for later and more advanced tree improvement. In contrast, the production orchards have as their objective the production of large quantities of seed with the maximum genetic gain from the very best clones. Too often the breeding orchard is given only minimal care. The production seed orchard is essentially a "dead end" insofar as being a major basis for future generations of breeding; the function of genetic conservation and breeding is carried out in the breeding orchard.

SEED CERTIFICATION

Ever since applied tree improvement became accepted, there has been increased interest and considerable emphasis toward certifying seed as has been done so successfully for agricultural crops. Seed certification in forestry is not new. As early as 1903, Doi reported on regulations for collecting and marketing *Cryptomeria* seed in Japan. Some recent certification plans have been put into practice, but there is no uniformity among the various methods. It seems that most people think that seed certification is a good idea. In a survey in the United States, 97% of the respondents favored seed certification. Of these, 93% wanted seed certification to be voluntary and not rigidly legislated (Horning, 1961; Cech et al., 1962). Most wanted a certification program to be administered by local groups that would have the legal authority to set standards, although Horning (1961), Rohmeder (1961), and Matthews (1964) felt that an international certification program was essential.

There are many meanings for seed certification and labeling, as have been discussed by Barber et al. (1962) and Barber (1969). The details of the various uses and methods of certification are not necessary to pursue in this book. However, certification requires a general supervision over the collection and handling of forest tree seeds in a uniform and consistent manner, and great care is needed (Barber, 1969; Barner and Koster, 1976). There is a tendency to emphasize the special importance of seed certification for the tropical and subtropical areas where exotics are used (Banks and Barrett, 1973), but it is also necessary for cool-weather areas where exotics are widely planted, such as in Europe and Canada (Presch and Stevenson, 1976).

Questions are often raised as to whether or not seed certification is really needed. It is becoming clear that with seed shortages, the great need for exotics, and the advent of genetically improved seed, that uniform and simple certification procedures are desirable. Many persons understand certification to mean what is correctly termed *labelling* in which the seed size, purity, germination, and other information about the seed is given. Others consider that there should be a minimum of *source certification;* that is, to have an accurate description of where the seed were obtained and, hopefully, from what quality of parent trees. Source certification is vital to all exotic forestry, and lack of such information has caused great frustration and losses in forestry throughout the world. To some foresters, *seed certification* means a statement about the *genetic quality* of the seed; this should be the ultimate certification goal. Quality certification is increasing as progeny tests mature and data become available on genetic performance. Quality certification is the most difficult of all because it depends on so many factors of product, size, crossing and test design, method of assessment, age, and other items. It is so complex that there may never be a consensus about just what should be required.

Regardless of definitions, the key to any type of forest tree seed certification program is to assure that the purchaser knows what is being bought. The objective of certification is not to legislate what can be sold; it is to make certain that the buyer receives what is being paid for. Certification methods are certain to multiply, both regionally and internationally. The real need is to avoid rigid and dictatorial certification schemes, while at the same time fulfilling the objective of assurance of genetic quality to the buyer.

LITERATURE CITED

Andersson, E. 1960. Froplantagen I Skogsbrukets Tjänst [Seed orchards in Swedish forestry] Särtyck ur Kungl. *Skogs-och Lantbruksakademiens Tidskrift* **99**(1–2):65–87.

Andersson, E. 1963. "Seed Stands and Seed Orchards in the Breeding of Conifers. 1st World Consul. on For. Gen. and Tree Impr., Stockholm, Sweden.

Banks, P. F., and Barrett, R. L. 1973. "Exotic Forest Tree Seed in Rhodesia." IUFRO, Symposium on Seed Processing, Bergen, Norway, Vol. II, pp. 1–8.

Barber, J. D., Callahan, R. Z., Wakeley, P. C., and Rudolf, P. O. 1962. More on tree seed certification and legislation. *Jour. For.* **60**(5):349–350, 352.

Barber, J. D., and Dorman, K. W. 1964. Clonal or seedling seed orchards? *Sil. Gen.* **13**(1–2):11–17.

Barber, J. D. 1969. "Control of Genetic Identity of Forest Reproductive Materials." 2nd World Consul. For. Tree Breed., Washington, D.C., pp. 7–16.

Barner, H., and Koster, R. 1976. "Terminology and Definitions to Be Used in Certification Schemes for Forest Reproductive Materials." XVI IUFRO World Congress, Oslo, Norway.

Barnes, B. V. 1964. "Self and Cross-pollination of Western White Pine: A Comparison of Height Growth of Progeny." U.S. Forest Service Research Note INT-22, pp. 1–3.

Barnes, R. D., and Mullin, L. J. 1974. "Flowering Phenology and Productivity in Clonal Seed Orchards of *Pinus patula, P. elliottii, P. taeda* and *P. kesiya* in Rhodesia." Forestry Research Paper No. 3, Rhodesia For. Comm.

Beers, W. L. 1974. "Industry's Analysis of Operational Problems and Research in Increasing Cone and Seed Yields." Colloquium: Seed Yields from Southern Pine Orchards, *Macon, Ga., pp. 86–96.*

Beland, J. W., and Jones, L. 1967. "Self-Incompatibility in Sycamore." 9th South. Conf. on For. Tree Impr., pp. 56–58.

Bengston, G. W., and Goddard, R. E. 1966. "Establishment, Culture and Protection of Slash and Loblolly Pine Seed Orchards. Some Tentative Recommendations. Proc. Southeast. Area For. Nur. Conf., Columbia, S.C., pp. 47–63.

Bergman, A. 1968. "Variation in Flowering and Its Effect on Seed Cost—A Study of Seed Orchards of Loblolly Pine." Technical Report No. 38, School of Forest Resources, North Carolina State University, Raleigh.

Bower, D. R., and Smith, J. L. 1961. "Partial Girdling Multiplies Shortleaf Cones." South. For. Notes No. 132, U.S. Forestry Service.

Bramlett, D. L. 1974. "Seed Potential and Seed Efficiency." Colloquium: Seed Yields from Southern Pine Seed Orchards, Macon, Ga., pp. 1–7.

Bramlett, D. L., Belcher, E. W., DeBarr, G. L., Hertel, G. D., Karrfalt, R. P., Lantz, C. W., Miller, T., Ware, K. D., and Yates, H. O. 1977. "Cone analysis of southern pines—A guidebook." Technical Report SE-13, U.S. Forestry Service.

Burrows, P. M. 1966. *A Theoretical Model for the Establishment of Seed Orchards from Plus Trees.* Biometrics Team, Agricultural Research Council of Central Africa, Southern Rhodesia.

Cameron, J. N., and Kube, P. D. 1980. "Management of Seedling Seed Orchards of *Eucalyptus regnans*—Selection, Strategy and Flowering Studies." Workshop on Gen. Impr. and Prod. of Fast Growing Trees, São Pedro, São Paulo, Brazil.

Cameron, R. S. 1981. "Toward Insect Pest Management in Southern Pine Seed Orchards." Texas Forest Service, Pub. No. 126.

Chapman, W. L. 1968. "Ideas Regarding Seed Orchard Management." Proc. Southeastern Area For. Nur. Conf., Stone Mountain, Ga., pp. 131–139.

Cech, F. C., Barber, J. C., and Zobel, B. J. 1962. Comments on "Who wants tree seed certification and why?" *Jour. For.* **60**(3):208–210.

Copes, D. L. 1973a. Genetics of graft rejection in Douglas fir. *Can. Jour. For. Res.* **4**(2):186–192.

Copes, D. L. 1973b. Effect of annual leader pruning on cone production and crown development of grafted Douglas fir. *Sil. Gen.* 22(5–6):167–173.

Copes, D. L. 1981. "Selection and Propagation of Highly Graft-Compatible Douglas-Fir Rootstocks—A Case History." U.S. Forest Service Research Note PNW 376.

Churchwell, B. 1972. "Hardwood Seed Orchard Management." Proc. Southeastern Area Nur. Conf., Greenville, Miss., pp. 84–87.

Davey, C. B. 1981. Seed orchard soil management. In *Tree Improvement Short Course,* North Carolina State University–Industry Cooperative Tree Improvement Program, School of Forest Resources, Raleigh, pp. 90–95.

DeBarr, G. L. 1971. "The Value of Insect Control in Seed Orchards: Some Economic and Biological Considerations." 11th Conf. South. For. Tree Impr., Atlanta, Ga., pp. 178–185.

DeBarr, G. L., and Williams J. A. 1971. "Nonlethal Thrips Damage to Slash Pine Flowers Reduces Seed Yields." U.S. Forest Service Research Note SE-160.

DeBarr, G. L. 1978. "Importance of the Seedbugs *Leptoglossus corculus* and *Tetyra bipunctata* and Their Control in Southern Pine Seed Orchards. Proc. Symposium on Flowering and Seed Development in Trees, Mississippi State University, pp. 330–341.

Dewers, R. R., and Moehring, D. M. 1970. Effect of soil water stress on initiation of ovulate primordia in loblolly pine. *For. Sci.* **16**(2):219–221.

Diekert, H. 1964. Einige Untersuchungen zur Selbsterilität und Inzucht bei Fichte und Lärche [Some investigations on self-sterility and inbreeding in spruce and larch]. *Sil. Gen.* **13**(3):77–86.

Doi, H. 1903. The enactment of the regulations on collection and marketing of *Cryptomeria* and hinoki cypress seeds. *Dainippion Saurin Kaihoo* [*Bull. Jap. For. Assoc.* **252**:37–42]. From *Abstr. Jap. Liter.* **1**(A), 1970.

Dorn, D. E., and Auchmoody, L. R. 1974. "Effects of Fertilization on Vegetative Growth and Early Flowering and Fruiting of Seed Orchard Black Cherry." 21st Northeast. Tree Impr. Conf., University of New Brunswick, pp. 6–18.

Dyer, W. G. 1964. "Seed Orchards and Seed Production Areas in Ontario." Proc. 9th Meeting of Comm. For. Tree Breeding in Canada, Part II, pp. 23–28.

Dyson, W. G., and Freeman, G. H. 1968. Seed orchard designs for sites with a constant prevailing wind. *Sil. Gen.* **17**(1):12–15.

Fashler, A. M. K., and Devitt, W. J. B. 1980. A practical solution to Douglas-fir seed orchard pollen contamination. *For. Chron.* **56**:237–241.

Faulkner, R. 1962. Seed stands in Britain and their better management. *Quart. Jour. For.* **56**(1):8–22.

Faulkner, R. 1975. "Seed Orchards." Forestry Comm. Bull. No. 54, Her Majesty's Stationary Office, London.

Feilberg, L., and Soegaard, B. 1975. "Historical Review of Seed Orchards." Forestry Comm. Bull. No. 54. Her Majesty's Stationary Office, London.

Fielding, J. M. 1964. Notes on a Monterey pine seed orchard on Tallaganda State Forest in New South Wales. *Aust. For.* **28**(3):203–206.

Franklin, E. C. 1969. "Inbreeding Depression in Metrical Traits of Loblolly Pine (*P. taeda*) as a Result of Self-pollination." Technical Report 40, School of Forest Resources, North Carolina State University, Raleigh.

Gallegos, C. M. 1978. "Criteria for Selecting Loblolly Pine (*Pinus taeda L.*) Seed Orchard Sites in the Southeastern United States." Proc. Symposium on Flowering and Seed Development in Trees, Mississippi State University, pp. 163–176.

Gallegos, C. M. 1981. "Flowering and Seed Production of *Pinus caribaea* var. *Hondurensis* (Results of a World-Wide Survey)." Symposium on General Improvement and Production of Fast-Growing Species, São Pedro, São Paulo, Brazil.

Gansel, C. R. 1973. "Should Slash Pine Seed Orchards Be Moved South for Early Flowering?" 12th South. For. Tree Impr. Conf., Baton Rouge, La., pp. 310–316.

Gansel, C. R. 1977. "Crown Shaping in a Slash Pine Seed Orchard." 14th South. For. Tree Impr. Conf., Gainesville, Fla., pp. 141–151.

Gerdes, B. C. 1959. "Some Thoughts on Douglas-Fir Seed Orchards and Their Establishment." Proc., Society American Forestry, San Francisco, Calif.

Giertych, M. M. 1965. Systematic lay-outs for seed orchards. *Sil. Gen.* **14**(3):91–94.

Goddard, R. E. 1964. Tree distribution in a seedling seed orchard following between and within family selection. *Sil. Gen.* **13**(1–2):17–21.

Greenwood, M. S. 1977. "Seed Orchard Fertilization: Optimizing Time and Rate of Ammonium Nitrate Application for Grafted Loblolly Pine" 14th South. For. Tree Imp. Conf., Gainesville, Fla., pp. 164–169.

Greenwood, M. S. 1981. Reproductive development in loblolly pine. II. The effect of age, gibberellin plus water stress and out-of-phase dormancy on long shoot growth behavior. *Am. Jour. Bot.* **68**(9):1184–1190.

Gregory, J. D. 1975. "Subsoiling to Stimulate Flowering and Cone Production and Ameliorate Soil Conditions in Loblolly Pine (*Pinus taeda*) Seed Orchards." Ph.D. thesis, North Carolina State University, Raleigh.

Gregory, J. D., Guiness, W. M., and Davey, C. B. 1982. Fertilization and Irrigation Stimulate Flowering and Seed Production in a Loblolly Pine Seed Orchard. *South. Jour. App. For.* 6:44–48.

Gregory, J. D., Guiness, W. M., and Davey, C. B. 1976. *Fertilization and Irrigation Stimulate Flowering and Seed Production in a Loblolly Pine Seed Orchard.* Soil Science Society of America, Houston, Tex.

Grigsby, H. 1966. "Irrigation and Fertilization of Seed Orchards." West Reg. Nur. Conf., Hot Springs, Ark., pp. 1–18.

Hansbrough, T., and Merrifield, R. G. 1963. "The Influence of Partial Girdling on Cone and Seed Production of Loblolly Pine." Louisiana State University Forestry Notes No. 52.

Harcharik, D. A. 1981. "The Timing and Economics of Irrigation in Loblolly Pine Seed Orchards." Ph.D. thesis, North Carolina State University, Raleigh.

Hatcher, A., and Weir, R. J. 1981. "Decision and Layout of Advanced Generation Seed Orchards." 16th South. For. Tree Impr. Conf., Blacksburg, Va., pp. 205–212.

Hattemer, H. H., Andersson, E., and Tamm, C. O. 1977. Effects of spacing and fertilization on four grafted clones of Scots pine. *Stud. For. Suec.* **141**:1–31.

Hedlin, A. F., Yates, H. O., Lovar, D. C., Ebel, B. H., Koerber, T. W., and Merkel, E. P. 1980. *Cone and Seed Insects of North American Conifers.* Canadian Forestry Service, Ottawa.

Heitmuller, H. H., and Melchior, G. H. 1960. Über die blühfördernde Wirkung des Wurzelschnitts, des Zweigkrümmens und des Strangulation auf japanischer Lärche (*Larix leptolepis*) [On the flower-promoting effects of root pruning, bending of branches and strangulation in Jap. larch]. *Sil. Gen.* **9**(3):65–72.

Horning, W. H. 1961. Society of American Foresters' report on a study of seed certification conducted by the committee on Forest Tree Improvement. *Jour. For.* **59**(9):656–661.

Jett, J. B., and Finger, G. 1973. "Stimulation of Flowering in Sweetgum." 12th South. For. Tree Impr. Conf., Baton Rouge, La., pp. 111–117.

Kellison, R. C. 1971. "Seed Orchard Management." 11th Congress on South. For. Tree Impr., Atlanta, Ga., pp. 166–172.

Klein, J. T. 1974. "A Jack Pine Seedling Seed Orchard Plantation of Unusual Design." 21st Northeast. For. Tree Impr. Conf., Fredricton, New Brunswick, pp. 55–65.

Koerber, T. W. 1978. "Tests of Bole Injected Systemic Insecticides for Control of Douglas-fir Cone Insects." Proc. Symposium on Flowering and Seed Development in Trees, Mississippi State University, pp. 323–329.

Kraus, J. 1974. "Seed Yield from Southern Pine Seed Orchards." Colloquium: Yield from Southern Pine Seed Orchards, Macon, Ga., pp. 1–100.

LaFarge, T., and Kraus, J. F. 1981. "Comparison of Progeny of a Loblolly Pine Seed Production Area with Progeny of Plus Tree Selections." 16th South. For. Tree Impr. Conf., Blacksburg, Va., pp. 302–310.

Lindgren, D. 1974. "Aspects on Suitable Number of Clones in a Seed Orchard." IUFRO Meeting, Stockholm, pp. 293–305.

Long, E. M., van Buijtenen J. P., and Robinson, J. F. 1974. "Cultural Practices in Southern Pine Seed Orchards." Colloquium: Seed Yield from Southern Pine Seed Orchards, Macon, Ga., pp. 73–85.

Matthews, J. D. 1964. Seed production and seed certification. *Unasylva* **18**(2–3):73–74, 104–118.

McElwee, R. L. 1970. "Radioactive Tracer Techniques for Pine Pollen Flight Studies and an Analysis of Short-Range Pollen Behavior." Ph.D. thesis, School of Forest Resources, North Carolina State University, Raleigh.

McKinley, C. R. 1975. "Growth of Loblolly Scion Material on Rootstocks of Known Genetic Origin." 13th South. For. Tree Imp. Conf., Raleigh, N.C., pp. 230–233.

Melchior, G. H. 1961. Versuche zur Ringelungsmethodik an Propflingen der europaischen Lärche (*Larix decidua*) und der japanischen Lärche (*Larix leptolepis*) [Experiments with girdling of grafts of European and Japanese larch]. *Sil. Gen.* **10**(4):107–109.

Miller, T., and Bramlett, D. L. 1978. "Damage to Reproductive Structures of Slash Pine by Two Seed-Borne Pathogens: *Diplodia gossypina* and *Fusarium monoliforme* var. *subglutinans*." Symposium on Flowering and Seed Development in Trees, Mississippi State University, pp. 347–356.

Miyake, N., and Okibe, A. 1967. Fundamental studies on seed gardens (orchards) of Japanese pine. 5. Effects of some fertilizers on the growth and cone crops of Akamatsu (*P. densiflora*). *Bull. Shimane Agr. Coll.* **15A-2**:101–112.

Moran, G. F., Bell, J. C., and Matheson, A. C. 1980. The genetic structure and levels of inbreeding in a *P. radiata* seed orchard. *Sil. Gen.* **29**(5–6):190–193.

Nanson, A. 1972. The provenance seedling seed orchard. *Sil. Gen.* **21**(6):243–248.

Neel, W. W., DeBarr, G. L., and Lambert, W. E. 1978. "Variability of Seedbug Damage to Pine Seed Following Different Methods of Applying Carbofuran Granules to a Slash Pine Seed Orchard." Symposium on Flowering and Seed Development in Trees, Mississippi State University, pp. 314–322.

Presch, R. F., and Stevenson, R. E. 1976. "Certification of Source-Identified Canadian Tree Seed under the O.E.C.D. Scheme." Can. For. Ser. For. Tech. Rept. No. 19.

Riemenschneider, D. E. 1977. "The Genetic and Economic Effect of Preliminary Cutting in the Seedling Orchard." Proc. 13th Lake States For. Tree Imp. Conf., St. Paul, Minn., (U.S. Forestry Service Technical Report NC-50, pp. 81–91.

Rohmeder, E. 1961. Probleme und Vorschläge internationaler Zertification des forstlichen Saatgutes [Problems and suggestions for an international certification of forest seed]. *Sonderdruck* **30**(9):253–255; **31**(8):219–221.

Schmidtling, R. C. 1972. "Importance of Fertilizer Timing on Flower Induction in Loblolly Pine." 2nd N. Amer. For. Biol. Workshop, Society of American Foresters, Oregon State University.

Schmidtling, R. C. 1978. "Southern Loblolly Pine Seed Orchards Produce More

Cone and Seed Than Do Northern Orchards." Symposium on Flowering and Seed Development in Trees, Mississippi State University, pp. 177–186.

Schmitt, D. M., and Perry, T. O. 1964. Self-sterility in sweetgum. *For. Sci.* **10**:302–305.

Schopmeyer, C. S. (Ed.). 1974. *Seeds of Woody Plants in the United States,* Agricultural Handbook No. 450. U.S. Forest Service, Washington, D.C.

Schultz, R. P., Wells, C. G., and Bengtson, G. W. 1975. "Soil and Tree Responses to Intensive Culture in a Slash Pine Clonal Orchard: 12 Year Results." U.S. Forest Service Research Paper SE-129.

Shibata, M. 1968. "Studies on Pruning and Shaping of Grafts in Seed Orchards." Trans. of 79th Mtg. Jap. For. Soc.

Slee, M. W., and Spidy, T. 1970. The incidence of graft incompatibility with related stock in *P. caribaea* v. *Hondurensis. Sil. Gen.* **19**(5–6):184–187.

Sniezko, R. A. 1981. "Genetic and Economic Consequences of Pollen Contamination in Seed Orchards." Proc. Southern Forest Tree Improvement Conf., Blacksburg, Va., pp. 225–233.

Sorenson, F. C., and Miles, R. S. 1974. Self-pollination effects on Douglas fir and ponderosa pine seeds and seedlings. *Sil. Gen.* **23**(5):135–138.

Sprague, J., Jett, J. B., and Zobel, B. 1978. "The Management of Southern Pine Seed Orchards to Increase Seed Production." Symposium on Flowering and Seed Development in Trees, Mississippi State University, pp. 145–162.

Squillace, A. E. 1967. Effectiveness of 400-foot isolation around a slash pine seed orchard. *Jour. For.* **65**(11):823–824.

Steinbrenner, E. C., Duffield, J. W., and Campbell, R. K. 1960. Increased cone production of young Douglas-fir following nitrogen and phosphorus fertilization. *Jour. For.* **58**(2):105–110.

Stern, K. 1959. The rate of inbreeding within the progenies of seed orchards. *Sil. Gen.* **8**(2):37–68.

Swofford, T. F. 1968. "Seed Orchard Management." Southeast. Area For. Nur. Conf., Stone Mountain, Ga., pp. 83–89.

Taft, K. A. 1961. "The Effect of Controlled Pollination and Honeybees on Seed Quality in Yellow Poplar (*Liriodendron tulipifera*) as Assessed by X-ray Photographs," Technical Report 13, School of Forestry, North Carolina State University, Raleigh.

Taft, K. A. 1966. "Cross and Self-Incompatibility and Natural Selfing in Yellow Poplar (*Liriodendron tulipifera*)." 6th World For. Cong., Madrid, Spain, pp. 1–11.

Taft, K. A. 1968. "Hardwood Seed Orchard Management." Southeastern Area For. Nur. Conf., Stone Mountain, Ga., pp. 90–91.

Talbert, J. T., and Heeren, R. D. 1979. Sex differences in green ash. *South. Jour. App. For.* **3**(4):173–174.

Thielges, B. 1975. "Forest Tree Improvement, the Third Decade." 24th Annual For. Symp., Louisiana State University, Baton Rouge, La.

Toda, R., Akasi, T., and Kikuti, H. 1963. Preventing upward curving of limbs of topped seed-trees by growth substance treatment. *Jour. Jap. For. Soc.* **45**(7):227–230.

Toda, R. 1964. Special issue of *Silvae Genetica* **13**(1) on vegetative and seedling seed orchards.

van Buijtenen, J. P., and Brown, C. L. 1962. "The Effect of Crown Pruning on Strobile Production of Loblolly Pine." For. Gen. Workshop, Macon, Ga.

van Buijtenen, J. P. 1968. "Seed Orchard Management." Proc. Southeastern Area For. Nur. Conf., Stone Mountain, Ga., pp. 123–127.

van Buijtenen, J. P. 1971. "Seed Orchard Design—Theory and Practice." 11th Conf. South. For. Tree Impr., Atlanta, Ga., pp. 197–206.

van Buijtenen, J. P. 1981. "Advanced Generation Tree Improvement." 18th Canadian Tree Impr. Assoc. Meet., Duncan, British Columbia," pp. 1–15.

Vande Linde, F. 1969. "Some Practical Aspects of Seed Orchard Management in the South." Proc. 10th South. Conf. For. Tree Impr., Houston, Tex., pp. 199–204.

van der Sijde, H. A. 1969. "Bending of Trees as a Standard Practice in Pine Seed Orchard Management in South Africa." 2nd World Cons. For. Tree Breed., Washington, D.C."

Wang, C.-W., Perry, T. O., and Johnson, A. G. 1960. Pollen dispersion of slash pine (*P. elliottii*) with special reference to seed orchard management. *Sil. Gen.* **9**(3):78–86.

Webster, S. 1971. "Nutrition of Seed Orchard Pine in Virginia." Ph.D. thesis, North Carolina State University, Raleigh.

Weir, R. J. 1975. "Cone and Seed Insects—Southern Pine Beetle: A Contrasting Impact on Forest Productivity." 13th South. For. Tree Impr. Conf., Raleigh, N.C., pp. 182–192.

Zobel, B. J., Barber, J., Brown, C. L., and Perry, T. O. 1958. Seed orchards; their concept and management. *Jour. For.* **56**:815–825.

Zobel, B. J., and McElwee, R. L. 1964. Seed orchards for the production of genetically improved seed. *Sil. Gen.* **13**(1–2):4–11.

CHAPTER 7

Use of Tree Improvement in Natural Forests and in Stand Improvement

The best way to effectively use the genetic differences among trees is through a plantation program. However, genetic manipulation can also be beneficial when forests are regenerated naturally. Also, it is of importance in such intermediate stand treatments as thinning, pruning, and fertilization. This is fortunate, because the bulk of the forests throughout the world are now, and will continue to be managed by using natural regeneration techniques.

Tree improvement principles can be used quite profitably both in regeneration of natural stands and in certain forest management activities that are applied to previously established stands. Although gains from the use of genetics will be less than when applied to a planting program, they can be obtained quickly because *the genetic manipulation can be applied immediately.* This is of great importance both in temperate climates and in the mixed species forests in the tropics and subtropics. In the efforts to use genetics in natural forests, most of the emphasis has been on the regeneration phase, and the real value of genetic manipulation in other aspects of silviculture has not been generally understood. In fact, even without serious genetic considerations, many of the best silvicultural practices have the effect of improving the genetic structure of the trees that are left to grow.

This chapter will cover some of the difficulties that are involved and some of the gains that can be made when applying genetic principles to natural regeneration or to previously established stands of trees. There is minimal research information that can be cited, because little has been done in this area, and economic studies are almost totally lacking. However, a knowledge of inheritance patterns and an idea of the economic worth of the characteristics involved will enable some realistic projections to be made.

NATURAL REGENERATION

Although most forest species are regenerated by seed, vegetative systems of regeneration, such as by root sprouts and stump sprouts, are also common, especially in the hardwoods. Vegetative reproduction also occurs in some conifers; two examples are redwood (*Sequoia sempervirens*) and *Cryptomeria*. It is easier to genetically manipulate natural stands when regeneration from selected seed trees can be employed. Complications arise when species sprout or have seeds that remain viable on the forest floor for many years.

Seed Regeneration

When seed trees or shelterwood systems of regeneration are used (as with the pines), genetic gain can be obtained by leaving the best trees as seed producers. The most important gain occurs because the progeny produced are well adapted to the site. Additionally, quality characteristics such as tree form can be improved by leaving the best phenotypes to produce seed, because these traits are usually highly heritable. Volume gains will be small, because growth characteristics are under only moderately additive genetic control, and most natural regeneration systems do not allow selection intensities great enough in the original stands to achieve substantial gains in growth. Resistance to many pests, such as fusiform rust in the southern pines, will be considerably improved. Goddard et al. (1975a) found that progeny from disease-free seed trees in stands heavily infected with fusiform rust produced seedlings that were considerably more disease resistant than those from the overall population.

When the best trees are removed, leaving the inferior ones to produce seed for the next generation, *dysgenic selection* will result. The most adverse cutting method within a species in even-aged stands is the *diameter limit cut* in which all trees over a given size are removed and the small diameter trees are left to grow and reproduce the stand (Trimble, 1971). The *diameter limit cut* is a type of dysgenic selection that is widely practiced throughout the world and results in succeeding generations of poorer-quality, slower-growing stands. A major dysgenic stand treatment occurs when special harvests are made to obtain certain high-value tree stems, such as poles or piling, or when the very best quality sawtimber trees are removed. An especially bad dysgenic practice occurs when only certain species are harvested, leaving the undesired species to occupy the site (Figure 7.1).

Even though there is now abundant proof that many of the poor-quality trees in even-aged stands are genetically inferior, some forest managers still claim that the lower-value and poor-quality trees can be left in a seed tree harvest or a shelterwood system because they can produce the needed quantities of seed to regenerate the site. Such trees are usually large crowned and frequently have been damaged by pests or adverse environments. Forest managers are understandably reluctant to leave high-quality and thus high-value trees because of the fear of loss from lightening, insects, or windthrow. The decision about which trees to leave as seed producers in natural regeneration systems is often made on the basis of their present value; this results in the most valuable trees being removed, leaving the crooked, deformed, and often, the diseased individuals.

Obviously, trees left to produce seed must have crown sizes large enough to produce sufficient quantities of seed. If trees are selected on growth characteristics as well as quality traits and occupy a dominant position in the stand canopy, their crowns will be of sufficient size and vigor to produce the seed needed.

Since such characteristics as disease resistance or straightness of tree bole are strongly inherited, a few generations of dysgenic selection can result in "minus-type" stands. Such a result is evident in the great increase in crookedness in the regeneration from stands that have been continuously harvested for poles or piling that removes the straightest trees. It has been stated that after many generations of continued removal of the best trees from a stand of *Pinus silvestris* in Germany, the initially good stand has degenerated to a forest of multistemmed bushes today.

One of the *most common misconceptions* by both foresters and laymen throughout the world is that *stands containing trees of various sizes will be uneven aged;* that is, trees of many different ages are growing intermixed. Generally, natural stands of both conifers and hardwoods are even aged, even though they

FIGURE 7.1

Throughout the world there is a reduction in the quality of forests because certain desirable species are preferentially logged, leaving less desirable trees. This practice is illustrated for Canada (*a*) where the conifers are removed leaving aspen residuals and sprouts. A similar situation is shown in the southern United States (*b*) where all pines have been removed, leaving low-quality hardwoods to claim the site.

contain trees of different sizes, or at most, they consist of two or three discrete age classes (Figure 7.2). When human beings selectively cut only a portion of the stand or when an overmature tropical hardwood stand is breaking up, truly uneven-aged stands may result; this also happens in the very tolerant temperate species such as beech (*Fagus*) and maple (*Acer*). The fallacy of considering stands with trees of different sizes as uneven aged has resulted in dysgenic selection in many areas of the world. The largest trees are harvested, with the incorrect assumption that the smaller trees are younger and are therefore genetically as good as the harvested trees. In even-aged stands, however, the smaller trees are the ones of poorest vigor and are more likely to be poorer genotypes, although differences may be accentuated as the result of competition. In these instances, harvesting only the larger and better trees is nothing more than a diameter limit cut in an even-aged stand.

One kind of seed regeneration that gives problems to the tree improver occurs in species whose seed can be stored in the forest floor for long periods of time. Leaving seed trees under these conditions is a waste of time, although it is done regularly. For example, seed of yellow poplar (*Liriodendron tulipifera*) can stay viable in the forest floor for many years (Clark and Boyce, 1964). Yet, there are

FIGURE 7.2

A common error is to assume that natural stands containing trees of all sizes are also all aged. This rarely is the case; most natural forest stands outside the tropics are even aged but all sized. This is true for both conifers or for hardwood stands such as the one illustrated.

laws in some states requiring that yellow poplars of a given size be left as seed trees to regenerate the area. Often, regeneration is very profuse with several hundred thousand seeds per hectare germinating. Most of these come from seed that had been stored in the forest duff for many years. Leaving high-quality seed trees does very little or nothing to improve the genetic quality of the new stand.

Dysgenic selection must be viewed in its true perspective. Despite removal of the best phenotypes, one or two generations of dysgenic selection will not result in future stands that are totally genetically degraded. Forest trees are very heterozygous, and after gene segregation and recombination some good genotypes will occur even in stands that have been heavily dysgenically selected for a few generations. The number of good trees from such a stand will be reduced, but in natural regeneration, in which perhaps only 1 out of 30 to 50 trees initially established will grow to constitute the dominant stand, the poorer-growing, noncompetitive, pest-susceptible genotypes will tend to be eliminated, and the resultant stand can be of satisfactory quality.

The ability of trees of poor phenotypic quality to produce good, young stands was used in an attempt to discredit the early forest genetic efforts. For example, many of the magnificent longleaf pine (*P. palustris*) stands in the southeastern United States were severely logged of all good trees, leaving only inferior, deformed individuals. Pole-sized reproduction from these stands was sometimes of high quality; these nice stands were pointed to with great enthusiasm by some foresters as evidence that selecting the good trees and leaving the poor trees to regenerate an area did not have a dysgenic effect on the new stand. This was correct for the first generation after high grading, where regeneration was commonly 30,000 to 100,000 stems per hectare. With the onset of competition, the number of trees was reduced to 1000 to 2000 of the best-growing stems per hectare, and these trees were of acceptable quality. However, if dysgenic selection is practiced for several generations, stand quality will be seriously diminished.

The effect of even one generation of dysgenic selection can be quite severe when seed are collected from inferior phenotypes and planted in nurseries for use in plantation programs. Under these conditions, nearly all seed have a chance to become established in plantations and to grow in the forest. Seedlings are fertilized, sprayed, and protected in the nursery and grown at spacings such that all trees, no matter how good or poor they are genetically, have a chance to grow and develop to a plantable size. Then, the seedlings are planted in the forest on site-prepared areas at wide spacings so they have little early competition. When competition from brush becomes too great, herbicides or mechanical methods are often used to release the planted trees. The good care of the young trees in the nursery and in the forests enables many genetically weak or inferior trees to survive that would be lost under a natural regeneration system in which competition is severe. Through nursery and plantation practices, people have developed a system that is very effective in preserving the weaker genotypes.

One of the most serious types of dysgenic selection is the harvesting of desired species from mixed stands, leaving only the undesired species (Figure 7.3). Vast changes in land productivity and timber quality have resulted and are still being

FIGURE 7.3

The best individuals of the best species are often logged from hardwood stands. Low-quality trees, such as the one shown, are often left to grow and to reproduce the new stand. The poor-quality residual trees occupy a lot of space but produce very little value.

produced by this policy. Such species-specific harvesting is almost universal. It is especially bad in the tropics, in the northeastern part of the United States, in central and eastern Canada and in the southern United States where there is a massive conversion from coniferous or mixed conifer hardwood forests to largely slow-growing, low-quality hardwood forests. Selection of certain key species is dictated by silviculture, policy, and harvesting: It must be considered here because of its effect on the genetic quality of the resultant stands. Selective-species harvesting is rarely considered in genetic terms, but in reality, the impact on the genetic constitution of the forest stand is tremendous; it is nearly as effective in causing gene complex losses as is clearcutting and converting forests to uses other than growing trees.

Selective-species logging is a classical and most serious problem in the tropical hardwood forests where only a few species of greatest value are removed. There is no easy solution to this problem. Although definitive results from long-term studies are generally lacking, observation indicates that a huge change in species frequencies, and consequently in gene complexes, is the result of this most common but adverse harvest system in the mixed tropical hardwoods. Genetic and economic losses from selective-species logging in the tropics is very severe because usually the species removed are the faster-growing, dominant, and intolerant species; these do not regenerate well following partial logging. Not only are good individual tree genotypes being lost, but whole species and gene complexes are being endangered.

Contrary to the general opinion, losses of good trees and whole species similar to the situation in the tropics are also occurring in the more temperate regions. As an example, about half the forest land in the southeastern United States contains primarily hardwood forests, most of which are composed of a mixture of valuable species. Too frequently these are selectively logged (dysgenically logged) of the best trees of the best species, resulting in massive loss of valuable gene complexes (Figure 7.3). As one example, in southern Alabama there was a species mix of cherrybark oak (*Quercus falcata* var. *pagodaefolia*), green ash (*Fraxinus pennsylvanica*) and hackberry (*Celtis occidentalis*), with other low-quality species. During logging, essentially all cherrybark oaks were removed as were all green ashes of merchantable quality. As a result, 20 years after logging, these once-beautiful mixed forests are now primarily composed of hackberry and low-quality hardwoods, including poor-quality ash. Cherrybark oak has essentially been removed from the forest, or only the poorest phenotypes are left. The result is the near loss of an outstanding species from the forest stand.

Another cause of adverse genetic change occurs when forests are broken up into farmland, leaving small scattered stands of trees that are separated by fields. This results in small, isolated breeding units in which inbreeding may occur; this result will be a degrading of the residual forest. Added to this is the usual practice of high grading the best trees from the farm woodlots. Forests of this scattered type are common throughout the tropics and in areas such as the small woodlots in the midwestern hardwood areas in the United States. In an assessment of black walnut, Beineke (1972) summarized the situation as follows: "By relying on natural regeneration in black walnut from scattered high-graded remnants in wood lots, small isolated breeding populations are developing . . . low vigor, slow growing, disease prone and poorly formed black walnut trees probably are being produced."

Although shifting agriculture, such as occurs in tropical and subtropical areas, is hardly a method of forest regeneration, it has the same effect because most of the farms are abandoned, often are grazed for a time, and eventually revert back into trees. Shifting agriculture is a severe cause for dysgenic selection in tropical areas. Usually, the abandoned farms are colonized by very fast-growing but poor-quality pioneer trees that can survive on the degraded soils with little humus and poor moisture regimes that are no longer suitable for the desired species. It is of special interest to assess the difference in colonizers between adjacent plots of abandoned farmland and a clear cut from the forest where the forest floor was left essentially intact.

Regeneration in Sprouting Species

Many species of hardwoods have sprouting as their major method of regeneration (see Figure 7.2). For example, in cutover forests, sweetgum (*Liquidambar styraciflua*) often has over 90% of its regeneration from stump or root sprouts. In general, the use of genetic principles in sprout-regenerated forest stands becomes very limited because genetic improvement will not occur when regeneration arises from sprouts from trees that are already established. Often, little can be done to improve the genetics of such stands other than later selective thinning. The small understory suppressed trees destroyed in the logging operation usually sprout more profusely than do those from the dominant overstory.

Sprouting species are especially difficult to manipulate in stands that have been badly degraded by previous selective logging. Following the selection harvest, the high-quality phenotypes that have been cut may sprout, but they are suppressed by the poor-quality trees remaining in the overstory and eventually die, or if they survive, they do not become part of the dominant stand. Any later attempt to regenerate the high-quality trees through removal of the remaining low-value trees will usually be met with failure, because the new stand will be composed largely of sprouts from the low-value trees. The only hope for genetic improvement in badly degraded sprout-regenerating stands is to destroy the sprouts, plant with good trees of the desired species, and then use sprouts from those good stumps for the following generations. This is a practice that is possible only when good silviculture is used and the desired species can be planted with reasonable success.

GENETIC IMPROVEMENT IN
PREVIOUSLY ESTABLISHED STANDS

Gains from tree improvement through its incorporation into intermediate silvicul-tural stand treatments has been greatly overlooked. The potential is great, and the incorporation of genetic principles into management activities is the fastest way of obtaining a significant amount of genetic improvement. Other than in regeneration activities, which automatically include harvesting, the best place to apply genetics in silviculture is in thinning, pruning, and less directly, in fertilizer operations. Thinning especially lends itself to the use of genetics.

Thinning

The most common error in forestry throughout the world is to establish, or leave, forest stands in an overdense or overstocked condition. Growing space is one of the major limiting factors in tree growth, and the genetic potential of a tree can never be realized fully when it is growing in an overstocked stand. Therefore, the silvicultural action of thinning is an essential tool if the full genetic benefits are to be obtained from improved strains of trees. There will always be argument about what are the proper spacings for different species, but the important thing is that the tree is free enough to grow in order to express its potential (Maki, 1969; Hofmann, 1974; Beck, 1975).

Aside from controlling spacing and competition, thinning can have other results that aid in forest management programs.

1. **Species control can be obtained**   Often, naturally regenerated forests contain large numbers of trees of undesirable species that take up needed growing space. A thinning operation can change the species composition in the desired direction (Roberge, 1975; McGarity, 1977). As one example, an organization in eastern Canada desired birch. The natural regeneration contained less than 20% birch, but following thinning over, 60% of the remaining forest consisted of birch. Similar, desirable species changes can be obtained in naturally regenerated conifer stands, especially in increasing the numbers of conifers over hardwoods or in increasing desired genera, such as spruce, over the less desired fir.

2. **Better trees can be left**   Thinning affords the opportunity to upgrade the quality of the residual stand above that of the original stand (Wahlenberg, 1952; Smith, 1967; Trimble, 1973; Little, 1974; Erdmann et al., 1975). The forester needs to know which characteristics can be changed by thinning and which are genetically set. For example, certain types of spiral twists in the tree bole are strongly inherited, and no thinning will help spirality in the remaining trees, although it can be directed toward leaving the straightest trees. Bole sweep is environmentally caused by uneven crowding, and therefore thinning can help alleviate this problem. Ramicorn branching is strongly inherited, and trees with this defect should be removed in thinning. Water sprouting and epicormic branching will occur differentially within certain hardwood species, and a knowledge of species susceptibility to sprouting and what conditions stimulate sprouting is essential as a guide to thinning (Kormanik, 1966; Books and Tubbs, 1970; Della-Bianca, 1973). Knowledge of the degree of resistance to insects and diseases should play a major role in guiding a thinning program.

Thinning of sprout clumps in hardwoods will become of increasing importance as forest practices become more intensive and shorter rotations with sprouting species are used (Bowersox and Ward, 1972). The genetics of coppicing is known, and the response to thinning of sprout clumps can be strongly specific (Beck, 1977) and individually controlled (Webb and Belanger, 1979).

Precommercial thinning of natural stands requires good genetic knowledge of which characteristics will respond and which will not. Quite often, the thinning is done too early, and the "winner" trees that are left may turn out to be "losers" some years later (Williams, 1974; Della-Bianca, 1975; Griswold, 1979). Such things as bole straightness in young trees are sufficiently correlated genetically with mature performance so that they can be judged early, but this is not true for growth rate. As better genetic stock becomes available, the need for precommercial thinning to improve quality traits will be diminished, and wider spacings will be used in plantations.

Thinning from below, or removing the smaller, less-vigorous trees, is acceptable to the geneticist, whereas thinning from above, or especially a thinning that removes only the best trees (as for poles or piling), is never acceptable either for continued growth of the stand or for regeneration. The comment is often made that removing the dominant trees leaves the intermediate and suppressed trees free to grow, but often the smaller trees are inherently slow growers and never respond satisfactorily to release. Certainly, the smaller, poor-quality trees should never be left to regenerate a new stand.

Pruning

Although there is less opportunity to use genetic principles in a pruning program than in thinning, the choice of trees to be pruned is important (Brown, 1965). A knowledge of the inheritance of bole form, limb characteristics, and growth rate is essential in choosing the trees that will bring the greatest return from pruning (Polge, 1969). For some species, like *P. patula,* a knowledge of limb conformation of individual trees and their reaction following pruning can be most useful (Harris, 1963; Minckler, 1967; Olischlager, 1969; Beineke, 1977). The key to all pruning is to time the operation so the selected trees will really be the dominants at harvest time. It is essential to prune only those individuals that are inherently fast growers and, as such, will retain their dominance within the stand.

Fertilization

A knowledge of fertilizer response by species is essential. Such information will not be available for natural stands or from plantations from randomly collected seed, but it can be most useful for plantations from seed orchards. Some species, like *P. taeda,* show little genotype × fertilizer interaction, so there usually is not a large differential response to fertilizers (Matziris and Zobel, 1976). Other species, like *P. elliottii,* often show large family-response differences to fertilization (Goddard et al., 1975b). When this occurs, it makes no sense to plant large acreages of plantations that do not respond to fertilization (or that may even react negatively to fertilization), and then to fertilize them as a standard operational procedure. Gains are possible by determining fertilizer response and then using the responders where fertilization will be used operationally. Family × fertilizer interactions have not been studied extensively in hardwoods, but because they often respond dramatically to small environmental changes, the potential for such interactions is great.

Another use of genetics that is related to fertilizers and nutrients is to use sources that are somewht tolerant to nutrient deficiencies. For example, in Chile, individual trees of *P. radiata* are occasionally found that grow very well in soils that are very deficient in boron and, there are indications that these trees can tolerate lower than normal levels of this substance. Studies have shown also that there is a differential tolerance to phosphate deficiencies (Burdon, 1969).

Fertilizer × genotype interaction will be covered more fully in the chapter dealing with genotype × environment interaction. It is sufficient to say that the differential genetic responses to fertilizers are great enough so that most forestry programs will not be completely successful without taking them into account, although outside the species and source levels this information cannot be well used in natural forests or plantations with unknown genetic backgrounds.

SUMMARY

The genetic approach has considerable potential in naturally regenerated forest stands when it is applied to the establishment of the new stand and follow-up silvicultural treatment. The most gain can be achieved through proper thinning. One major problem is that it is difficult to quantify the gains and to pinpoint economic results from various activities. As time progresses, the use of genetic principles in natural regeneration will be of considerably greater value than is currently believed; this is especially true in managing the tropical hardwoods. Similar use of genetic principles will be made in improvement of plantations established from seed of unknown genetic sources. Much of the gain will not be improvement as such, but it will be prevention of losses by practices such as dysgenic selection or selective logging.

LITERATURE CITED

Beck, D. E. 1975. "Board-Foot and Diameter Growth of Yellow-Poplar after Thinning." U.S. Forest Service Research Paper SE-123.

Beck, D. E. 1977. "Growth and Development of Thinned versus Unthinned Yellow-Poplar Sprout Clumps." U.S. Forest Service Research Paper SE-173.

Beineke, W. F. 1972. "Recent Changes in the Population Structure of Black Walnut." 8th Central States For. Tree Impr. Conf., Columbia, Mo., pp. 43–46.

Beineke, W. F. 1977. Corrective pruning of black walnut for timber form. *For. Nat. Res.* **76**:1–5.

Books, D. J., and Tubbs, C. H. 1970. "Relation of Light to Epicormic Sprouting in Sugar Maple." U.S. Forest Service Research Note NC-93.

Bowersox, T. W., and Ward, W. W. 1972. Long-term response of yellow-poplar to improvement cuttings. *Jour. For.* **70**(8):479–481.

Brown, G. S. 1965. The yield of clearwood from pruning: Some results with radiata pine. *Common For. Rev.* **44**(3):197–221.

Burdon, R. 1969. "Clonal Replication Trial in *Pinus Radiata.*" Forest Research Institute, New Zealand Forest Service, Wellington, New Zealand.

Clark, F. B., and Boyce, S. G. 1964. Yellow poplar seed remains viable in the forest litter. *Jour. For.* **62**:564–567.

Della-Bianca, L. 1973. Screening some stand variables for post-thinning effect on epicormic sprouting in evenaged yellow-poplar. *For. Sci.* **18**(2):155–158.

Della-Bianca, L. 1975. "Precommercial Thinning in Sapling Hardwood Stands." 3rd Annual Hardwood Symp., Cashiers, N.C., pp. 129–133.

Erdmann, G. G., Godman, R. M., and Oberg, R. R. 1975. "Crown Release Accelerates Diameter Growth and Crown Development of Yellow Birch Saplings." U.S. Forest Service Research Paper NC-117.

Goddard, R. E., Schmidt, R. A., and Vande Linde, F. 1975a. Effect of differential selection pressure on fusiform rust resistance in phenotypic selections of slash pine. *Phytopathology* **65**(3):336–338.

Goddard, R., Zobel, B., and Hollis, C. 1975b. "Response of Southern Pines to Varied Nutrition." Physiological Genetics Conference, Edinburgh, Scotland, Tree Physiology and Yield Improvement, pp. 449–462.

Griswold, H. C. 1979. "An Analysis of Precommercial Thinning After Ten Growing Seasons." Research Note No. 75, Georgia Kraft Co., Rome, Ga.

Harris, J. M. 1963. "The Effect of Pruning on Resinification of Knots in Radiata Pine. New Zealand Forest Research Note No. 34.

Hofmann, J. G. 1974. "Thinning in Short Rotation Plantation Forests—Will It Come to Pass? 1974 Annual Meeting, TAPPI, pp. 189–193.

Kormanik, P. P. 1966. "Epicormic Branching and Sprouting in Hardwoods; a New Look at an Old Problem." Symposium on Hardwoods of the Piedmont and Coastal Plain, Georgia Forest Research Council, Macon, Ga., pp. 21–24.

Little, N. G. 1974. "Analysis of Pre-commercial Thinning." Georgia Kraft Co., Research Note No. 44, Rome, Ga.

Maki, T. E. 1969. "Major Considerations in Thinning Southern Pines." IUFRO Conference, Stockholm, Sweden.

Matziris, D., and Zobel, B. 1976. Effects of fertilization on growth and quality characteristics of loblolly pine. *For. Ecol. Management* **1**(1):21–30.

McGarity, R. W. 1977. "Ten-Year Results of Thinning and Clearcutting in a Muck Swamp Timber Type." International Paper Co. Technical Report No. 38.

Minckler, L. S. 1967. Release and pruning can improve growth and quality of white oak. *Jour. For.* **65**(9):654.

Olischlager, K. 1969. "Studies on the Increase in Value of Spruce After Pruning." Universität zu Göttingen in Hann., Münden, Germany.

Polge, H. 1969. Densité de plantation et élagage de branches vivantes—ou pourquoi, quand et comment élaguer? [Density of planting and pruning of live branches—or why, when and how to prune]. *Silviculture* 21:451–465.

Roberge, M. R. 1975. Effect of thinning on the production of high-quality wood in a Quebec northern hardwood stand. *Can. Jour. For. Res.* **5**(1):139–145.

Smith, L. F. 1967. "Effects of Spacing and Site on the Growth and Yield of Planted Slash Pine." U.s. Forest Service Research Note SO-63.

Trimble, G. R. 1971. "Diameter Limit Cutting in Appalachian Hardwoods: Boon or Bane?" U.S. Forest Service Research Paper NE-208.

Trimble, G. R. 1973. "Response to Crop-Tree Release by 7-Year-Old Stems of Yellow-Poplar and Black Cherry." U.S. Forest Service Research Paper NO-253.

Wahlenberg, W. G. 1952. Thinning yellow-poplar in second-growth upland hardwood stands. *Jour. For.* **50**(9):671–676.

Webb, C. D., and Belanger, R. P. 1979. "Inheritance of Sprout Growth in American Sycamore (*Platanus occidentalis*)." 15th South. For. Tree Imp. Conf., State College, Miss., pp. 171–175.

Williams, R. A. 1974. "Precommercial Thinning." Symp. on Management of Young Pines, Alexandria, La., pp. 72–74.

CHAPTER 8

Genetic Testing Programs

MATING DESIGNS
Incomplete Pedigree Designs
Open-Pollinated Mating
Polycross (Pollen Mix) Designs
Complete Pedigree Designs
Nested Design
Factorial Design
Single-Pair Mating
Full Diallel
Half Diallel
Partial Diallel
Multiple-Population Systems
EXPERIMENTAL DESIGNS
Types of Plots
Distribution of Plots
Completely Random Design
Randomized Complete-Block Design
Other Designs
Testing Procedures
Choosing Test Sites
Replication and Plot Layout
Nursery Procedures
Site Preparation
Documentation
Test Maintenance
Thinning
ANALYSIS OF GENETIC TESTS
The Analysis of Variance
Nature of the Analysis
Individual-Tree Heritability Calculations
Family Heritability
Heritability from Parent–Offspring Regression
Genetic Correlations
GENOTYPE X ENVIRONMENT INTERACTION
Impacts of *GE* **Interaction**
Species and Provenance Interaction
Family and Clone x Environment Interaction
Interaction and Forest Management
LITERATURE CITED

Trees chosen for use in seed orchards or for breeding purposes are usually selected because they appear to be superior; that is, they have a good phenotype. Once trees are selected, their genetic worth is tested by mating them in some fashion, and the offspring are then established in genetic tests. Genetic testing is mandatory for any aggressive and successful tree improvement program. It lays the foundation for genetic decisions involving management of seed orchards and provides the material and information that will be the basis for advanced-generation tree improvement efforts. It is essential, therefore, that extreme care be taken to insure that the genetic testing program is designed and implemented so that maximum gains can be achieved in both the short and long term. Genetic testing is one of the most expensive aspects of a tree improvement program, but the profitability of tree improvement efforts will be directly related to the quality of the genetic testing program.

There can be several different objectives of genetic testing, but it must be emphasized that no one testing design will be the best to meet all objectives. Therefore, care must be taken to use the proper design; this can only be done when the objectives of the testing program are clear. Objectives of genetic testing include the following:

1. **Progeny testing** The best way to evaluate the genetic worth of selected parents is to grow their progeny in a way that allows estimation of the parental breeding values. This enables one to separate parents whose phenotypic superiority may have resulted from growing in a good environment from those that are superior because they have a good genotype. If the parents who are progeny tested have already been established in production seed orchards, the undesirable ones can be removed from the orchard by roguing. Dramatic genetic differences often appear among progeny from different parents. For example, the slow-growing and fast-growing *Eucalyptus grandis* progeny shown in Figure 8.1 both have parents that were selected because they had superior phenotypes. However, they obviously had different genotypes.

2. **Estimation of variance components and heritability** The choice of which traits to emphasize in a tree improvement program is highly dependent upon the degree of their inheritance. Such a choice can only be made when the tree breeder has assessed the relative contribution of genetics and of environment to the total variation. Only after this is known is it possible to devise the most efficient method of selection and breeding.

3. **Production of a base population for the following generations of selection and breeding** Perhaps the most important function of the genetic test in the long term is that it provides a source of material from which selections for the following generation can be obtained. The opportunity for improvement in advanced-generation seed orchards is dependent on the types of mating and

FIGURE 8.1

Radical family differences often appear in genetic tests when the parents are progeny tested. The slow-growing *Eucalyptus grandis* family on the left and the fast-growing family's on the right both come from phenotypically superior parents. This progeny test resulted in the parent of the slow-growing family's being removed from the breeding program.

testing schemes followed in earlier generations. It is not possible to develop a long-term tree improvement program without suitable, well-planned genetic tests.

4. **Demonstration or estimation of genetic gain**   The profitability of a tree improvement program depends upon the tree breeder's ability to create populations that have better characteristics than did the unimproved populations. The only way to assess accurately the progress made in a tree improvement program is to compare the relative performance of the improved and unimproved stock in the same test.

No single genetic testing scheme will be best to satisfy all of the preceding objectives. The ideal is to design tests specifically to accomplish one objective; however, because of cost and manpower restrictions, it is often necessary to try to satisfy several objectives in the same test. This presents a real challenge to the tree breeder. The design used should be most successful for the most important function of the test, but as much information as possible will be obtained for the other functions. Careful consideration of the objectives of testing, their relative importance, and the most suitable design require the most sophisticated skills of the tree breeder in developing an efficient tree improvement program.

Once the objectives of a testing program have been defined, two additional major decisions must be made. These are the choice of the mating design that is to be used to create the progeny population and the experimental design that is to be employed when the test is established in the field. In general, the mating design will determine the *type* of information that will be derived from the testing program, and the field test design will determine the *quality* of information that is obtained.

This chapter will cover various types of mating and field test designs used in tree improvement; certain concepts of methods used to analyze the tests will also be introduced. Additionally, because of its importance to any testing program, the nature and importance of genotype–environment interaction will be covered in some detail. In actuality, genotype–environment interactions are pertinent to several chapters in this book, but their relevance to the design and implementation of genetic tests makes presentation of this topic especially relevant here.

MATING DESIGNS

Numerous mating designs have been proposed for forest trees. These have been discussed in detail by several authors, including Burdon and Shelbourne (1971) and van Buijtenen (1976). For convenience, mating designs can be divided into two general classes: (1) *incomplete pedigree designs,* in which only one parent is known for any given progeny, and (2) *complete pedigree designs,* in which both parents are known to the breeder.

Incomplete Pedigree Designs

Open-Pollinated Mating   The easiest and least expensive means of creating a progeny population is to use open- or wind-pollinated offspring from selected parents. The method is simple, consisting of collecting open-pollinated seed from the parent trees that are to be tested. Seed may be collected from parent trees growing in natural stands or plantations and can also come from genetic tests, or from genotypes established in a seed orchard.

Open-pollinated seed are useful in fulfilling several breeding objectives. They can serve a progeny-testing function by giving estimates of parental general combining ability that is necessary to rogue genetically poor parents from production seed orchards. If seed orchard parents are to be progeny tested using open-pollinated seed from seed collected in the orchard, it is best not to collect seed until the orchard is in heavy seed production. Pollination patterns characteristic of young seed orchards are often very different from those of a fully productive orchard because at a young age only a few clones produce most of the pollen; therefore seed collected from young orchards may have only a limited number of pollen parents. In fact, observation by several loblolly pine seed orchard managers has indicated that in young orchards about 80% of the seed is produced by 20% of the clones, and it is only after 10 to 12 years that a somewhat more equal distribution occurs. Estimates of breeding values using open-pollinated progenies from young orchards will be biased when nonrandom pollination patterns exist. For example, if only one clone in the orchard was producing all of the pollen, estimates of the general combining abilities of the parents being tested would be totally confounded with the specific combining abilities of crosses between those parents and the single pollen parent. Several examples could be cited of nonuniform seed production. For example, Bergman (1968) reported, for one seed orchard, that 50% of the seed produced had one of the clones in the orchard as either the male or female parent.

Open-pollinated tests can also provide estimates of additive genetic variance and heritability values for the population being tested. Because only one parent is known, estimates of nonadditive genetic variance cannot be obtained.

Open-pollinated tests are of limited utility for future generations of selection. If progeny are grown from seed collected in the seed orchard, the breeder will not know whether the individuals selected were related through a common male parent. The risk of inbreeding when selections from such tests are used in advanced-generation seed orchards is a severe limitation to using the open-pollinated design. Advanced-generation selection can be made from tests involving open-pollinated progeny if seed was collected from widely spaced trees growing in natural stands or unimproved plantations, without fear that selections are related through a common male parentage. However, a drawback to use of this kind of open-pollinated material is that the pollen parent is unselected, which lowers gain from selection efforts.

Polycross (Pollen Mix) Designs   In a polycross design, each female parent is crossed with a mix of pollen from a number of male parents. Generally, a considerable number of pollens are included in the mix to insure that female parents are pollinated by a representative sample of other parents. Several variations of the polycross design have been proposed (Burdon and Shelbourne, 1971).

Like the open-pollinated design, the polycross design can be used to efficiently estimate additive genetic variances, heritabilities, and breeding values of the female parents involved. However, because the male parent's identity is unknown, estimates of nonadditive variance and specific combining abilities are not possible. Also, there is considerable danger that the estimates of breeding values obtained may be biased due to nonrandom pollination by the pollens included in the mix. Research is needed to determine if this bias is large enough to appreciably affect breeding value estimates. As with the open-pollinated design, selection efforts using a pollen mix are usually limited because the male parent is unknown, and the proportion of outstanding trees may be very largely biased in favor of one or two good general combiners that occurred by chance in the pollen mix.

Complete Pedigree Designs

For any mating design, the maximum number of unrelated families that can be created is one half the total number of parents in the breeding population, assuming that all parents are unrelated. Only when equal numbers of parents are used as males and as females are the maximum number of unrelated families created. When crosses are shown, the convention is to list the female parent first.

Nested Design   The *nested design,* which is also known as the *hierarchial mating design,* is a scheme in which groups of parents of one sex (in the case of monoecious species, sex is "designated") are mated to members of the other sex (Figure 8.2). Therefore, the progeny are composed of full-sib families that have both parents in common, and half sibs that have one parent in common.

The nested design has been used extensively in agricultural crops. In forest trees, the best example of its use is the Loblolly Pine Heritability Study, which was established jointly by the International Paper Company, the National Science Foundation, the National Institutes of Health, and North Carolina State University to determine inheritance patterns in an unimproved loblolly pine population. A summary of results from the study through its first ten years has been given by Stonecypher et al. (1973).

Use of the nested design allows the tree breeder to estimate both additive and nonadditive genetic variances and heritabilities; this is the test objective that is best served by the design. It has some disadvantages, however. As shown in Figure 8.2, estimates of general combining ability can be obtained only for

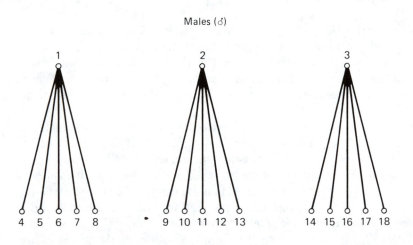

Males (♂)

Females (♀)

FIGURE 8.2

The nested, or hierarchial, mating design is a control-pollinated design that has been used in forest trees. In the diagram, Males 1, 2, and 3 are used as the rarer sex, and each is mated to five different female parents.

members of the rarer sex, because members of the more common sex are used in only a single cross. The number of unrelated selections that can be made among the progenies is limited by the number of parents used as the rarer sex.

Factorial Design   The *factorial design* is one in which members of one sex (in case of monoecious species, sex is "designated") are crossed in all combinations with several members of the other sex (Figure 8.3). The most common way this design has been employed in forestry is when four to six parents are designated as testers and crossed to all other parents in the population being tested. The factorial design is therefore also known as the *tester design*.

The factorial design is very useful for progeny-testing purposes, since breeding values can be estimated for all genotypes in the population being tested. It also allows a reasonable estimation of variance components and heritabilities. Specific combining ability can also be estimated for the actual crosses that are made. A disadvantage of the tester design is that the number of unrelated progeny that can serve as parents for the next generation is limited by the number of parents used as testers. If five testers were used (as in Figure 8.3) only five unrelated families can be produced, regardless of the number of parents used in the crossing program.

A derivation of the factorial mating scheme is the *disconnected factorial design*. With this approach, the breeding population is divided into several sets of parents,

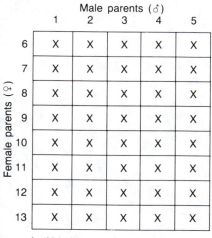

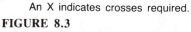

An X indicates crosses required.

FIGURE 8.3

With the factorial mating design, parents of one sex are crossed to several members of the other sex. The factorial design is sometimes called the *tester design* when, as shown, a few members of one sex are designated as testers and are crossed to all other parents in the population.

and a factorial mating design is employed within each set. For example, the 18 parents shown in Figure 8.3 could be divided into three groups of six parents and factorials involving three parents as males and three parents as females could be used in each group. This is shown diagramatically in Figure 8.4. Disconnected diallels have the advantage of maximizing (if equal numbers of males and females are used) the number of unrelated families that are created while maintaining at acceptable levels the number of crosses that must be made. Therefore, the disconnected factorial is an appropriate mating design if selection is the primary objective of testing. It is not as efficient as the tester design for progeny testing because individuals in different sets are mated to different parents, and estimates of general combining ability will be biased to the extent that the genetic quality of parents differs from set to set. This becomes less of a problem as more parents are included in each set.

Single-Pair Mating   With the *single-pair mating* scheme, each parent is mated to one other member of the population (Figure 8.5); this creates the maximum number of unrelated families in each generation with a minimum number of crosses. Its use allows maintenance of a large, effective population size.

Single-pair matings are not suitable for roguing seed orchards. Since each parent is involved in only one cross, it is not possible to estimate general

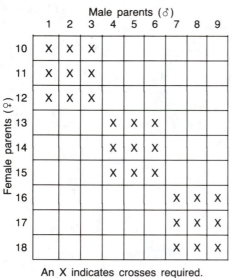

An X indicates crosses required.

FIGURE 8.4

In the disconnected factorial mating design, the breeding population is divided into small sets, and factorial matings are used in each set. Shown are three disconnected factorials that might be used for a breeding population of 18 parents.

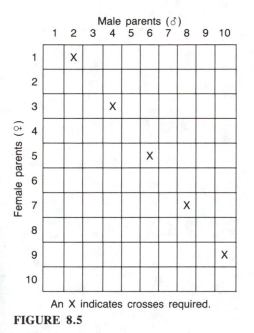

An X indicates crosses required.

FIGURE 8.5

The single-pair mating scheme requires that each parent be mated to one other parent in the population. In the diagram, 10 parents have been mated in single-pair fashion, requiring five crosses. In the example shown, it is assumed that the species is monoecious and that parents can be used as either males or females in the crossing scheme.

combining ability that is needed as the basis for roguing. Similarly, the design cannot be used to estimate additive and nonadditive variances. If general combining abilities of the parents involved have previously been estimated in other types of genetic tests, single-pair matings with genetically proven parents will be useful in a breeding program to produce populations for advanced-generation selection.

Full Diallel   The most comprehensive mating system available is the *full diallel* in which each parent is crossed to all others in every combination, including reciprocal crosses. *Reciprocal crosses* are two matings involving two parents in which, in the first mating, the first parent is used as a female, the second as the male; whereas in the reciprocal cross, the second parent is used as a female and the first parent is used as a male. Reciprocal crosses can be diagrammed as $A ♀ × B ♂, B ♀ × A ♂$. A full diallel involving 10 parents is shown in Figure 8.6.

The full diallel will yield information on general combining ability of all parents, specific combining ability of all crosses, and variance component infor-

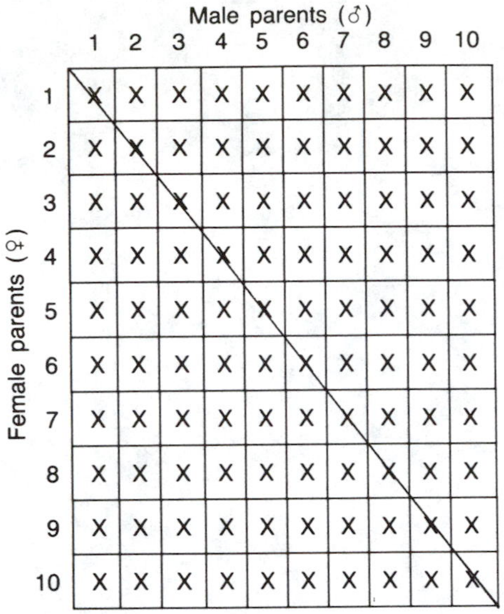

An X indicates crosses required.

FIGURE 8.6

The full-diallel crossing scheme requires that all parents be crossed to each other in every combination, including selfs. The 10-parent full diallel shown requires 100 crosses.

mation; it also creates the maximum number of unrelated families available for future selection. Thus, it might appear to be the ideal mating design. However, a major disadvantage of the full diallel is that the large number of crosses required makes it expensive and time consuming, especially when large numbers of parents are involved. Furthermore, there often are not enough reproductive structures available to carry out a full diallel. For example, a full diallel involving $n = 10$ parents (as in Figure 8.6) would require $n^2 = 10^2 = 100$ separate crosses. A full diallel involving 200 parents, which is a more realistic breeding population size, would require 40,000 crosses. As a result, several modifications of the full-diallel system have been proposed that are planned to decrease the numbers of crosses required but also to maintain many of the desirable attributes of the full diallel. Obviously, diallels can not be used with dioecious species because the design requires that individuals be used as both male and female parents.

Half Diallel *Half diallels* are similar in design to full diallels, except that reciprocal crosses (and usually selfs) are not made. A half diallel involving 10 parents, excluding selfs, would require the crosses shown above or below the diagonal line in Figure 8.6.

Half diallels yield nearly as much information as full diallels but at less than half the cost and effort. However, a large number of crosses are still required for half diallels when many parents are involved. In a 200-parent half diallel, excluding selfs, $n(n-1)/2 = 19,900$ crosses would be required. As a result, large half diallels are rarely used in tree improvement programs other than for studies that are very basic in nature.

Partial Diallel   Several other most useful modifications of the diallel have been developed; these are usually called *partial diallels*. One type of partial diallel, named the *sytematic or progressive mating scheme* by Wright (1976), is shown in Figure 8.7. With this design, crosses are made that fall in particular diagonals.

Male parents ($\male$)

Female parents ($\female$)	1	2	3	4	5	6	7	8	9	10	11	12	13	14	15	16	17	18
1		X	X					X	X	X							X	X
2			X	X					X	X	X							X
3				X	X					X	X	X						
4					X	X					X	X	X					
5						X	X					X	X	X				
6							X	X					X	X	X			
7								X	X					X	X	X		
8									X	X					X	X	X	
9										X	X					X	X	X
10	X										X	X					X	X
11	X	X										X	X					X
12	X	X	X										X	X				
13		X	X	X										X	X			
14			X	X	X										X	X		
15				X	X	X										X	X	
16					X	X	X										X	X
17						X	X	X										X
18							X	X	X									

An X indicates crosses required.

FIGURE 8.7

A systematic or progressive diallel is a modification of a full diallel in which crosses are made that fall on particular diagonals. A systematic diallel with 18 parents involving each parent in 10 crosses is illustrated.

Diagonals are chosen so that no one parent is involved in more than a few crosses. This design has many of the advantages of a full diallel or half diallel in that it creates the maximum number of unrelated crosses, allows estimates of general combining ability for each parent, estimates of additive and nonadditive variances, and estimates of specific combining ability for a part of the possible combinations.

A second type of diallel modification is the *disconnected diallel* scheme, which

Male parents ($\male$)

	1	2	3	4	5	6	7	8	9	10	11	12	13	14	15	16	17	18
1		X	X	X	X	X												
2			X	X	X	X												
3				X	X	X												
4					X	X												
5						X												
6																		
7								X	X	X	X	X						
8									X	X	X	X						
9										X	X	X						
10											X	X						
11												X						
12																		
13														X	X	X	X	X
14															X	X	X	X
15																X	X	X
16																	X	X
17																		X
18																		

Female parents ($\female$)

An X indicates crosses required.

FIGURE 8.8

The disconnected-diallel mating scheme is a modification of the full diallel scheme in which the population is divided into groups and diallel or half-diallel matings are done in each group. The three six-parent disconnected half diallels shown would require 45 crosses.

is shown in Figure 8.8. With this design, parents are divided into small 5 to 10 tree groups, and diallel or half-diallel matings are done within each group. For example, in Figure 8.8, the 18-parent breeding population has been divided into three groups of six individuals, and each group of six parents has mated in half diallels, excluding selfs. The disconnected diallel maintains most of the advantages of more complete diallels but greatly reduces the number of crosses required. For example, a full diallel using 18 parents requires $n^2 = 18^2 = 324$ crosses; a half diallel, excluding selfs, with the same parents, would involve $n(n - 1)/2 = 153$ crosses. The three six-parent disconnected half diallels involve $n(n - 1)/2 = (6 \times 5)/2 = 15$ crosses per diallel, or $3 \times 15 = 45$ crosses for the entire 18-parent population.

Multiple-Population Systems

One type of breeding scheme that is increasing in popularity among tree breeders is that of multiple-population breeding. This is actually not a mating scheme per se, because it can involve one or several of the mating schemes discussed previously. With the multiple-population design, a large breeding population is divided into a number of smaller breeding groups. Crossing is done only within each group, and crosses are never made between individuals in different groups other than in production seed orchards. Once established, breeding groups maintain their genetic integrity and selected individuals always remain in the same breeding groups, regardless of whether they are first-, second-, or more advanced-generation selections. Inbreeding quickly becomes commonplace within groups because of the small group sizes. Breeding groups usually contain up to 25 individuals. Specific mating designs within groups can be chosen to meet the needs and constraints of the breeder.

One use of multiple population breeding is "sublining," a system designed to permanently avoid inbreeding in production seed orchards (van Buijtenen, 1976; Burdon et al., 1977; Namkoong, 1977). With this system, numerous small breeding groups are formed. Although inbreeding soon occurs within groups, production seed orchards are established that use only the best individual from each breeding group. To assure a suitable genetic base for the orchard, a number of groups are required. Using only one tree from each group in the production orchard assures outcrossing to produce the commercial seed crop.

Sublining breeding systems have been developed for several important tree species, including black walnut (McKeand, 1982) and loblolly pine (van Buijtenen and Lowe, 1979). An example of the sublining breeding scheme proposed for black walnut that involves 16 breeding groups, each with 25 parents, is shown in Figure 8.9. There are several problems involved with using sublining. First is the difficulty of selecting the best individual tree from within each group since varied amounts of related matings have occurred. Also, the inbred trees used in the production seed orchard of some species may be slow growing and produce few reproductive structures.

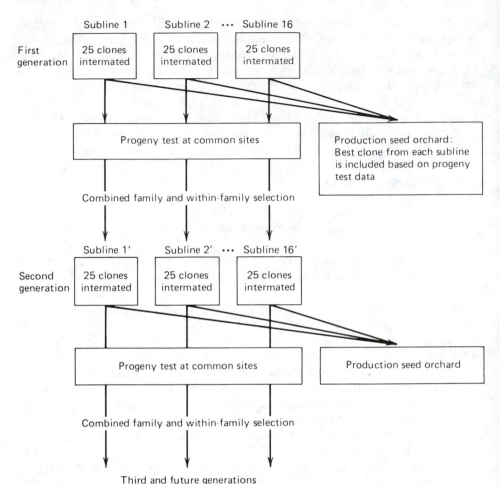

FIGURE 8.9

Sublining has been proposed as a breeding system for several forest tree species. The scheme shown for black walnut involves 16 sublines, each of which is composed of 25 clones. (from McKeand and Beineke, 1980).

EXPERIMENTAL DESIGNS

Once the desired progeny have been created, the next step in genetic testing is to establish the seedlots in the nursery, greenhouse, or field so that genetic and environmental effects on progeny performance can be assessed. Many different types of experimental designs are available for use by the forest tree breeder. However, only those that are in wide use and are relatively efficient on a large scale under field conditions will be discussed here. Excellent sources of information on design and analysis of experiments can be found in basic statistical textbooks, such as those by Cochran and Cox (1957), Steel and Torrie (1960), and Snedecor and Cochran (1967). Although the following discussion will center mainly on testing of seedlots in which one or both parents are known, the same general considerations apply to provenance testing and even to species trials.

Types of Plots

A *plot* can be defined as a group of trees of a single family, provenance, or species that vary in size from one to several hundred trees that are treated as a unit in a tree-breeding experiment. In most experiments, trees within a plot are planted adjacent to one another; that is, they are planted contiguously. More than one, and usually several, plots of a given seedlot must be planted in an experiment to estimate genetic and environmental effects on progeny performance.

Plot shapes vary, depending on the goals and resources of the researcher. They may consist of single trees, rows of trees, or square or rectangular blocks of trees. When differences among provenances or species are being tested, it is usually best when large block plots are used, because in this way competitive effects between species or provenances that will not be planted together operationally can be minimized. Large block plots allow seedlots that may differ widely in growth rates to fully express their genetic potential without being suppressed by others throughout the length of the experiment. If large differences in growth patterns exist and single-tree or row plots are used, seedlots that are the largest and most vigorous when tree-to-tree competition begins in the test may suppress and eventually even kill adjacent, less vigorous seedlots, and the freely growing seedlots will develop with minimal competition. This results in exaggerated differences among seedlots when the slower-starting one has no chance to express its genetic potential. In the hypothetical situation shown in Figure 8.10, use of row or single-tree plots might make it appear that Species A would be the best at rotation age because it is larger when competition started than the initially slower-growing Species B that would actually be larger at rotation age if it were not suppressed. When the goal is to measure yield of genetic entries per unit area, large-block plots are the only types to use, regardless of whether or not the seedlots are species, provenances, open-, or control-pollinated families, or clones. When block plots are used, the outside row of trees in each plot is usually considered a border row, and measurements are made only on the interior trees. The rule to follow is to use block plots for tests of materials that will be grown together operationally. For example, rarely would provenances (or species) be interplanted; they would be used as discreet units, whereas seedlings from individual clones from a seed orchard would normally be planted together.

Row plots are commonly used to test genetic differences among open- or control-pollinated families. In some instances, single-tree plots are used; statistically, single-tree plots often are the most efficient types of plots. However, when single-tree plots are used, death of even a single tree can severely complicate the statistical analysis. Additionally, if each tree is not labeled in the field, errors of

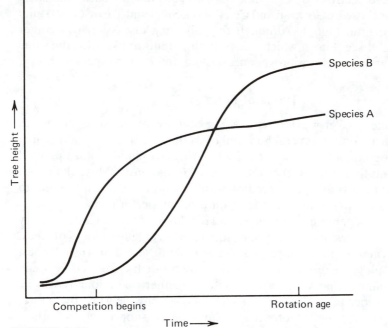

FIGURE 8.10

Block plots are preferred in tree improvement experiments when large differences in growth curves exist among the seedlots being tested. In the hypothetical situation shown, Species A and Species B have quite different growth curves. Use of single-tree or row plots in a test of these two species might result in Species A's being erroneously declared the winner at rotation age because it entered competition with a growth advantage over Species B. Species B would not have had the opportunity to express its genetic potential.

identification are often made that can ruin an entire experiment. Single-tree plots are most appropriate in situations in which each tree is given much care, and in which measurements are taken before competition and subsequent mortality begins, as for example, in Christmas tree studies.

Distribution of Plots

Several different methods have been developed to best distribute plots in an experiment; only a few have found common usage in forestry.

Completely Random Design In a *completely random design,* two or more plots of the seedlots to be tested are distributed at random throughout the test area. It is important that the seedlots be distributed at random so that each seedlot can encounter the range of environmental conditions on the test site. Completely random designs are easy to analyze statistically, and they work well when the test area is very uniform. Because test sites for forest trees are rarely uniform, most tree breeders prefer the randomized complete-block design (that is discussed next) over completely random designs.

Randomized Complete-Block Design   Each family or seedlot is represented by a single plot in a test subunit called a *replication* or *block,* when a *randomized complete-block design* is used. Several replications are established for each test, with the plots being randomly distributed within each replication. The purposes of blocking are to adjust for different environmental conditions on the test site statistically, and to ensure that members of a seedlot are exposed to the full range of environments within the test area. Differences that occur among replications indicate the degree of environmental variation and can be adjusted for statistically so that the genetic differences can be assessed.

Randomized complete-block designs are used more frequently than any other type of design in tree improvement work. Installation of tests using this design are relatively straightforward, and analysis is relatively simple. Statistical precision is greater than for the completely random design because environmental differences (differences among blocks) can be separated from effects due to family and within-family genetic variation.

Other Designs   Several other test designs are used by tree breeders that are more precise statistically, but they are not easy to implement, maintain, and analyze in field situations where controls are difficut. These more sophisticated designs are generally used in special situations, such as in nursery beds, greenhouses, or in special experiments. In a *latin square* design, the test site is divided into rows and columns, and one plot of each seedlot is planted in each row and in each column. In *balanced incomplete-block designs* or *lattice designs,* replications of equal size are established, but each replication contains fewer than the total number of seedlots involved in the test. The major benefit of incomplete-block designs is a reduction in replication size, which makes the test more precise because of reduction in environmental variation within replications. Loss of trees and plots is particularly difficult to handle in the sophisticated tests.

One type of sophisticated design that deserves special mention is the *split-plot design* that is often employed when it is necessary to test two or more types of seedlots in one test (for example, provenances and families within provenances). It is also used where environmental treatments are to be superimposed on the experiment (e.g., to determine if families respond differently to fertilizer treatments). In a split-plot design, the treatment or seedlot type that requires large areas (fertilizer treatments or provenances) are established in major plots that are randomly distributed in each replication. Within each major plot, subplots of each family or other sources of genetic variation are planted—again in random distribution. Split-plot designs are generally not difficult to install or analyze when only two or three factors are to be tested.

Testing Procedures

Regardless of the type of experimental design used, the success of a genetic test depends to a great extent on the breeder's ability to properly install the test in the field and to keep it intact throughout the life of the experiment. Many of the factors involved in proper field installation are as much an art as a science and are often nonstatistical in nature. Even though an experimental design is potentially statistically precise, if it is improperly installed or maintained, results will not be satisfactory. There is no way to adjust for a poorly installed test. In field forestry experiments, losses of individual trees and sometimes parts or whole plots are not unusual. The many years during which the experiments must be kept and the environmental or pest catastrophes that can occur, all result in tests that rarely come to completion as established.

There are a number of factors involved that will ensure proper test installation and maintenance. Because most genetic tests employ randomized complete-block designs, the discussion that follows will concentrate on proper use of that design.

Choosing Test Sites   An elementary but very important rule in choosing test locations is that genetic tests should be located on sites that are representative of the lands that are to be reforested with the improved planting stock that is being assessed. For example, if a large majority of the land to be planted is of an organic soil type, it would not be suitable to plant genetic tests on mineral soils that might be more easily accessible. Rare site types and unusual environments should be avoided as test sites, unless the improved seedlings or vegetative propagules are to be grown on such sites in the operational program.

It is impossible to find sites for genetic testing that have no environmental variation. However, since the purpose of a genetic test is to distinguish between performance that is caused by genetic and environmental factors, tests should be established on land that is as uniform as possible within replication with respect to topography, drainage, and soil factors. Differences between replications can be handled statistically, but there is no way to separate genetic differences from environmentally caused variation within a replication.

One of the most common errors is to choose and prepare a test site that is too small or that has a shape that will not accommodate the test. Untold frustrations occur when it is time to plant to find then that the test will not fit on the land available. The experimentor should be sure to precalculate the amount of area needed. This includes borders, roads, and fire breaks. Also, the experimentor should make a trial to see whether or not the test fits. Preferably, this should be done both in the office and in the field. If the topography of the site is nonuniform, extra land may be needed so that replications can be properly located.

Replication and Plot Layout   A general rule for layout of randomized complete-block designs is that replications should be oriented so that their long axis is perpendicular to the environmental gradient. The progeny plots (family row plots) should be oriented so that their long axis is parallel to the environmental gradient. The objective is to minimize environmental variation within each replication and to have each family sample what environmental variation does exist within the replication. For example, if a test is to be established in sloping terrain and family row plots are to be used, there should be a minimum amount of elevation change within each replication, and family row plots should run up and down the hill (Figure 8.11). When poorly drained test sites have been ditched, family rows should run perpendicular to the drainage ditches. On bedded sites, family rows should run across beds, and they should be perpendicular to windrows when these are present.

Abrupt, isolated site changes in an area allocated for a replication can often be avoided by splitting the replication or through the use of filler rows. For example, if a small gully runs through a replication, filler rows, which are not measured, should be planted in the gully and in areas immediately surrounding the gully. Although small "splits" in replications are occasionally acceptable, it is absolutely mandatory that all sections of a replication be planted on similar sites as near to one another as possible. Examples of correct and incorrect ways to split replications are shown in Figure 8.12

The number of replications to be used in a test depends on the variability of the plant material being tested, the uniformity of the test site, the precision needed for evaluating seedlot differences, the amount of plant material that is available for the test, and the size of the plots. The question is often raised whether it is better to have more replications with small plots, or fewer replications with larger plots.

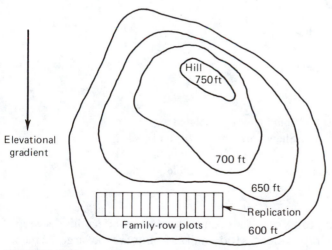

FIGURE 8.11

A generalized scheme for genetic test layout where a randomized complete-block design is used on sloping terrain is shown. The replication extends lengthwise along the contour, or across the elevational gradient. Family-row plots extend along the elevational gradient.

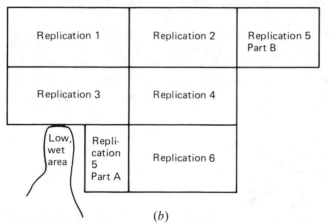

FIGURE 8.12

Sometimes replications must be "split" to avoid isolated extreme environmental conditions. Shown are correct (*a*) and incorrect (*b*) ways to split replications. The rule is to establish sections of replications as near to one another as possible.

Cost of establishment, uniformity of the site, and other factors affect this decision. Commonly, the number of replications is too small and plots are too large. In most instances, 4 to 6 replications are adequate if sites are reasonably uniform. Not much additional information is gained by using more than 10 replications. When there is a question, the choice should be to use smaller plots with more replications. Rarely are less than 3 replications acceptable. The success of a field plot layout does not depend so much on its size but on the uniformity of the environment within a replication and on how well the plots sample the environment within the replication.

Nursery Procedures Successful field establishment of genetic tests depends upon the production of quality plants. Maximum care must be used to insure that hardy trees are produced for outplanting. All testlots must be exposed to the same uniform growing conditions in the nursery or greenhouse. If a randomized complete-block design is to be used, it is best that the test be sown in the nursery or greenhouse in the same manner that it will be planted in the field. Then it should be planted in the field using the same pattern. This method is especially useful when measurements are to be made at young ages in the field because environmental effects associated with plant production in the nursery or greenhouse then will show up largely as replication (environmental) effects in the field and will not confound estimates of genetic differences among seedlots. Differences in seedlots that are caused by large environmental effects in the nursery can persist for a number of years in the field, and these differences make genetic tests imprecise. Most organizations that do not take measurements at early years in the field do not replicate in the nursery if nursery beds are uniform. This is because the small nursery-caused differences will disappear by the time when later measurements are made.

It is mandatory that genetic tests be fully documented when they are being grown for field establishment; this documentation includes a map of the nursery bed or greenhouse layout. The plot locations need to be indicated accurately and referenced to a specific starting point. It is amazing how often poor nursery monumentation is made and later lost; for example, using a pen marker on wooden stakes will lead to disaster. Loss of seedlot identity will require remaking the crosses with a resultant serious loss of time, money, and badly needed information. What is most usual is that the seed orchard will require roguing at the time of the earliest progeny measurement. To delay this measurement can seriously reduce the value of the tree improvement program.

Site Preparation Site preparation prescriptions will vary depending upon geographic areas, species, and the organization establishing the test. In all instances, it should maximize test area uniformity, seedling survival, and insure adequate tree growth following establishment. It is highly important to get the seedlings established and growing well so that correct answers can be obtained from the tests as soon as possible. Site preparation and care should be similar to that used operationally; however, extra care is sometimes justified to obtain earlier results, although experiments should be made to determine if family ranks change when the extra care is given to the test.

Documentation Each genetic test should be permanently monumented with stakes and tags in the field, and maps should be made of the field layout and kept in two separate locations for safety. It is essential to indicate where measurements start; that is, which end of the row, or which corner of the block. Overlooking this simple rule has caused confusion about, or actual loss of, many experiments. The maps showing replications and family identities will be a part of an establishment report that should also include access routes and the general location of the test. In addition, it should include the date, weather, site preparation, planting condition.

planting method used, and anything else that might affect the survival and growth of the test plantation. Checks in the field should be made periodically, especially before measurement, to ensure that the test is still fully monumented. Numerous errors can result during measurement of tests because of poor field monumentation and lack of field maps. There are even instances in which vandals removed the tags, or even worse, in which they actually switched tags. Monumentation and documentation must be done carefully.

Test Maintenance   Close attention must be paid to genetic tests after planting if they are to survive and yield accurate information. What is too often done is that tests are planted and then neglected until measurement time arrives. Tests should be monitored closely after planting to ensure that seedlings are not threatened by competition and to ensure that volunteer seedlings or sprouts of the same species that are being tested have been removed. Pest control is essential. This ranges from the periodic application of pesticides to fencing in cases in which animals like deer, elk, or cattle are a problem. One certain fact is that a deat or deformed seedling can give no information on the genetic growth potential or quality of the test material. It is obvious, of course, that if the test is to determine resistance to insects or diseases, then control measures should not be applied.

Thinning   Thinning practices for genetic tests will vary, depending upon initial spacing, age of final assessment, the objectives of testing, and the thinning practices used in operational plantations. In cases in which thinning is a planned silvicultural treatment, special test designs have been developed to accommodate this test treatment (Libby and Cockerham, 1981).

ANALYSIS OF GENETIC TESTS

Choice and implementation of the appropriate mating and experimental designs are only the first steps in a genetic testing program. The value of the tests to a tree improvement program depends upon proper analysis and interpretation of the measurements obtained. Genetic test analyses usually are relatively straight forward and require only a very basic knowledge of statistics. When mating or experimental designs are complex, or when tests become extremely unbalanced, analysis may then require an in-depth knowledge of statistics, data processing, and much experience. In this book, only the most basic concepts are covered. Those who wish to follow genetic analyses in more detail are referred to quantitative genetics and breeding texts, such as those by Becker (1975), Falconer (1960), or Hallauer and Miranda (1981).

The Analysis of Variance

Nature of the Analysis   Most genetic experiments are analyzed by what is known as an analysis of variance, which is a statistical method by which the total variation in an experiment can be partitioned into different sources. In a genetics experi-

TABLE 8.1

An Analysis of Variance for a Half-Sib Genetic Test Planted in a Randomized Complete-Block Design[a]

Source of Variation	Degrees of Freedom	Mean Squares	Expected Mean Squares
Replications	R-1	MS_4	$\sigma_W^2 + T\sigma_{RF}^2 + TF\sigma_R^2$
Families	F-1	MS_3	$\sigma_W^2 + T\sigma_{RF}^2 + TR\sigma_F^2$
Families × replications	$(F$-1$)(R$-1$)$	MS_2	$\sigma_W^2 + T\sigma_{RF}^2$
Trees within plots	$RF(T$-1$)$	MS_1	σ_W^2

[a]F, R, and T refer to the number of families, replications, and trees per family-replication plot. σ_W^2, σ_{RF}^2, and σ_F^2 and σ_R^2 are the within-plot, replication × family, family, and replication variance components, respectively.

ment, an analysis of variance allows the tree breeder to partition the observed variation into genetic and environmental components, and when suitable, to assess their interactions.

A simple analysis of variance that is appropriate for an experiment involving open-pollinated families or families representing a polycross mating scheme that are planted in the field in a randomized complete-block design is shown in Table 8.1. Several terms used in analysis of variance warrant explanation. The term *source of variation* is self-explanatory; it simply denotes which part of the total test variation is being accounted for in that line of the analysis. *Degrees of freedom* indicates the number of independently variable classes, whereas *mean squares* for any particular source denotes all of the variation that has contributed to the observed differences for that effect. The term *expected mean squares* denotes relative contributions of each type of variance to each mean square. For example, in Table 8.1 the family mean square is composed of variations from trees within plots, interactions of replications and families as well as family variation. The mean squares are used to test for the statistical significance of an effect, but further manipulation of the analysis is needed to estimate the variance components for within-plot (σ_W^2) replication × family (σ_{RF}^2), family (σ_F^2), and replication (σ_R^2). The component σ_W^2 is estimated directly from the mean square for trees within plots, MS_1. The variance component for the interaction of replications shown by mean squares in Table 8.1 and families (σ_{RF}^2) can be estimated from the within-plots mean square MS_1 and from the replications × families mean square MS_2 in the following way:

$$\sigma_{RF}^2 = \frac{MS_2 - MS_1}{T} = \frac{(\sigma_W^2 + T\sigma_{RF}^2) - \sigma_W^2}{T}$$

The family component of variance is estimated from the family mean square MS_3 and the replication × family mean square MS_2:

$$\sigma_F^2 = \frac{MS_3 - MS_2}{TR} = \frac{(\sigma_W^2 + T\sigma_{RF}^2 + TP\sigma_F^2) - (\sigma_W^2 + T\sigma_{RF}^2)}{TR}$$

The component of variance for replications (σ_R^2) would be estimated in the same way. These variance components may be used to calculate heritabilities; this will be discussed in the next two sections.

The coefficients T, R, and F (denoting number of trees/plot, number of replications, and number of families) that appear in the expected mean squares can be arranged in simple multiplicative fashion only if the data are completely balanced; that is, all families occur in all replications, and the same number of trees occur in each plot. If data are unbalanced, the coefficients for the expected mean squares will change somewhat. If only a few trees are missing, the change may not be large enough to cause concern, although this depends on the distribution of missing trees. For example, loss of entire plots is a matter of some concern, and it complicates the analysis.

An analysis of variance for a test using a factorial mating design with randomized complete-block field design is given in Table 8.2. It is assumed that there is a completely balanced set of data. Variance components are derived in the same manner as for the half-sib family test. Sources of variation due to females and male families (with variance components σ_F^2 and σ_M^2, respectively) are equivalent to half-sib families (σ_F^2 in Table 8.1). It is important to emphasize again that the coefficients for the expected mean squares hold only if data are completely balanced, that is, all males are crossed to all females, there are no plots missing, and all plots have the same number of measurable trees. In reality, this is often not the case, especially for the more complicated mating designs. Where data are considerably unbalanced, complicated techniques must be used to derive the coefficients for the expected mean squares. Methods of hand calculation can be found in many advanced statistical tests. Many organizations now have computers with the capability of easily making the complex calculations.

Analyses of variance for other mating designs are in most respects similar to the ones for half-sib family tests (open-pollinated or polycross mating designs) and tests utilizing factorial mating schemes. It should be obvious from the preceding discussion of the two designs that complex mating designs require more complicated analytical techniques. When computers are available, the complexity of

TABLE 8.2

An Analysis of Variance of a Test Employing a Factorial Mating Design and a Randomized Complete-Block Field Design[a]

Source	Degrees of Freedom	Mean Squares	Expected Mean Square EMS
Replications	$R1$		
Males	$M-1$	MS_5	$\sigma_W^2 + T\sigma_{MFR}^2 + TS\sigma_M^2 + TSF\sigma_M^2$
Females	$F-1$	MS_4	$\sigma_W^2 + T\sigma_{MFR}^2 + TS\sigma_{MF}^2 + TSM\sigma_F^2$
Males × females	$(F-1)(M-1)$	MS_3	$\sigma_W^2 + T\sigma_{MFR}^2 + TS\sigma_{MF}^2$
Pooled error	$(FM-1)(R-1)$	MS_2	$\sigma_W^2 + T\sigma_{MFR}^2$
Within plots	$FMR\,(T-1)$	MS_1	σ_W^2

[a]Pooled error represents variance due to replication × males, replications × females, and replications × males × females.

the analysis required should rarely be a limiting factor in the choice of a mating design. Rather, the constraints reside in the objectives of the breeder, and the need to proceed with the breeding and testing program as rapidly and efficiently as possible.

Analyses for other mating schemes have not been discussed here. An excellent reference that details results of an experiment using the nested mating scheme can be found in Stonecypher et al. (1973). An example of use of the diallel mating scheme is the work with longleaf pine (*Pinus palustris*) by Snyder and Namkoong (1978).

Individual-Tree Heritability Calculations   Variance components such as those derived from the analysis shown in Table 8.1 can be used to estimate heritability. This involves equating the *statistical* components of variance to their *genetic* counterparts, such as the components for additive genetic variance (σ_A^2), nonadditive genetic variance (σ_{NA}^2), and phenotypic variance (σ_P^2).

For the analysis in Table 8.1, where half-sib families are being tested, the family component of variances (σ_F^2) is equal to one fourth of the additive genetic variance (σ_A^2). With the testing scheme used in this example, the phenotypic variance (σ_P^2) is estimated by the sum of the within-plot variance component (σ_W^2), the replication × family variance component (σ_{RF}^2), and the family component (σ_F^2). Symbolically,

$$\sigma_P^2 = \sigma_W^2 + \sigma_{RF}^2 + \sigma_F^2$$

Individual-tree narrow-sense heritability, the ratio of additive genetic variance to phenotypic variance, is calculated as

$$h^2 = \frac{\sigma_A^2}{\sigma_P^2} = \frac{4\sigma_F^2}{\sigma_W^2 + \sigma_{RF}^2 + \sigma_F^2}$$

Knowledge of the genetic meaning of the family component σ_F^2 allows one to interpret the meaning of the other components. The component σ_{RF}^2 results from the failure of families to behave the same way relative to each other in different replications, and σ_W^2 is composed of the remainder of the genetic variation plus environmental variation within plots. Thus,

$$\sigma_W^2 = 3/4\sigma_A^2 + \sigma_{NA}^2 + \sigma_E^2$$

where

$$\sigma_A^2 = \text{additive genetic variance}$$
$$\sigma_{NA}^2 = \text{nonadditive genetic variance}$$
$$\sigma_E^2 = \text{environmental variance}$$

The mating design used for the example in Table 8.1 cannot be used to calculate *broad-sense heritability* (H^2), the ratio of all of the genetic variation to the phenotypic variation, because none of the statistical components of variance gives an estimate of the nonadditive genetic variance σ_{NA}^2.

The component $\sigma_F^2 = 1/4\ \sigma_A^2$ holds for the example in Table 8.1, only if the families were half sibs. If the families had been unrelated full-sib families, the family component σ_F^2 would equal one half the additive variance plus one fourth the nonaddititive variation ($\sigma_F^2 = 1/2\ \sigma_A^2 + 1/4\sigma_{NA}^2$). Therefore, if the nonadditive variance was of importance for a particular trait, there would be no way to estimate narrow-sense heritability from a test utilizing a single-pair mating scheme because the estimate of additive genetic variation would be confounded with that of the nonadditive genetic variance.

For the factorial mating design analysis presented in Table 8.2, the variance component for males and females (σ_M^2 and σ_F^2) are equal to one fourth of the additive genetic variance. With this mating scheme, the variance component σ_{FM}^2 estimates one quarter of the nonadditive variance. Narrow-sense individual tree heritability can be calculated as

$$h^2 = \frac{2(\sigma_F^2 + \sigma_M^2)}{\sigma_M^2 + \sigma_F^2 + \sigma_{MF}^2 + \sigma_{MFR}^2 + \sigma_W^2}$$

Since the variance component σ_{FM}^2 gives an estimate of the nonadditive variance, this mating design can be used to estimate broad-sense heritability (H^2), the ratio of all of the genetic variation to phenotypic variation. This is calculated as

$$H^2 = \frac{\sigma_G^2}{\sigma_P^2} = \frac{2(\sigma_F^2 + \sigma_m^2) + 4\sigma_{MF}^2}{\sigma_m^2 + \sigma_F^2 + \sigma_{MF}^2 + \sigma_S^2 + \sigma_W^2}$$

Family Heritability   In addition to individual-tree heritability, tree breeders are often concerned with heritabilities of family means, which become important when selection can be practiced on families as well as on individuals, as is done in advanced-generation tree improvement when selections are made from genetic tests. These may be calculated directly from the components of variance and a knowledge of numbers of trees per plot, replications, and families. For example, the heritability of family means for the half-sib test example given in Table 8.1 would be calculated as

$$h_F^2 = \frac{\sigma_F^2}{\dfrac{\sigma_W^2}{TR} + \dfrac{\sigma_{FR}^2}{T} + \sigma_F^2}$$

Variance components and coefficients are as defined in Table 8.1. Family heritabilities are usually higher than individual tree heritabilities because they are based on averages estimated with a sample of many progenies. The effects of environmental factors within the test are thus averaged out for the family mean.

Heritability from Parent–Offspring Regression   There are other methods of calculating heritabilities in addition to sib analysis. When measurements have been made on both parents and progeny, it is possible to calculate heritability for a trait through regression techniques that relate progeny performance to parental values. Essentially, regression is a statistical technique that fits a line relating two groups of variables. The regression equation may be written as

$$Y = bX + e$$

where

Y = average of progeny values
b = regression coefficient (slope of line)
X = parent value
e = error (lack of fit of values to the line)

The regression coefficient b may be estimated as

$$b = \frac{\Sigma(X_i - \bar{X})(Y_i - \bar{Y})}{\Sigma(X_i - \bar{X})^2}$$

where

X_i = individual parent values
$\bar{X}$ = average of parental values
Y_i = progeny family averages
$\bar{Y}$ = average of all progeny

Where half-sib families are involved (i.e., measurements have been made on only one parent), the coefficient b is equal to one half the narrow-sense heritability. When full-sib families are measured, and progeny values are regressed on the average value of the two parents (midparent value), then b equals narrow-sense heritability.

Heritabilities estimated with parent–offspring regression techniques are usually much lower than those estimated through sib analyses of genetic tests. There are at least two possible reasons for this. First, parents of the progeny often do not occur in a uniform environment but are scattered throughout very large, environmentally variable stands, or they may be growing in many different stands. This is especially true when selected parents have been obtained from many different areas. The environmental variation affecting the parent trees lowers heritability. Second, measurements usually occur on parents and progeny of different ages; this will lower heritability if the traits that are being measured are most variable on trees of differing ages. Traits that are under very strong genetic control and are slightly influenced by local environments, such as wood specific gravity, may show similar heritabilities regardless of the method of estimation.

Genetic Correlations

Genetic correlations among traits are useful and of interest to tree improvers because they indicate the degree to which one trait will change as a result of a change in another trait. Conversely, they play a role in determining the degree to which indirect selection, or selection for one trait in the hopes of improving another trait, will be successful.

Analyses similar to those in Tables 8.1 and 8.2 can be used to estimate genetic correlations among traits. The major departure is that, rather than computing an analysis of *variance*, an analysis of *covariance* is performed. This involves calculating mean cross products rather than mean squares. Subsequently, the mean cross products are used to determine the components of covariance. For completely balanced data, the coefficients for the expected mean cross products are exactly the same as those for the expected mean squares. Genetic correlations may be estimated as

$$r_G = \frac{\sigma_{F_{XY}}}{\sqrt{\sigma_{F_Y}^2} \; \sqrt{\sigma_{F_X}^2}}$$

where

r_G = genetic correlation
$\sigma_{F_{XY}}$ = half-sib component of covariance for traits X and Y
$\sigma_{F_X}^2$ = half-sib component of variance for trait X
$\sigma_{F_Y}^2$ = half-sib component of variance for trait Y

GENOTYPE × ENVIRONMENT INTERACTION

A special concern in tree improvement and genetic testing relates to genotype × environment interaction. Simply defined, *genotype × environment interaction* means that the relative performance of clones, families, provenances, or species differ when they are grown in different environments. This may involve an actual change in rank, which is the most important type of interaction to the tree breeder or may simply involve a change in variation from one environment to the other, with no change in rank. A graphical presentation of genotype × environment interaction involving rank changes is shown in Figure 8.13, where Family 1 is the better performer in Environment A, while Family 2 is the superior family in Environment B.

Considerable variability can result in a genetic testing program when there is an interaction because genotypes respond differently to differing environments. It is

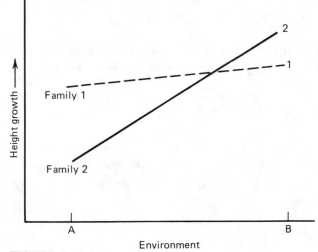

FIGURE 8.13

Genotype × environment interaction occurs when genotypes rank differently relative to one another in different environments. In the hypothetical situation shown above, Family 1 is the better grower in Environment A; in Environment B there is a change of rank, and Family 2 becomes superior.

a complicating factor that must be recognized and used. It can be of great benefit when maximum gains are desired in specific environments, but the interaction can become a formidable barrier when broadly adapted strains suited to several different environments are sought.

Genotype × environment interaction may be expressed symbolically by an extension of the model relating phenotypic performance to genetic and environmental effects. The model is extended to read

$$P = G + E + GE$$

where

P = phenotypic value
G = the genetic effect
E = an environmental effect
GE = an effect due to genotype × environment interaction

The *only* way to account for the *GE* interaction effect is to plant genetic tests in more than one environment. When tests are established in one location, the interaction term is *hidden* in the genetic (*G*) effect. This has two major consequences if there is appreciable *GE* interaction.

1.   Heritability may be overestimated. The appropriate formula for narrow-sense heritability when tests are planted at multiple locations is

$$h^2 = \frac{\sigma_A^2}{\sigma_P^2} \qquad \frac{\sigma_A^2}{\sigma_A^2 + \sigma_{NA}^2 + \sigma_{GE}^2 + \sigma_E^2}$$

where

σ_{GE}^2 = variance due to genotype × environment interaction and other variance components are as defined in earlier sections of this chapter

However, in the presence of *GE* interaction and with information from only one environment, we have

$$h^2 = \frac{(\sigma_A^2 + \sigma_{GE}^2)}{\sigma_A^2 + \sigma_{NA}^2 + \sigma_{GE}^2 + \sigma_E^2}$$

The variance components for additive genetic variance and *GE* interaction are confounded and cannot be separated when tests are established in only one location. This will result in an overestimation of gain from selection when the improved material is planted in a nontest environment.

2.   Suboptimal genotypes may be used in environments that are different from the test environment if the *GE* interaction results in rank changes. This is because the genotypes chosen as *best* in the test environment are not the best genotypes to use in other environments. The end result will be less genetic gain in the nontest environment.

Because of the possible presence of *GE* interaction, it is always advisable, and sometimes mandatory, that genetic tests be established in multiple environments. Environments may consist of different locations, different years, or different site preparation or management treatments.

Two methods of reducing genotype × environment interaction have been employed in breeding programs. One alternative is to stratify areas within a breeding region into subregions with similar environmental conditions. Similarities can be determined from such macroenvironmental factors as temperature regimes, physiographic provinces, or soils or may be indicated by experience and the performances of the genotypes themselves. Different breeding programs are then conducted for each subregion.

Stratification of environments can be useful when breeding programs cover large areas, but it is inefficient when the subregions become too small. In addition, stratification is not effective for unpredictable environmental factors such as large year-to-year climatic changes. Tree breeders are thus often faced with a significant *GE* interaction within subregions.

A second alternative is to select for genotypes that perform well and that show little interaction over a wide variety of environmental conditions. This approach has received much attention from breeders in recent years, and several different statistics estimating general adaptability (stability) have been developed (Finlay and Wilkinson, 1963; Eberhardt and Russell, 1966) that have been used in tree improvement programs. For example, Morgenstern and Teich (1969) used genotypic stability parameters in an investigation of genotype × environment interaction of jack pine provenances. Stability parameters can often be used in conjunction with environmental stratification to reduce genotype × environment interactions within subregions.

There will be no attempt to deal with the statistical aspects of *GE* interactions here. Appropriate statistical methods to analyze experiments planted at multiple locations are an extension of the methods given in the previous section and may be found in the work of Cochran and Cox (1950). Subsequent discussion of *GE* interactions will focus on the general impact of *GE* interactions in tree improvement and forest management activities.

Impacts of *GE* Interaction

For most large-scale tree improvement programs, the major objective is to develop widely adapted strains of trees that can be used over many environments. This requires genotypes that perform well in different environments. When ignored, large production losses in operational forestry can result from *GE* interaction. Losses may be of two types: death and reduced growth or quality. The former is easily recognized, but it often requires close observation to assess the importance of the latter. It must be emphasized that genotypes may interact for both growth and quality characteristics. Quality aspects are sometimes given little attention, but in some instances they can be more interactive than growth traits.

There are many causes for interaction, but it is generally conceded that most of them are more closely related to edaphic than to climatic factors (Shelbourne, 1972; Goddard, 1977), unless very small geographic areas are being considered, or where climatic variables change appreciably over short distances, as in mountainous terrain. Large genetic or environmental differences do not always result in genotype × environment interaction. For example, none was found for 3-year height growth or fusiform rust resistance in slash pine (Squillace, 1969), whereas Wright (1973) reported genetic and site effects to be stronger than genotype–site interaction in 8- to 12-year-old provenance and progeny tests of 11 forest tree species in the north central United States.

Species and Provenance Interaction

Many examples of the effect of interaction resulting in growth loss and quality for species and provenances have been cited by Binet (1963), King (1965), and others. Much of the genetic effort in forestry has been concerned with the determination of gross genotype × environment interactions, and several books could be written on the results, showing the wide adaptability of some seed sources and the required specific environmental conditions for others (Suassuna, 1977). A classic example of the interaction effect at the population level was shown by Fuentes (1971) who collected seeds of loblolly pine from both wet and dry sites and backplanted the resultant seedlings in both site conditions. He found a strong interaction for both tree diameter and height between planting site and seed source.

Determination of the environmental conditions to which a seed source is adapted *is the first essential step* if a forestry program is to be successful. Identification of an adapted strain for the test area does not completely solve the problem, because there is still the danger of *offsite* planting as a forestry program expands to include different sites. Much of the importance of genotype × environment interaction on a species or a source within-species level has been covered in the earlier disussions on seed source and provenance, and it will not be repeated here. But it *is absolutely necessary to assess the adaptability and possible interaction of any species and source in the environments in which they will be grown.*

Family and Clone × Environment Interaction

Family or clone × environment interaction has been documented within a number of species. Burdon (1971) found a very striking clone × site interaction for frequency of branch clusters, stem crookedness, and tree vigor in *P. radiata*. This mainly reflected the inability of certain clones to perform well in soils that were low in phosphorus. The danger of using families of *P. radiata* indiscriminately in production forestry has been illustrated by Fielding (1968), based on his research results that showed the failure of the progeny of one of the parents tested on dry sites but good performance of the same progeny on good sites. For the same species, Pederick (1972) found that one of 28 families had much more interaction in survival and growth than did the others. No interaction among progenies of slash and loblolly pines were found for fusiform rust and test location, but a strong one was found for crown form (Kraus, 1970). In a larger geographic area encompassing the area where Kraus worked, Powers and Zobel (1978) reported a family interaction with rust resistance.

It is obvious that the genotype × environment response is difficult to assess. There may be a significant interaction for growth, whereas there is none for form—or there may be one for adaptability and little for growth when the tree survives. For example, an occasional genotype of the most southern source of loblolly pine from Florida survives and grows very well under severe conditions of drought in Texas, whereas overall survival of the families from that source is only 5 to 10% of the local Texas families. The trees from the Florida source that *do* survive under Texas conditions grow more rapidly than the local Texas race, thus maintaining the same relative growth pattern as they do when the two sources are planted in Florida.

Problems of the practicing forester are compounded when there is genotype × environment interaction of both pest and host, such as that which occurs with fusiform rust resistance in slash and loblolly pines. It is known that there is genetic variation within both the rust and pine host and that both interact with the environment. Selected slash pine families have been identified as having good growth performances throughout a broad geographic area. In a section of that area, the tested trees may have good resistance to fusiform rust, but they may be highly susceptible to the disease at another location. Some selected families of loblolly pine respond similarly to slash pine in growth and disease resistance, but others seem to be resistant wherever the species is grown. It appears that a genotype × disease interaction is evident within loblolly pine.

Some characteristics, such as wood specific gravity, do not show much genotype × environment interaction on a family or individual level. Often, the average values change dramatically with environment, but the individual high- or low-gravity genotypes maintain their relative position regardless of the average. Changes of wood specific gravity caused by environmental differences are evident for loblolly pine, but the changes are small and ranking by families is essentially constant (Chuntanaparb, 1973). However, the wood properties of families of other species are far less stable than that of the southern pines when they are planted in an exotic environment. Although such species as *P. caribaea* from Honduras generally have their wood drastically altered when grown in differing environments, not all individuals respond in the same way to the changed conditions. This difference in interaction will enable the development of trees of this species that have usable wood for sites where most trees produce a low-value product.

Interaction and Forest Management

Humans change the environment drastically when intensive forest management practices are employed. Often overlooked is the fact that the new environment may not be well suited to the type of trees that were growing on the original site. Of more concern, however, is the fact that the changed environment of the improved site may not be suited to the genetically improved stock that was developed to grow on the unimproved site.

Fertilization greatly affects the environment and more is known about the genotype × fertilizer interaction than about other management interactions (Goddard et al., 1975; Matziris and Zobel, 1976; Roberds et al., 1976). Interactions are generally lacking when fertilizers are applied at nominal rates, but whether a reaction will occur is highly dependent upon species and family. Slash pine appears to have a greater genotype × fertilizer interaction than does loblolly pine (Goddard et al., 1975). Interaction appears to be greater with nitrogen than with phosphorus fertilization, but definite phosphorus × family interactions have been reported for both slash and radiata pine (Shelbourne, 1972; Jahromi et al., 1976). Burdon (1971) found that most of the clone × site interactions for vigor in *P. radiata* arises from clonal differences in their ability to tolerate low available phosphorus and that soil fertility is more likely to account for site effects than is climate.

It was initially suspected that it might be necessary to develop special seed orchards with clones that are particularly responsive to fertilization and others that would contain genotypes that were most suited to grow without fertilization. To determine the validity of this assumption, three of the six replications of each of the early progeny tests in the North Carolina State Cooperative were fertilized. Results obtained 8 years after test establishment showed that most of the clones that performed best with fertilization also performed best when grown without supplemental nutrients. The major difference was that the poorer-growing clones responded somewhat more to fertilizers than did those that grew fastest without fertilizers. However, rankings did not change.

Considerable interaction appears to be present with differing hardwood genotypes, but it is hard to classify. Growth of planted hardwoods without fertilization (except on the very best sites) is generally so poor that family differentiation is essentially meaningless. On sites too poor to be considered normally for hardwood planting in the southern United States, an occasional family will grow quite well when fertilized, whereas most other families have completely unacceptable growth rates.

An area in need of intensive study is the interaction of genotypes with site preparation, cultivation, or in the tropical areas, cleaning. The results indicate that with increasing intensity of site preparation and cultural measures there is greater fusiform rust incidence on loblolly and slash pines (Miller, 1977). However, it is uncertain whether there is interaction between inherently rapid tree growth and fusiform rust infection. Experience indicates that interactions between tree growth and intensity of site preparation will be greater with hardwoods than with pines, but this has not been quantified.

Experience gained from 4000 acres of genetic tests of the southern pines shows that moderate site preparation and test maintenance do not change the relative family rankings. That finding has resulted in the intensified care of genetic tests, so that reliable measurements of family performance can be obtained several years earlier than those without the intensive care.

One of the greatest needs in operational forest mangement is to determine the magnitude of the interaction of genetically improved families with site preparation, cultivation, or both. When intensive site preparation is used, is followed by cultivation, and especially when fertilization is also applied, the potential for meaningful interactions is greatly increased. Such tests are complex and expensive, but the effects of interaction can be so important that great effort needs to be made to determine its presence.

Perhaps the most serious error in forest management is to ignore genotype × environment interaction. It makes little sense, for example, to plant a genotype that does not respond to fertilization in an operation where fertilization is standard. It is economically indefensible to plant a family on land that has been intensively site prepared if it does not respond favorably. On a more gross basis, it is absolutely essential to use species or sources within species that are suitable to the new environment when exotic plantations are established. To ignore the potential effect of genotype × environment interaction often spells disaster in operational forest management.

LITERATURE CITED

Becker, W. A. 1975. *Manual of Quantitative Genetics*. Student Book Corporation, Washington State University, Pullman, Wash.

Bergman, A. 1968. "Variation in Flowering and Its Effect on Seed Cost— A Study in Seed Orchards of Loblolly Pine." Tech. Rept. 38, School of For. Res., North Carolina State University, Raleigh.

Binet, F. 1963. An instance of interaction of genotype and environment at the population level. *Gen. Today* 1:1–47.

Burdon, R. 1971. Clonal repeatabilities and clone-site interactions in *Pinus radiata*. *Sil. Gen.* 20(1–2):33–39.

Burdon, R. D., and Shelbourne, C. J. A. 1971. Breeding populations for recurrent selection: Conflicts and possible solutions. *N. Z. Jour. For. Sci.* 1:174–193.

Burdon, R. D., Shelbourne, C. J. A., and Wilcox, M. D. 1977. "Advanced-Generation Strategies." 3rd World Cons. on For. Tree Breed, Canberra, Australia.

Chuntanaparb, L. 1973. "Inheritance of Wood and Growth Characteristics and Their Relationships in Loblolly Pine (*Pinus taeda*)." Ph.D. Thesis, School of Forest Resources, North Carolina State University, Raleigh.

Cochran, W. G., and Cox, G. M. 1950. *Experimental Designs*. John Wiley & Sons, New York.

Eberhart, S. A., and Russell, W. A. 1966. Stability parameters for comparing varieties. *Crop. Sci.* 6:36–40.

Falconer, D. S. 1960. *Introduction to Quantitative Genetics*. Ronald Press, New York.

Fielding, J. 1968. Genotype-site interaction in *Pinus radiata*. *Newsletter* 1:14–15. (Research Working Group No. 1, Res. Comm., Aust. For Council.)

Finlay, K. W., and Wilkinson, G. N. 1963. The analysis of adaptation in a planting breeding programme. *Aust. J. Agri. Res.* 14:742–754.

Fuentes, J. 1971. "Interaction Between Planting Site and Seed Source of Loblolly Pine." 7th World For. Cong., Buenos Aires, Argentina.

Goddard, R. 1977. "Genotype × Environment Interaction in Slash Pine." 3rd World Cons. on For. Tree Breeding, Canberra, Australia.

Goddard, R., Zobel, B., and Hollis, C. 1975. Response of southern pines to varied nutrition. In *Tree Physiology and Yield Improvement* (M. G. A. Cannell and F. L. Last, eds), pp. 449–462. Academic Press, New York.

Hallauer, A. R., and Miranda, J. B. 1981. *Quantitative Genetics in Maize Breeding*. Iowa State University Press, Ames.

Jahromi, S., Smith, W., and Goddard, R. 1976. Genotype-×-fertilizer interactions in slash pine: Variation in phosphate (33_p) incorporation. *For. Sci.* **22**:21–30.

King, J. 1965. Seed source-×-environment interactions in Scotch pines. *Sil. Gen.* **14**:141–148.

Kraus, J. 1970 "Progeny-×-Planting Location Interactions in Five-Year-Old Slash and Loblolly Pine Tests in Georgia." 1st North American For. Biol. Workshop, Michigan State University, East Lansing.

Libby, W. J. and Cockerham, C. C. 1981. Random non-contiguous plots in interlocking field design layouts. *Sil. Gen.* **29**:183–190.

Matziris, D., and Zobel, B. 1976. Effects of fertilization on growth and quality characteristics of loblolly pine. *For. Ecol. Man* **1**:21–30.

McKeand, S. E., and Beineke, 1980. Sublining for half-sib breeding populations for forest trees. *Sil. Gen.* **29**(1):14–17.

Miller, T. 1977. "Fusiform Rust Management Strategies in Concept: Site Preparation." Proc. Management of Fusiform Rust in Southern Pines Symp., University of Florida, Gainsville, pp. 110–115.

Morgenstern, E. K., and Teich, A. H. 1969. Phenotypic stability of height growth of jack pine provenances. *Can. J. Genet. Cytol.* **11**:110–117.

Namkoong, G. 1977. "Choosing Strategies for the Future." 3rd World Consul. on For. Tree Breeding, Canberra, Australia.

Pederick, L. A. 1972. "Genotype–Environment Interactions Calculated for Height Growth of Young *radiata* Pine Families at Three Locations." 3rd Mtg. Res. Comm. of the Australian For. Council, Mt. Gambier, South Australia.

Powers, H., and Zobel, B. 1978. Progeny of specific loblolly pine clones vary in fusiform rust resistance according to seed orchard of origin. *For. Sci.* **24**(2):227–230.

Roberds, J., Namkoong, G., and Davey, C. 1976. Family variation in growth response of loblolly pine to fertilizing with urea. *For. Sci.* **22**(3):291–299.

Shelbourne, C. 1972. "Genotype-Environment Interaction, Its Study and Its Implications in Forest Tree Improvement." IUFRO Genetics–SABRAO Joint Symposia, Tokyo.

Snedecor, G. W., and Cochran, W. G. 1967. *Statistical Methods.* Iowa State University Press, Ames.

Snyder, E. B., and Namkoong, G. 1978. "Inheritance in a Diallel Crossing Experiment with Longleaf Pine." U.S. Forest Service Research Paper 50-140, Southern Forest Experiment Station.

Squillace, A. 1969. Field experiences on the kinds and sizes of genotype–environment interaction. *Sil. Gen.* **18**:195–197.

Steel, R. G. D., and Torrie, J. H. 1960. *Principles and Procedures of Statistics.* McGraw-Hill, New York.

Stonecypher, R. W., Zobel, B. J., and Blair, R. L. 1973. "Inheritance Patterns of Loblolly Pines from a Nonselected Natural Population." Tech. Bull. No. 220, North Carolina State University, Raleigh.

Suassuna, J. 1977. A cultura do Pinus—uma perspectiva e uma preocupacão [The culture of pine—A perspective and a concern]. *Brazil. Flor.* **29**(8):27–36.

van Buijtenen, J. P. 1976. "Mating Designs." Proc. IUFRO Joint Meeting Genetic Working Parties on Advanced-Generation Breeding, Bordeaux, France, pp. 11–20.

van Buijtenen, J. P., and Lowe, W. J. 1979. "The Use of Breeding Groups in Advanced-Generation Breeding." Proc. 15th Southern For. Tree Imp. Conf., Starkville, Miss., pp. 59–65.

Wright, J. 1973. Genotype–environment interaction in the north-central United States. *For. Sci.* **19**:113–123.

Wright, J. W. 1976. *Introduction to Forest Genetics.* Academic Press, New York.

CHAPTER 9

Selection and Breeding for Resistance to Diseases, Insects, and Adverse Environments

One of the most important objectives of tree improvement is to reduce damage by disease and inspect pests and to produce strains of trees that are particularly well suited to grow in adverse environments. Success can spell the difference between profitable forestry or failure to produce an economically viable crop. No matter what benefits might result from tree improvement, only limited returns can be obtained from forest operations unless the trees produced are free to develop without excessive damage or mortality caused by pests or severe environments.

In pathology and in entomology, strong and sometimes controversial stands have been taken with respect to the use of the words *tolerance, resistance,* and *immunity*. In this book the word *resistance* will be used to indicate the ability of trees to grow and develop normally even when attacked by pests or when subjected to adverse environments. The word *tolerance* is preferred by some because many people assume that resistance indicates a total lack of damage. However, the latter is properly called *immunity*. There is no expectation of breeding immune strains of trees that *will not* be affected at all by pests or adverse environments; only trees can be developed that can tolerate pests so that they will be more productive. Some pathologists use the term *tolerance* to indicate specifically the degree to which a tree can grow with a pest or in an adverse environment and still retain its economic value.

As forestry becomes more intensive and as requirements for agricultural products increase, foresters are finding their operations shifted from the best sites to areas that currently are marginal or submarginal for growing economic crops of trees. If forestry is to expand—even if it is to hold its own—trees must be profitably grown on some sites that formerly were considered to be so poor that they were of little value for forest production. Many forest geneticists are now concentrating more on developing trees that are suitable for marginal sites than on the improvement of trees that are already adapted to grow on good sites.

When combined with good forest management, breeding for resistance to disease and insect pests and for better adaptability to adverse environments has proven to be very feasible, and some remarkable achievements already have been recorded. For example, the whole cost of large-scale breeding programs in the southern pine region of the United States will probably be recovered from increased production resulting from selection and breeding for resistance to just one disease—fusiform rust (*Cronartium quercuum f. sp. fusiforme*). This development will enable the successful management of pines on many millions of acres on sites on which disease is so prevalent that they previously were not profitable for pine forestry operations.

Disease and insect pests are always a problem in forest enterprises. As forest management becomes more intensive, losses will appear to become more prevalent. This *apparent increase* is partially the result of closer observation by the forester as well as a greater concern about pests and their effects on intensive forest management. In this situation, disease and insect pest attacks that were formerly overlooked or were considered to be minor nuisances suddenly become important. A good example is the impact of spider mites on the southern pines. In the past, this pest was hardly noticed in natural stands or plantations, but now it is considered to be serious on grafts in seed orchards and to some extent on young trees in genetic tests.

Sometimes the apparently greater incidence of pests is a *real* increase resulting from more intensive forest management. A good example is fusiform rust on *P. taeda;* this disease becomes progressively more serious when forest management practices such as site preparation are intensified or when nitrogen fertilizers are used (Miller, 1977).

Another reason for an increase in pests under intensive forest management is the establishment of trees with restricted genotypes over large contiguous areas. Although this concern is sometimes overemphasized, it is important in tree improvement programs. Large plantings of a single species do not present a serious hazard or loss unless the genotypes used are so uniform that a true monoculture results. This danger was emphasized by Heybroek (1980) who showed the advantages of mixing genotypes to discourage pest attack. The mixing of genotypes should be done consciously but can only be effective after much knowledge has been obtained about both the host and the parasite. There is a special danger of damage by pests when vegetative propagules of restricted genetic backgrounds are used for operational planting. Vegetative propagation in itself is not hazardous if a broad enough genetic base is assured through the use of several different genotypes.

The true amount of genetic diversity in forest trees is generally underestimated. Tree populations are almost always highly heterogeneous genetically, both among stands and trees within stands. Recent studies have indicated that forest trees have the greatest variability of all plants (Conkle, 1979) or, in fact, of any organism. Much tolerance to pests and adverse conditions would be expected, because a tree is perennial must survive many years and reproduce under varied growing conditions and pest attacks within a year and from year to year. Even though a forest may be established with all the individuals originating from a few parents from a seed orchard, the heterozygosity of the parent trees and genetic recombinations will result in a heterogeneous stand of trees.

Tree species differ greatly in their abilities to withstand differing pests or environments. Even the ramets of a clone of a vegetatively reproduced plantation (where each individual has the same genotype) may contain several resistance genes. The danger from clonal plantings of the same genotype occurs because all ramets of the clone will have the same set of resistance genes. When these are overcome or exceeded, catastrophic losses can result because then all individuals within the plantation are susceptible to the pest or severe environment.

In an intensive tree improvement program, as in all forestry activities, there is *always* a trade-off between *the gain achievable and the risk encountered to obtain greater gains.* This is especially evident when pests or adverse environments are involved. Decisions must constantly be made that are relative to the option of obtaining greater gain from the use of better trees along with more intensive management of the forests, or the use of similar genotypes, and the possibly increased danger of pest attacks or loss to adverse environments. Only when resistant (or tolerant) strains of trees are used, can forestry come close to the optimal production of desired products and thus make its full contribution to society. Full realization of gains from the use of resistant trees can only occur if forest management and tree improvement activities are synchronized with pest control. This type of activity is particularly crucial for the success of exotic forestry.

In forest trees there is more latitude with respect to how much genetic uniformity can be tolerated to obtain greater gains that is also relative to the possible greater danger of losses from pests or adverse environments than in most crops. This is because of the opportunity for the geneticist to breed for narrowing the genetic base for economic characteristics, while at the same time broadening it for adaptability and pest resistance. This is possible because the genetic systems controlling these characteristics are usually independent. For example, it is feasible to develop straight trees that are tolerant to cold, drought, or to excess moisture. The opportunity to combine the two types of characteristics independently is a great boon to the forest tree improver.

There usually is considerable genetic resistance to most pests and adverse environments in forest trees, although some problems are much easier to breed against than are others. Zsuffa (1975) feels that the use of genetics is a powerful tool for controlling forest tree diseases and insects. He points out that most genetic studies have been directed toward the host trees, but that the genetics of the pest should also be studied. A good example is in fusiform rust on loblolly and slash pines about which Dinus et al. (1975) have discussed the differences in breeding strategies that are made necessary by variability in the pathogen. There are instances when the host species shows limited variability, such as for red pine (*P. resinosa*); the value of breeding for pest resistance in this species has been questioned (Nicholls, 1979).

The subjects of pest resistance and greater adaptability to adverse environments are huge and complex. Several books and numerous symposia have been produced that were related to each. In this chapter, it will be possible to cite a few of the many studies and breeding achievements. Because of its importance, more effort is being expended each year on pest tolerance (resistance) or on the adaptability aspects of tree improvement. There is no question about the importance and need for breeding for resistance to pests and adverse environments in forest trees (Zobel, 1980b).

ENEMIES OF THE FOREST—GENERAL CONCEPTS

Introduction

There are many kinds of forest enemies, ranging from insects and diseases to man. For convenience, they will be divided into the four categories of *diseases, insects, environmental,* and *miscellaneous. Miscellaneous* includes all types of things, such as animals, parasites, mistletoe, and so forth. Some problems encountered in forest management for which resistance can be developed do not fit neatly into any category; that is, they do harm as part of the environment but are usually caused *artificially,* frequently by people. Their effects and damage patterns may be similar to those caused by pests; therefore, breeding can be done for greater adaptability and resistance to them. Air pollution and acid rain are examples of this category of pseudopest. It is evident that the preceding is an arbitrary and not totally acceptable categorization but will be helpful in discussing enemies of the forest.

The fear has been expressed that "pests will defeat us" and if action is taken to develop resistance, the situation will only become worse; some persons feel it would be best to do nothing. Such negative attitudes are misguided because any forest, managed or not, will be attcked by pests that must be controlled in one way or another if optimal yields are to be obtained. Like any crop, when intensive forestry is applied, "new" pests are found or suddenly become important, often not really because they are new, but because they had not been closely observed previously. Intensive site preparation, cultivation, thinning, and fertilization sometimes do result in trees becoming more pest susceptible, but other times the opposite result occurs in which intensively cultivated trees become more pest resistant. A good example of the latter is pine bark beetles; the healthy, well-tended stands are more resistant to insects than are the offsite, the overaged, or the overdense forests.

THE NEED FOR RESISTANT TREES

Any breeding program with forest trees is long term and expensive. Why, therefore, should one spend a lot of time and money on controlling pests through developing resistance rather than using silvicultural, chemical, or natural predator controls? Whether or not breeding for pest resistance should be done depends upon the availability and suitability of other methods of pest control.

The concept will be illustrated later, using fusiform rust and *Fomes* as examples. One would not generally try to develop a pine to be resistant to fusiform rust in the nursery; fungicide sprays are so simple, successful, and economical in the nursery that spraying is the preferred control. But in forest plantations the rust cannot be successfully controlled using chemicals. It has been suggested that actions such as reducing the alternate host can be applied and may be helpful (Squillace and Wilhite, 1977); no known practical or safe method is available, however, for efficient control of the oak host in large forest plantings. Therefore, if fusiform rust on pine is to be controlled successfully in large plantations, the only feasible method is to breed for resistance. The pine host shows large variability in susceptibility to the disease, and inheritance of resistance is of such a magnitude that it is relatively easy to develop useful resistant strains (Kinloch and Zoerb, 1971). The pest also varies genetically; although this complicates the problem of breeding for resistance, it still appears to be possible to make good gains.

Let us continue with the fusiform rust example. The improved silvicultural methods of better site preparation, fertilization, and cultivation currently used to improve tree growth and yield in the southern pines are all conducive to increased rust infection (Miller, 1977). The need for fusiform rust control and thus resistance breeding has increased as forest management has become "better." Many persons are convinced that without the use of genetically resistant species or strains in the "hot spot" rust areas in the southeastern United States the potentially greater growth resulting from improved forest management will be more than offset by the greater losses from the increased fusiform rust attack that results from intensive forest management.

In contrast to fusiform rust, *Fomes annosus* root rot[1] appears to be an important widespread pest against which breeding for resistance will not be easy or of much value. Although some progress has been made in breeding for resistance to this fungus, the major control mechanisms are through silvicultural and species manipulation For example, *Fomes* has generally been kept under reasonable control by good silviculture and stand management, including the use of borax and competing fungi on the cut stump and thinning during the time of reduced spore flight.

Although the development of resistant strains is expensive and takes special skills and much time, it has the advantage that, once obtained, resistance is relatively permanent, and more resistance can be added. The long-term cost of disease management is usually less with breeding than it is with direct controls because costs of the latter reoccur frequently, sometimes several times throughout one rotation. A good example is *Dothistroma* on radiata pine. Several studies have indicated that there is a reasonable resistance to the disease (Carson, 1977). Despite this, some organizations have chosen to control *Dothistroma* by spraying with a copper fungicide during the most susceptible period of the plantation's life, rather than to include resistance to the pest in breeding programs. If only one generation of planting is considered, there is no doubt that resistance breeding would be less efficient than sprays for *Dothistroma* control. For several generations, however, this probably is not so. Resistance breeding is currently being reconsidered.

BREEDING FOR PEST RESISTANCE

No tree breeder expects to eliminate totally losses due to diseases, insects, or to environmental factors. The *objective is to reduce damage to a tolerable level.* Some foresters speak about producing trees that are immune to pests, a goal that is usually not achievable and usually not even desirable, as will be explained later. What is wanted are trees that can live with the pest and still produce a quality product that approaches the maximum yield under the given site and climatic conditions involved.

---

[1]This is now known as *Heterobasidium annosum*.

One reason that is sometimes given against breeding for resistance in forest trees is that more virulent strains may soon evolve that can overcome the resistance of the previously resistant trees. This concern arises because insect- or disease-causing organisms are also genetically variable and subject to selection. A change in the genetic structure of the trees being used could result in selection for increased virulence in the pest.

The potential for development of supervirulent strains is a real and serious consideration but has often been overemphasized. Because of several outstanding examples of pests overcoming resistance, like rusts on wheat, the attitude often is that overcoming forest tree resistance by new or virulent strains of the pest will be a regular, rapid, and normal occurrence. An experimental demonstration of this possibility for fusiform rust on slash pine has been provided by Snow and Griggs (1980). They took rust spores from seven families that were moderately resistant to fusiform rust on slash pine and, after passage through the alternate oak host, they placed them back on the same seven families. One rust source proved to be more virulent on one family than the general rust for the area, but no rust source was more virulent on all families.

In a study on loblolly pine, Powers et al. (1978) did not find much increased virulence when rust spores were obtained from infected individuals from the tolerant families. In a study by Carson (1982), considerably increased virulence was found on the resistant families from which spores were obtained. In addition, the greater virulence extended to other resistant families. However, Walkinshaw and Bey (1981), working on slash pine, found that some isolates from random galls were as virulent as isolates derived from galls on a resistant family. Hattemer (1972) states that the danger of damage from more virulent strains can be greatly reduced in long-lived, heterozygous forest trees when proper precautions are taken and when resistance is obtained from a number of genes rather than a few.

The development of more virulent pests would be more critical to the forester if management methods and the kind of trees planted on a given site remained static over consecutive rotations—with the same genetic stock's being used for succeeding generations of plantations. But forest tree improvement is not static and has a special advantage because of the long rotation ages. This enables *succeeding crops to be genetically different from the previous crop* planted on a given site. The trees planted later on a given site should be more genetically improved and therefore will be different from those that had been planted there previously, if progressive breeding programs are being followed. During the period required for the crop to mature, new and improved strains of trees usually will have been developed, although this potential is less with shorter rotations. Therefore, the fear of a supervirulent pest evolving directly from the first crop to destroy succeeding crops is a concern, but is not likely to occur when aggressive development programs are pursued. The adaptation to resistant strains and evolving of pests commonly happens on some annual crops in which the same genotypes are used, cycle after cycle. Much breeding effort is concerned with overcoming the increased virulence. Specially effective is the use of multilines that combines resistant genes into the crop. This should not occur as frequently or quickly on long-lived perennials, if an active, ongoing breeding program is underway, so that the new planting stock will differ from that originally planted in a given area.

A major danger is that once virulent strains of the pest have developed, they may spread rapidly in the managed forest. Such rapid development of the pest is more likely when the forest has a uniformly high resistance than when some individuals are still susceptible. When both types of trees are present, the tendency toward maintaining a natural equilibrium is increased, thus reducing the pressure for establishment of mutants. Totally resistant (immune) trees will rarely be developed because of the biological, time, and cost restraints involved. Thus, the most desirable goal for the breeder is resistance rather than immunity.

If optimal gains through tree improvement are to be obtained, a reduced genetic base for the economically important characteristics may result. On the surface, this would appear to develop in the direction of a monoculture that would be ideal for the spread of pests. But greater pest attacks will not occur if care is taken in the development of the tree improvement program. As mentioned previously, but reemphasized here, it is most fortunate that in forest trees, almost all the economically important traits, such as tree straightness or wood quality, are genetically independent from the characteristics of resistance to pests or adverse environments. In general, the most important concept is that *one can breed for economic characteristics such as straightness or wood qualities, while at the same time one can breed for broad adaptability to pests and adverse environments*. This is not known, or is overlooked, by many foresters and laymen. With few exceptions, these two sets of characteristics are controlled by multiple genes and usually are inherited independently from one another and thus are not strongly correlated.

Often, the magnitude and complexity of controls required in forest stands, whether they be chemical or biological, make it impossible to use anything efficiently other than a genetic resistance breeding program (Figure 9.1). Biological control is also a desirable option, but as of this date, with a few exceptions, successful large-scale biological control of forest pests is not currently a method that is useful on a commercial scale. However, in forest trees the success of biological control of pests is of the greatest importance and will enable foresters to avoid or decrease dependence on chemical methods of pest management.

PESTS AND INTENSIVE FOREST MANAGEMENT

As forest management becomes more intensive, foresters must adapt to and reduce the increased pest attacks. The high costs of intensive management and the high potential value of the products make it mandatory to prevent pests from seriously reducing the volume and quality, and thus the value of the forest products that are grown. It is not possible to live with the pests or "let the pests have their share." Historically, growing forest trees has yielded a low return on the investment, and pest attacks can easily turn what would otherwise be a profitble venture into a losing enterprise. As the demand for forest products becoms greater, it is of paramount importance to society to have each forested acre more productive

FIGURE 9.1

Certain pests can only be controlled through use of genetic methods. A good example is fusiform rust in southern pine plantations; shown is a stand in which more than 50% of the trees have been destroyed by the disease. There is no known silvicultural control that is *economically suitable* for this disease in forest conditions, although it is very satisfactorily controlled in the nursery by sprays.

since land on which forest trees can be grown for wood products is becoming less available. Because the best forest lands are being taken for agriculture, forest operations are shifting toward the more marginal sites, resulting in the trees' being grown under greater stress. They then become more susceptible to pest attack, and this further increases the need for pest resistance.

As has been mentioned previously, certain pests are increased by activities such as fertilization, site preparation, or thining. Also, the biological and physical changes in the environment resulting from short-rotation forestry can alter the forest's susceptibility to pests (McNabb et al., 1980). McNabb and co-workers emphasize the importance of stand density and have coined the term *spatial resistance* to deal with changes in the susceptibility of different stands to pests.

PESTS AND EXOTICS

The relationship between pests and exotics was briefly discussed in Chapter 3. Because of the importance and complexity of this subject, it will be explored in greater depth here.

Exotics are generally planted as large blocks of single species, with the seed for the plantings sometimes being from either restricted souces or from small numbers of parents. In almost all instances, the exotic is not well adapted to its new environment. The result is that frequently the trees in the new forests are growing under considerable stress. Much flagrant offsite planting has been done with exotics and is still being done, especially with some *P. radiata, P. caribaea,* and certain *Eucalyptus* plantings. These outstanding forest trees have often been placed in quite unsuitable environments, making them more susceptible to pests.

As stated in Chapter 3, there are not too many "absolutes" in biology, but *one that comes the closest to certainty is that exotic plantations will be attacked by pests of one kind or another.* The situation is especially insidious because the exotic often grows well while it iş pest free in its early years in the new environment. Too often this leads to euphoria and false projections about its production potentials (Martinsson, 1979). But it never fails—it may take 2 years or it may take 10— pests of some type, often very destructive ones, will become established in exotic plantations. A recognition of the potential of exotic planting's being seriously affected by pests is critical to those involved in exotic forestry. Many times pest attacks will result from the severe stresses caused by conditions that often occur where exotics are grown. The poor physiological condition of the exotic trees in these conditions enhances the spread and damage by pests that previously may have been unknown or that were considered to be of only minor importance or a nuisance.

It is of great concern for the future of healthy and profitable forests to recognize the magnitude and seriousness of losses from pest attacks on exotic forest tree plantings. Massive programs have failed from such pest attacks; examples are the large *P. radiata* plantations destroyed by *Dothistroma* in several regions of the world, including Brazil, Zimbabwe (Rhodesia), and east central Africa. The loss of confidence in forestry that has occurred as a result of destruction by this disease could have been prevented, and millions of dollars could have been saved if those who established the exotic *P. radiata* had heeded the advice of a few pathologists who warned that *Dothistroma* probably would become severe in these radiata pine plantations planted in regions where there are warm and moist summers. Many persons have been puzzled about how *Dothistroma,* which is found on *P. radiata* in its indigenous range in California (where the disease is considered to be primarily a nuisance), has managed to become established in such widely separated regions throughout the world. The important fact, however, is that *Dothistroma has spread* to these areas, or was already there but not recognized, and there is no reason to believe it will not also spread to other areas having environments that are suitable for development of the disease.

Sometimes the pest is harmful, not because it kills the exotic, but because it deforms the tree, making it of value only for low-quality products. A good example of this is the so-called cypress stem canker on *Cupressus* planted in Colombia, Kenya, and other areas. Stem deformation is so severe that the only suitable use for the tree is for fiber products, and the good-quality solid-wood products for which many of the plantations were established cannot be produced.

HOW TO BREED AGAINST PESTS

There is no simple answer about how to proceed in a breeding program against pests, and no general program can be outlined. The choice of method depends on the kind of pest, on the variability of the host, the variability of the pest, and their interactions. Whatever choice is made, genetic diversity within the tree population must be maintained, because populations without diversity are poor risks, with danger of major losses (Schmidt, 1978). A knowledge of the following is necessary to construct a breeding program against pests:

1. The economic worth of the host.
2. The potential economic losses from the pest.
3. The biology and genetic variation within the host species.
4. The biology and genetic variation of the pest.
5. The interaction of the environment with tolerance of the host and virulence of the pest (Nelson, 1980).
6. The interactions between the pest and the host.

The preceding information is just a start. The difficult problem about which breeding method to use has had much thought given to it (Borlaug, 1966; Gerhold et al., 1966).

The concept of *integrated pest management* is simple but most difficult to achieve in silviculture (Waters and Cowling, 1976). This integrated approach is becoming routine with other organisms but has been used too infrequently in forestry. To apply integrated pest management, which certainly must be the final goal, the pest and host populations and the environment must be looked at as a dynamic ecosystem. Models have been developed that will allow the bringing together of the many variables in a way that most efficiently manages pests. A major objective of the tree breeder, along with the pathologists, entomologists, and economists, should be to help uncover the information necessary for a successful integrated pest management program.

Part of pest management relates to deployment of the tolerant stock, such as where the resistant trees should be used and how intensively they should be used. These questions, along with how much diversity is needed, must all be addressed once resistant trees have been developed. Too often the emphasis is on developing the tolerant material with little thought about how best to use it.

SELECTING RESISTANT TREES

Although the same general principles apply in selecting trees to use in a pest resistance breeding program as for those involving other characteristics, there are several special concepts that must be kept in mind. The most important of these is that selection of trees must be made in badly infected stands if results from mass selection are to be successful. If trees are chosen in a forest stand that has only been lightly attacked, the chance of an "escape" is very great. An *escape* is a tree that has been very lightly attacked or not attacked by the pest. When such trees are chosen, the breeding program will fail because the progeny of the selected trees often do not show special resistance. It is this situation that has resulted in the attitude by some that breeding for pest resistance is not successful. It generally will be the most successful the more severely the stand has been attacked by the pest. Put another way, selection intensity for mass selection efforts is greatest if disease-free individuals are chosen in heavily infected stands.

Another important consideration is the age of the trees screened for desirable pest-resistant characteristics. Susceptibility to pest attack sometimes changes greatly with the age of the host, because resistance genes have been turned on or off, resulting in morphological or physiological changes with age that make the tree more susceptible or resistant. The age restriction is of particular importance when working with those pests to which trees are susceptible only at older ages; for example, some of the bark beetles rarely attack young trees.

A third important consideration that applies when selections are to be made in genetic tests is to place the genetic tests so that the trees are subjected to at least moderate levels of attack by the pest. If this is not done, family and individual tree separations will not be good, and results from the tests will be inconclusive or of no value. It has been shown with several species of trees and pests that the greatest discrimination among families can be obtained if there is an intermediate amount of infection in the test. This is in contrast to situations when mass selection is to be practiced in which the greatest gain may be obtained by selection in highly infected stands.

DISEASES

General

What is a disease? There are many differing definitions. For example, Ford-Robertson (1971) defines *disease* as "harmful deviation from normal functioning of physiologicl processes generally pathogenic in origin." Webster (see Webster and McKechnie, 1980) uses "any departure from health" or "a particular destructive process in the body with a specific cause and characteristic symptoms," among a number of other definitions. Disease is not easy to define simply. For purposes of this volume, the definition of disease will be "abnormal physiology of an organism that has a specific cause." It will be immediately clear that in using this definition, air pollution and similar agents cause diseases along with fungi, algae, viruses, or other agents that are usually considered to be the causative agents for disease. However, air pollution and other related environmental causes of "disease" will be treated separately in this chapter.

Many books have been written about disease resistance in forest trees; generally, these refer to fungal organisms. Breeding for resistance to diseases is the most difficult aspect of breeding forest trees (Heimberger, 1962). Stern (1972) stresses the importance of the coevolution of host and parasite and how natural selection may lead to a balance between these. Since breeding can upset the balance, this factor must be considered in estimating genetic gain. To complicate the situation, diseases like blister rust on white pines or fusiform rust on the southern pines have alternate hosts, so the tree breeder must work with a complex genetic system involving both the host(s) and the parasite (Day, 1972). Although great gains have been made in breeding for resistance to diseases of forest trees, the basic biological foundations are usually poorly known. This fact, of course, does not negate the ability to produce and use disease-resistant trees, but it does make advanced-generation breeding methodology very difficult.

Genetically, most disease resistance in forest trees is complex, and it is not determined by a simple Mendelian dominant–recessive system. Fusiform rust and white pine blister rust are examples in which tolerance may be inherited through a complex, quantitative system (Bingham, 1963; Blair 1970). One good example of a simply inherited resistance was for *Thuja*. As reported by Soegaard (1969), the leaf-rust disease is controlled by one pair of dominant and recessive genes. The genetic complexity of many host–parasite systems with the frequently added problem of alternate hosts makes breeding for disease resistance unusually difficult.

Comparisons are often made between the relative difficulty of breeding for resistance against disease and against insects. Most breeders prefer working with diseases because the pest organism is usually less mobile and usually does not exhibit the wild fluctuations in population numbers that are common with insects. Artificial inoculation of a disease on the host plant often is easier to achieve than is the forced feeding of an insect. Overall, testing for resistance to insects is complicated. Diseases usually are present as *endemic* populations (i.e., they are at a normal, balanced level), so they are always present in the forest and are available to work with. Often, populations of insects go through extreme cycles in which they are *epidemic* (i.e., a buildup, often rapid, to highly abnormal and generally injurious levels). This is followed by periods in which the insect becomes so scarce that it becomes difficult to work with. Of course, the preceding is not a generality; diseases also go through endemic and epidemic cycles. However, the general experience of most tree improvement workers is that, all things being equal, the chances of successfully breeding for resistance will be greater with most diseases than with most insects.

Breeding Disease-Resistant Trees

Despite the difficulties, breeding for disease resistance has progressed well, and there have been some remarkable improvements, such as in *Albizzia* to *Fusarium* (Toole and Hepting, 1949), or pine to fusiform rust (Zobel, 1980a). Some poor results have been obtained when trees that were not infected (escapes) have been considered to be resistant because of microclimatic or other factors (Riker and

Patton, 1961). The importance of the effects of the interaction of environment and the disease has been stressed by Schreiner (1963) and many others. He illustrates these effects on clones in the genus *Populus,* which he considers to be one of the most disease-prone genera in forestry but one in which large gains in resistance will be possible.

Generally, those diseases whose symptoms are easily observed and that occur early in the life of the tree are the easiest to use in a breeding program (Heimburger, 1962). These include the rusts, the canker and gall diseases, and most leaf diseases (Bingham, 1963). Those diseases that are not easily visible or that are not evident until the tree reaches an advanced age are hard to breed against. Root diseases and heart rot organisms fit into this category. For example, it is much easier to breed for resistance to a leaf disease such as *Melampsora* rust (Schreiner, 1959; Chiba, 1964), or for a canker disease such as *Cronartium quercuum f. sp. fusiforme* (fusiform rust) (Zobel, 1980a) than for a disease like *Fomes annosus* (*Heterobasidium annosum*) (Dimitri and Frohlich, 1971; Kuhlman, 1972).

Such authors as Callaham (1966) recognize that there is pseudoresistance as well as true genetic or inherent resistance. The former refers to the apparent resistance of potentially susceptible plants that may be caused by age, environment, cultural conditions, or other factors. Both must be clearly distinguished if a resistance breeding program is to be successful. It is also important to develop resistance that is effective over several developmental stages of the host rather than for only one.

There are many types of resistance, ranking from resistance at the species level to resistance of individual trees within families. Differences among species are well recognized (Powers, 1975), but the magnitude of differences in resistance among sources within species has also been great. For example, Stephen (1973) found great differences in susceptibility to *Rhadbocline pseudotsugae* among sources of Douglas fir with the southern sources often being the most susceptible. Geographic variation in the needle cast disease of jack pine was considerable (King and Nienstaedt, 1965), and 29 different sources of jack pine showed great differences in resistance to needle cast disease and stem rusts (Martinsson, 1980). In numerous studies on the southern pines in the United States, geographic trends of resistance to fusiform rust have been found (for example, Wells and Switzer, 1975). Thielges and Adams (1975) also reported large differences in *Melampsora* rust resistance among provenances of cottonwood. For Scots pine in Norway, Dietrichson (1968) was able to find geographic differences with the northern sources being more tolerant of *Scleroderis*; he could also relate disease resistance to cold resistance. Trees that were more frost resistant were also less susceptible to disease because of the high dry matter content of their foliage, early growth cessation, and initially rapid shoot elongation. Similar geographic patterns in resistance to the canker disease of sycamore were found (Coggeshall et al., 1981). In addition to family differences, Coggeshall and coworkers found the southern sources of sycamore to be the most resistant.

Most studies on disease resistance have been made for families as well as for individual trees within families. The literature on this is voluminous. In his summary paper, Bjorkman (1964) states that the general combining ability for disease resistance for many diseases of forest trees is high. High general combining ability was reported for fusiform rust resistance by Kinloch and Kellman (1965), Blair (1970), and several others (Figure 9.2). Even for the root disease *Phytophthora cinnamomi*, which seems to affect many genera of trees throughout the world, there appears to be genetic variation in resistance. For example, Bryan (1965) found good evidence for inheritance of tolerance to littleleaf disease in shortleaf pine that is caused by *Phytophthora*. He found that certain mother trees produced outstandingly resistant progeny. Similar results were found for parents used to form aspen hybrids, in which some parents produced progeny that were very resistant to Hypoxylon canker (Einspahr et al., 1979). In loblolly pine, Powers and Zobel (1978) reported that resistance to fusiform rust varied considerably from progeny from one seed orchard to another. Most of the difference was related to resistance of the specific clones used within the orchards, although some was related to differences in geograhic origin of seed orchard clones. Most reported results on disease resistance have been on the basis of overall family performance. Filer and Randall (1978), for example, found that families of sweetgum varied from 7 to 74% in resistance to *Botryosphaeria ribis*.

Experience has shown that, especially for the rust and canker diseases, breeding for resistance by selecting within the wild populations has been generally successful (Figure 9.3), although there have been specific failures in breeding for disease resistance, such as for American chestnut blight. Nearly a half century of selection and hybridization have resulted in very little improvement in the chestnut genus (MacDonald et al., 1962). The potentials of improvement through use of variants (hypovirulent strains) that will destroy the capacity of the virulent strain of the pathogen have been described for chestnut blight (Horsfall and Cowling, 1980). The potentials are exciting.

Hybrid development has been used successfully to produce tolerant trees in the chestnuts. One of many successes using hybrids has been the shortleaf × loblolly crosses and backcrosses against fusiform rust. LaFarge and Kraus (1980) and others have found that the desired resistance can be maintained along with desired growth and form by choice of parents and backcrossing. The approach for using hybridization to develop disease resistance in individuals has been detailed by Powers and Duncan (1976).

Although forced attacks under artificial environments give useful information about resistance, they are often more severe than would normally occur by pests in the forest, and assessment of results must take this into account (Callaham, 1966; Dinus, 1969). The simplest kind of forced attack is to use methods of screening in the laboratory or greenhouse. This has been done on a large scale for fusiform rust, and it has been described by Phelps (1977). Artificial screening is justified when a suitable correlation can be established between laboratory and field performance. Methods of inoculation, incubation, testing, and analysis can all be crucial in relating test results to field results. For example, Walkinshaw et al. (1980) greatly improved the correlation of laboratory to field infections by changing the method in which fusiform rust was assessed and scored.

FIGURE 9.2

Inheritance of resistance to diseases is often strong by individuals. Shown (*b*) is a row of trees from fusiform-resistant parents; (*a*) shows a row from fusiform-susceptible parents. These two rows of trees were growing adjacent to one another.

What Causes Disease Resistance?

One could hardly ask a question that would have more answers than "What causes disease resistance?" A main problem is that often more than one type of resistance is present in a poulation of a given species. This compelxity and the difficulties posed to the breeder have been mentioned by Miller et al. (1976). The ideal is to combine all, or several, kinds of resistance to a single tree, but forest tree breeding has not developed to the stage of determining how much or how easily this can be done.

There are many differing ideas that have been offered to explain why trees differ from one another in disease resistance. These have been summarized by Hare (1966) in a review article on the physiology of resistance to fungal diseases. He made three general classifications that are (1) exclusions; (2) growth restriction after entry by methods such as "walling off"; and (3) destruction of the pathogen after entry. Hare makes the rather strong point that *exclusion* probably is of limited importance and that the pathogen usually enters both the susceptible and nonsusceptible tissue. He suggests that what happens after entry will determine the seriousness of the disease. Specific possible mechanisms for resistance have been mentioned by Bjorkman (1964); these include the moisture content of the host tissue, the nutrient status of the plant, the pH, the osmotic pressure

(b)

FIGURE 9.2 (*continued*)

within the cell, and other factors. To this can be added the presence or absence of resin acids, phenols, or other substances within the plant that might be toxic to the fungus (Forrest, 1980). A much-discussed resistance method is walling off the diseased tissue by the formation of a wound periderm or cork cells, which is triggered by the balance of auxin content and kinins (Boyer, 1966). The structure of the leaf and stomatal surface also can influence disease resistance. For example, Patton and Spear (1980) reported that wax on the needles hinders infection by white pine blister rust along with inhibitory substances in the stomatal subcham-

FIGURE 9.3

The diseases most easily worked with in resistance breeding are the rust and canker diseases, which are typified by the galls of fusiform rust illustrated. Symptoms are easy to see, are usually observable early, and a large genetic component of resistance is frequently evident.

ber. Elgersma (1980) reports that internal-wood anatomical features restrict the spread of Dutch elm disease.

The terpenes (or terpenoid compounds) often appear to give resistance to disease (Bridgen and Hanover, 1980). The resin acids and monoterpenes seem to be very effective. An area in which only recent progress is evident is the possible resistance to certain wood rotting and root rot organisms. This appears to result from a compartmentalization, a walling-off of the infected tissues from the healthy tissues (Shigo, 1980). Considerable genetic variability in resistance to root rot appears to be present among trees (Ladeitschikova, 1980).

One type of resistance that has puzzled foresters for many years is recovery from disease after infection. Although the mechanisms, which are often related to compartmentalization, are imperfectly known, many foresters recognize that trees sometimes recover after disease. Several authors, such as Boyer (1964; for white pine blister rust), claim that recovery from disease has a genetic basis; certain families recover much better than others. We definitely agree with this concept and have watched trees badly infected with fusiform rust recover, heal over the diseased area, and grow to a normal life span.

Gains from Resistance Breeding

No matter what the mechanisms of resistance are, the results from breeding for resistance to diseases are that useful strains of forest trees are now becoming available on a large scale. For example, many seed orchards have been established with fusiform rust-tolerant parents by the North Carolina State–Industry Tree Improvement Cooperative. Seed are available in large quantities with many thousands of hectares being planted in the disease "hot spots" in the southern United States. Other organizations are doing the same thing. Another example is the very successful breeding against several leaf diseases in the poplars (Chiba, 1964; Einspahr et al., 1979; and many others).

Space does not permit us to outline all the gains achieved by breeding for disease resistance. Large areas of forestland, which were formerly considered to be marginal or submarginal, have been made into economic forests by use of disease-resistant trees (Zobel et al., 1971; Zobel and Zoerb, 1977) (Figure 9.4).

FIGURE 9.4

Shown are two rows of loblolly pine progeny from different parents in a clonal seed orchard. The row on the left from one clone has been very heavily attacked by fusiform rust, whereas the one on the right has been only lightly infected. Such great genetic differences enable a profitable forest enterprise even where the disease incidence is large.

Predicted and achieved gains have been significant and of great economic value (Bingham, 1967; Porterfield, 1973; Rockwood and Goddard, 1973; and many others). *Breeding* for *disease resistance* is *economically profitable* and a *necessity* for *most tree improvement programs* (Zobel, 1980b).

INSECTS

General

Although damage to forest trees by insects is sometimes catastrophic, much less progress has been made in developing insect-resistant strains of forest trees than has been achieved for diseases. There are many reasons for this, among which are the mobility of the insect, the lack of ability to predict where and when an attack will occur, lack of knowledge of genetics of the insect, lack of knowledge about what causes resistance, and in some instances, lack of ability to produce "forced attacks" as needed for controlled genetic studies (Connola and Belskafuer, 1976). Furthermore, some insects can be relatively easily controlled silviculturally, and some attack at only one phase in the life cycle of the host. Epidemic insect buildups can be very rapid and are highly dependent on environmental fluctuations.

Despite the preceding, genetic gains can be made in developing resistance to insects, and much more research and study is needed in this area. A good example is the recent report by Trial (1980) in which he made an in-depth assessment of the history of attacks by the spruce budworm (*Choristoneura fumiferana*), a member of one of the most destructive groups of insects in forestry. This insect has caused huge losses in forestry and has affected the future and fate of communities, companies, and even governments. Little breeding for resistance to spruce budworm has been attempted (Figure 9.5).

Somehow the idea has developed that resistance of forest trees to insects is not great and that other methods of control should be the only ones used. This is puzzling, and we do not generally agree with that concept. In his report, Soegaard (1964) makes it clear that both conifers and broad-leaved forest trees have shown considerable resistance to insect attacks. Attempts are being made to determine this, even for the very difficult bark beetles (Waring and Pitman, 1980). Henson et al. (1970), using resistance to sawflies (*Nediprion*) as an example, are of the opinion that trees can be bred with comparatively low susceptibility to insect damage. Generally, this can be accomplished along with selection for growth and form. In our opinion, the use of chemical sprays should be viewed as an interim method of keeping the forest intact until resistant strains and–or biological control of the insects have been achieved. This is because of the potentially deleterious effects of pesticides on the environment. Some insects appear to be so general in their feeding habits that it will be difficult to breed for resistance. For example, the gypsy moth (*Lymantria dispar* or *Porthetria dispar*) eats foliage from all but a few species of hardwoods and even eats conifers during epidemics.

FIGURE 9.5

It is difficult to breed for resistance to some pests. Damage to spruce and fir by the spruce budworm in Newfoundland, Canada, is shown. Little resistance breeding has been attempted, although some trees are less severely attacked than are others.

Generally, three types of resistance to insects can be recognized: (1) *nonpreference,* in which the insect is not attracted to, or is repelled from, feeding and ovipositing on a tree; (2) *antibiosis,* in which the insect is killed, injured, or prevented from completing its normal life cycle after feeding on a tree; and (3) *tolerance,* in which the tree recovers from insect attack by a population approximately equal to that which will damage a normal susceptible tree (Gerhold, 1962). In their bulletin, Henson et al. (1970) proposed the term *susceptibility* as a measure of how much the tree will be attacked by the insect and *vulnerability* as a measure of how much damage will be done by the insect.

Many reasons have been given for resistance to insects. One of the most common is the amount and quality of the resin production; it has been suggested that some trees "pitch out" the attacking beetles. Some resin vapors are toxic to the insect, or they inhibit feeding of adult beetles. Trees within a species can have greatly variable resins (Smith, 1966). Physical factors, such as bark thickness, also seem to be important as do many other kinds of factors. Waring and Pitman (1980) hypothesized that the amount of carbohydrate reserves in the host controls host resistance in bark beetles. They feel that tree vigor is related to carbohydrate production and thus affects beetle resistance.

One group of insects that causes very heavy damage to the tree improvement effort is the group that attacks seeds, cones, and flowers. These can be harmful to natural regeneration and devastating for seed orchards. Even though genetic variation is very evident among clones (Sartor and Neel, 1971), a breeding program for tolerance to insects solely for seed orchards is not a feasible alternative because other methods can be used to control the insects more efficiently in this special situation. Areas are small and seed orchards are very intensively managed and isolated. Much work has already been done in insect control for seed orchards, usually by use of *systemic* insecticides that are combined with good management. The systemics enter the tissues of the tree, and the insect is either killed or repelled by the chemicals that have been established in the plant tissues. If applied correctly, the systemic has little adverse effect on the general environment. Very intensive studies have been made on the use of systemics in seed orchards with some outstanding results (Drew and Wylie, 1980; van Buijtenen, 1981).

Resistance to Insects

Variation in susceptibility to insects has been known and recognized for numerous species of trees. For example, in two major groups of pines, differences in damage were found when the trees were attacked by shoot moths (*Rhyacionia*). These insects were widespread, they occur on many species, and they are persistent pests that cause much damage and loss of quality. Some species have been found that show some resistance (Holst, 1963).

Just as for diseases, resistance to insects can be related to individual families or the geographic origins of the host tree. For example, Batzer reported in 1961 great differences in white pine weevil attack, depending on the geographic source of jack pine; the indigenous source was the least damaged. Considerable work has been accomplished on resistance to white pine weevil (Gerhold, 1962; Garrett, 1970), but the results have been variable. For example, Connola and Belskafuer (1976) have emphasized that lightly weeviled trees may be heavily weeviled in other environments, especially under caging experiments. However, some usable resistance has been found; one of several studies reporting this is that of Heimburger and Sullivan (1972). Even the extensive tip moth and webworm attacks on slash and loblolly pine showed some differences with source of seed (Hertel and Benjamin, 1975).

Resistance on an individual-tree basis is perhaps the most important in control of insect damage, although only limited studies have been done on this aspect of resistance to insects. Work has been published on resistance to white pine weevil (Garrett, 1970), but the results have not been conclusive. Heimburger (1963) and Connola and Belskafuer (1976) have emphasized the great effect of environment in selecting and testing for individual-tree resistance. Success was reported in selecting black pines (*P. thunbergii*) whose progeny showed many scars but a low rate of gall formation when attacked by the pine gall midge. This insect, which is found in many regions of the world, does not injure certain tree genotypes.

The aphids are another group of serious insects that cause widespread damage. On Douglas fir, Meinartowicz and Szmidt (1978) found populations infected from 0 to 94%, with those sources of Douglas fir from east of the Cascade Mountains being most resistant. They concluded that the differences appear to be under genetic control.

Not all studies on insect resistance have been on conifers. Good work has been accomplished on the poplars; additionally, differences in susceptibility among willows to the cottonweed leaf beetle (*Chrysomela scripta*) have been reported by Randall (1971). Studies have shown that black locust (*Robinia pseudoacacia*) trees have exhibited variability in resistance to the locust borer (*Cullene robiniae*) (Soegaard, 1964).

Insects are particularly destructive to some tropical hardwoods. Luckily, resistance has sometimes been found, such as that against the shoot borer *Hypsypla* in *Toona ciliata* (Grijpma, 1976). Much study is needed in this area because insects, both the leaf eaters and shoot borers, are a major problem when tropical hardwoods are grown in plantations. Differences in insect resistance among species and sources within species are evident in the eucalypts, but only a limited amount of work has been done on individual-tree resistance to beetles and larvae that so commonly attack members of this genus. One of the great needs in tree improvement is to obtain better information on breeding for resistance to insects.

Just as for diseases, both the hosts for insects and the insects themselves vary genetically. For example, Stock et al. (1979) found genetic differences between *Dendroctonus pseudotsugae* from Idaho and coastal Oregon. Apparently, very destructive races have been formed. In the white pine weevil, Heimburger (1963) states that "the weevil is genetically a very versatile organism." It attacks several widely different species of pine, spruce, and other species. The host–insect interactions can be very important, as has been reported for *Carya* by Harris (1980).

One group of pests that does widespread damage in the tropical and subtropical regions is the leaf-cutting ants. as far as we know, no resistant testing has been carried out with respect to the leaf cutters, although the ants have definite species preferences. It is possible that selection and breeding for tolerance to ant attack could be helpful. In Venezuela, it has been noticed that within the species *P. caribaea*, the ants prefer and first defoliate those trees with finer needles; they attack the trees with heavy needles last. The problem that faces breeding for tolerance to ant attack is that when food becomes scarce, the ants seem to feed on anything that is green, and they would probably also destroy the resistant trees, even though the attack might be delayed or lessened by a resistant strain of trees.

MISCELLANEOUS PESTS

General

There are a host of pests other than diseases and insects that attack forest trees. Some of these have been studied sufficiently to show that genetic resistance by the host is present, whereas others show no resistance at all. It was mentioned earlier that humans often can be considered to be severe pests. We do not like to say "something can't be done," but the nearest thing to impossibility is to develop resistance in forest trees to people as pests. The "people problem" has strong social and economic, as well as biological foundations. However, the subject of humans as pests will not be considered here. Damage by them can be great, but control is most difficult.

The other pests will be briefly listed and referenced with statements about the potential for resistance breeding. Some pests that appear to be of minor importance at the present time could become of major importance as forestry becomes more intensive and human population pressures become greater.

Some Different Pests

A newly recognized and somewhat frightening loss has been caused by wood nematodes, especially in Japan. Ohba (1980) reports heavy losses caused by nematodes but found enough resistant trees to begin a breeding program and the early results are encouraging.

Another pest that causes extensive damage is mistletoe. Foresters commonly associate it with poor sites and overaged stands, but it sometimes is very frequent in young stands. When mistletoe attacks plantations, it can have a very adverse effect. Several studies have been made on the resistance potential to this pest. Persons such as Roth (1978) have found some resistance to mistletoe in conifers, although it is not as great as the resistance to many other pests. Frochot et al. (1978) have reported considerable species difference in poplars that are resistant to attack by mistletoe. They found, for example, that *P. trichocarpa* is very susceptible and *P. nigra* is quite resistant. Intermediate types between the two had intermediate infections. Hawksworth and Wiens (1966) have reported on witches broom and hosts for members of the genus *Phoradendron*. Different patterns occur within and between host and parasite species, but it is most difficult to determine if true resistance is present.

In addition to mistletoe, other parasitic plants cause trouble. However, no studies are known relative to resistance against these insidious plants. They are mentioned here, however, because of their effect on tree improvement that is related to progeny testing. If not recognized, they can seriously disrupt progeny testing, and even cause incorrect selection because trees that are not parasitized grow much faster than those that are. If the tree is noninfected because it is resistant, that would be fine, but often the nonparasitized trees are merely escapes, and this will cause the selection process to be imprecise in choosing trees with genetic superiority. There are several groups of parasitic weeds that infect forest trees. Two examples are *Agalinis purpurea,* parasitic on sycamore, sweetgum, and loblolly pine (Musselman et al., 1978), and *Seymeria cassioedes,* on pines in the southeastern United States (Fitzgerald et al., 1977).

Animals frequently become serious pests. Bear, deer, elk, rabbits, and "mountain beaver" in the western United States and elsewhere can be most destructive to young forests (Marquis, 1974). Little resistance breeding has been done for these, except against deer and rabbits (Radwan, 1972). It is evident that certain families are greatly preferred over others; also, nursery-grown seedlings are usually much preferred over naturally regenerated seedlings. No serious breeding for animal tolerance has been done because control has been sought by reducing the animal populations or by the use of repellants. One breeding scheme that would help reduce animal depredations would be to develop trees that start growth rapidly so that they could grow quickly out the animal damage zone. This characteristic has been used as a criterion for selection in a number of tree-breeding programs.

One type of damage that is difficult to categorize as a pest results when a tree species "poisons" the growth environment, either for itself or for other species (Fisher, 1980). The term used to describe this phenomenon is *allelopathy* that is defined as "the influence of plants, other than microorganisms, upon each other, arising from the products of their metabolism." A simpler definition is given in Webster (Webster and McKetchnie, 1980): Allelopathy is "the reputed influence of one living plant upon another due to secretion of toxic substances." Some species poison the environment; therefore they cannot be grown in pure plantations or with other species. Allelopathy is suspected among tropical forest tree species; the most well-known species in the temperate area is *Juglans nigra* (walnut) (Gabriel, 1975), which produces the toxic substance *juglone*. When, for example, pines are planted where walnuts are grown, they often die or are stunted; similar allelopathic tendencies have been reported for cherrybark oak (DeBell, 1971). Many species, such as those of the genus *Eucalyptus* and even some pines, appear to have allelopathic effects because of changes they cause in the environment. Many weed species are allelopathic, resulting in more excessive damage than that caused by simple competition. No within-species breeding programs have been initiated to overcome allelopathy, although choice of species and how they will be grown often depends on a knowledge of allelopathy. Some of the barriers to growing some tropical hardwoods in pure plantations may be of allelopathic origin.

AIR POLLUTION AND ACID RAIN

General

Most plants obtain a part of their essential nutrients from the air: carbon dioxide for photosynthesis, nitrogen and sulfur for synthesis of proteins, oxygen for respiration, and many of the major and minor mineral elements (Witwer and Bukowac, 1969). Uptake of nutrients from the atmosphere is especially important in forests, because nutrients from other sources are scarce and fertilization is still an infrequent management practice.

But these uptake processes also make plants susceptible to injury by air pollutants. Most toxic substances occur as gases or fine aerosol particles (smoke or smog) that diffuse readily through the open stomata of trees and kill the leaves. In some areas, especially around metal smelters, damage can be severe, and no forests will grow until resistant trees are available, or until the source of pollution is decreased.

Air pollutants can cause death, deformation, or growth loss. Some of these effects are very obvious (Berry, 1961). The most insidious, widespread, and overall serious loss is from a reduction in growth. This is often not noticed by foresters and is accepted without question.

In some parts of the world, resistance of forest trees to air pollutants has been studied, and enough variation in resistance to air pollutants is available to permit selection and breeding of more tolerant individuals, strains, or even species. But such breeding is difficult and tricky and has not been followed with suitable vigor. Most of the emphasis has been on decreasing the sources of pollution rather than on developing trees that will grow normally in polluted air.

Fume Damage

Many different chemicals are released into the atmosphere by factories, automobiles, volcanoes, and many other sources. Many different gases can damage forests; the most prevalent and damaging are ozone, sulfur dioxide, oxides of nitrogen, and the fluorides, but a host of others are sometimes involved. Losses, especially in growth, are increasing, and when one estimates the area of forestland affected by air pollution, the amount of loss of forest products is staggering (Figure 9.6). The loss in one species (*P. ponderosa*) was reported by Cobb et al. (1970). They emphasized not only the direct loss to pollutants but the secondary problems from diseases and insects that attack the pollutant-weakened trees.

Many publications have dealt with breeding for resistance to air pollutants; one example is Patton (1981) who reported on the effects of ozone and sulfur dioxide on the growth and wood of poplar hybrids. The subject has been dealt with in a symposium (Bialobok 1980), and interest in breeding for resistance to air pollutants seems to be increasing. Bialobok criticizes the simple mass-selection approach, and states that the genetic basis for resistance to air pollution must be better understood.

Breeding for resistance to air pollutants is just as complicated as is breeding for disease resistance. Trees resistant to ozone may be susceptible to sulfur dioxide or vice versa. Trees that are resistant to both ozone and sulfur dioxide separately may succumb or be injured when subjected to both of them together. Whether and how much trees are affected depends on age, time of year, and the physiological condition of the tree (Berry, 1973). There is no area of forest tree breeding in which greater emphasis is needed. Strains of trees with reasonable tolerance to air pollutants are needed in many parts of the world.

Acid Rain

Another possible threat to forest trees is acid rain; the word *possible* is used because the evidence about the balance between beneficial and harmful effects of acidic and acidifying materials in the air is still not clear. Sulfur and nitrogen oxides are toxic gases. They can kill individual trees or whole forests, but when

FIGURE 9.6

Fume damage can be severe, when the resistant trees are healthy, the sensitive ones are dead. Often the results from fume damage are a loss in growth. Shown is a row from a sensitive parent that has grown only half as much as a resistant family row behind it. When grown in a clean environment, both families grew at the same rate.

some substances are transported over long distances, they are transformed chemically in the moist air to sulfuric and nitric acids. These acids can be deposited on soils and trees. In water solution the acids dissociate (break up) into ions—H^+, NH_4^+, NO_3^-, and SO_4^{2-}—that are sometimes injurious and sometimes beneficial for plants, depending on their nutrient status, concentration, and the physiological condition of the trees.

The question of the effects of acid rain on forests is so new, the variability among forest trees and soils is so great, and the growth responses are so long term that the actual effects on forest growth have not yet been quantified (Cowling, 1979; Cowling and Davey, 1981; Hornbeck, 1981).

Several different mechanisms of adverse effects caused by acid rain have been suggested (Tamm and Cowling, 1977). Some of these ideas have been verified by experiments—but always in greenhouse or field tests with simulated acid rain. These tests show that acid rain *can* damage forest trees under some conditions, but no direct evidence has yet shown that acid rain *does* damage forests.

The ideas that have been put forward include the following: erosion of protective waxes on leaves (Shriner, 1976) and killing of feeder roots by acid-mobilized aluminum (Ulrich et al., 1980), both of which would predispose trees to drought. Ozone damage has been shown to be greater in plants that also receive simulated acid rain than in plants with normal rain. Older needles on loblolly pine turn brown prematurely when exposed to the simulated acid rain of pH 3.2 (Shriner, 1976). Leaching of nutrients from foliage and from soils by simulated acid rain has been discussed by Wood and Bormann (1974).

No selection or breeding work has yet been done to determine the resistance to acid rain. Once the major mechanisms of acid rain damage have been demonstrated, it should be possible to find and use variation in resistance in the same way as with other "pests."

Some authors (Lee and Weber, 1979) feel that acid rain effects will be most severe at the regeneration stage in the life of the forest. They found that seed germination on the forest floor was much reduced by acid rain for some species and that root development of the seedlings was inhibited.

Breeding for aspects of nutrient deficiencies like those that would result from acid rain has been tried and has been successful (van Buijtenen and Isbell, 1970). The solution is to have source control of this pollutant so the situation will not continue to worsen. A limit will be reached beyond which breeding can no longer assure reasonable health and growth of trees established on soils changed by acid rain. This is an area needing intensive and urgent research.

ADVERSE ENVIRONMENTS

Perhaps the most successful of all tree-breeding efforts has been to produce trees that are better suited to adverse environments. As the human population expands, the need for more land to grow food crops will continue to increase greatly. There is already a growing trend to convert the best forest sites to agricultural uses. As the need for more forest products becomes greater, one way to produce more wood is to grow trees economically on sites that are now considered to be marginal or submarginal for forest production.

Adverse environments may be caused by conditions that are too dry, too wet, too hot, or too cold for normal tree growth. Adverse environments may be caused by such other factors as nutrient-deficient soils, hard wind, or excess salts in the soil. All of these have been encountered in tree improvement efforts and considerable success has been achieved in breeding for resistance to them. For example, Monk and Wiebe (1961) found tolerance differences to salts in woody ornamental plants. Although no breeding has been done on an individual-tree basis, tolerance of various sources within species to wind and ice damage is sometimes striking (Kerr, 1972; Williston, 1974). It is not uncommon in seed orchards that have been damaged by ice to find a few clones that are much more severely damaged than the others.

Great gains from tree improvement have been obtained from cold- and drought-resistance breeding. There are numerous publications listing results for both conifers and hardwoods. For example, Kriebel (1963) was able to select drought-tolerant strains of sugar maple whereas van Buijtenen et al. (1976) reported upon drought-hardy pines. Drought-hardy strains of forest trees are relatively common. Drought resistance occurs both by species and by individuals within species. Much gain can be obtained through provenance selection (Ferrell and Woodward, 1966); this is true for most "adverse environment" characteristics and is usually the first approach to be followed after species have been chosen for use in adverse environments.

Cold tolerance is a most important characteristic and one that has had a great breeding emphasis. For example, in Douglas-fir, Szöny and Nagy (1968) have reported on the relationship between frost resistance and growth. A key charac-teristic in expanding the range of *Eucalyptus* is to develop cold-tolerant strains (Boden, 1958; Hunt and Zobel, 1978). Some excellent work has been done on juvenile selection of the eucalypts to frost resistance (Marien, 1980). The ability to withstand cold is of great value in the species that grow in cold climates, and much work has also been done in this area, some of it many years ago (Bates, 1930). Great progress has been made in breeding for cold resistance in the Scandanavian countries. As one of many examples, Schummann and Hoffman (1968) tested 1-year-old spruce seedlings for frost resistance, and a number of workers report a good correlation between the dry-matter content of the needles and cold resist-ance. Sometimes little genetic variation is found, such as for radiata pine where no differences in cold tolerance by stand origin were found (Hood and Libby, 1980).

If one is to breed intelligently for resistance to something such as drought, it is important to know the possible causes for resistance. In their report, van Buijtenen et al. (1976) found that drought resistance was determined by only a few avoidance-or-tolerance mechanisms. They listed the avoidance mechanisms as follows:

1. Stomatal control with the drought-hardy seedlings transpiring rapidly when water was available but conserving water under stress. (A similar pattern has been reported for some species of *Eucalyptus*.)

2. Root morphology with the drought-hardy trees having deeper, more fibrous root systems.

3. Needle morphology in which the drought-hardy trees have smaller, deeper stomatal pits.

4. Number of stomata per unit needle length.

The physiological effects of drought and flooding are often similar, with flooding causing poor root development that then makes trees less efficient in nutrient and water uptake so that they become very susceptible when droughts do occur (Kormanik and McAlpine, 1971). Just as for flooding and drought, the physiology of resistance to cold and to drought also seems to be similar. This has been discussed by Shirley (1937), Pisek and Larcher (1954), and Schönbach et al. (1966).

Although many more pages could be written about resistance to adverse environments, little more needs to be said here to make clear the gains to be obtained from breeding trees that are suitable for such conditions. This activity is of vital importance, and it is an integral part of developing land races. Much of the effort of the tree improver in the future will be expended on intensified breeding for resistance to adverse environments. The applied phases of the breeding efforts have far outstripped the needed fundamental information. Without continued progress in *both* the applied and fundamental phases of this type of breeding, forestry will not progress in the future as much as it should.

LITERATURE CITED

Bates, C. G. 1930. The frost hardiness of geographic strains of Norway pine. *Jour. For.* **29**:327–333.

Batzer, H. O. 1961, "Jack Pine from Lake States Seed Sources Differ in Susceptibility to Attack by the White-Pine Weevil." Technical Note No. 595, Lake States Forest Experiment Station.

Berry, C. R. 1961. "White Pine Emergence Tipburn, A Physiogenic Disturbance." U.S. Forest Forest Service, Southeastern Experiment Station Paper No. 130.

Berry, C. R. 1973. The differential sensitivity of eastern white pine to three types of air pollution. *Can. Jour. For. Res.* **3**(4):543–547.

Bialobok, S. 1980. "Forest Genetics and Air Pollution Stress." Symp. Effects of Air Pollutants on Mediterranean and Temperate Forest Ecosystems, Riverside, Calif., pp. 100–102.

Bingham, R. T. 1963. "Problems and Progress in Improvement of Rust Resistance of North American Trees." First World Con. For. Gen. and Tree Impr., Stockholm, Sweden.

Bingham, R. T. 1967. "Economical and Reliable Estimates of General Combining Ability for Blister Rust Resistance Obtained with Mixed-Pollen Crosses." U.S. Forest Service Research Note INT-60.

Bjorkman, E. 1964. Breeding for resistance to disease in forest trees. *Unasylva* **18**(2–3):73–81.

Blair, R. L. 1970. "Quantitative Inheritance of Resistance to Fusiform Rust in Loblolly Pine." Ph.D. thesis, North Carolina State University, Raleigh.

Boden, R. W. 1958. Differential frost resistance within one *Eucalyptus* species. *Aust. J. Sci.* **2**(3):84–86.

Borlaug, N. E. 1966. Basic concepts which influence the choice of methods for use in breeding for disease resistance in cross-pollinated and self-pollinated crop plants. In *Breeding Pest-Resistant Trees*, pp. 327–348, Pergamon Press, Oxford, England.

Boyer, M. G. 1964. "The Incidence of Apparent Recovery from Blister Rust in White Pine Seedlings from Resistant Parents." 9th Comm. For. Tree Breed. in Canada, Part II.

Boyer, M. G. 1966. Auxin in relation to stem resistance in white pine blister rust. In *Breeding Pest-Resistant Trees*, pp. 179–184, NATO and NSF Adv. Study Inst. on Gen. Impr. for Dis. and Insect Res. of For. Trees, Pergamon Press, Oxford, England.

Bridgen, M. R., and Hanover, J. W. 1980. "Biochemical Aspects in Resistance Breeding. Indirect Selection of Pest Resistance Using Terpenoid Compounds." Workshop on Genetics of Host–Parasite Inter. in For., Wageningen, Holland.

Bryan, W. C. 1965. "Testing Shortleaf Pine Seedlings for Resistance to Infection by *Phytophthora cinnamomi*." U.S. Forest Service Research Note SE-50.

Callaham, R. Z. 1966. "Tree Breeding for Pest Resistance." Sexto Congresso Forestal Mundial, Madrid.

Carson, M. J. 1977. "Breeding for Resistance to *Dothistroma pini*. Breeding *Pinus radiata*." IUFRO Working Party Newsletter No. 1., pp. 2–4.

Carson, M. 1982. "Breeding for Resistance to Fusiform Rust in Loblolly Pine." Ph.D. thesis, North Carolina State University, Raleigh.

Chiba, O. 1964. "Studies on the Variation and Susceptibility and the Nature of Resistance of Poplars to the Leaf Rust Caused by *Melampsora laricipopulina*." Bull. Govt. For. Expt. Stat., Japan, No. 166, pp. 85–157.

Cobb, F. W., and Stark, R. W. 1970. Decline and mortality of smog-injured ponderosa pine. *Jour. For.* **68**(3):147–149.

Coggeshall, M. V., Land, S. B., Ammon, V. D., Cooper, D. T., and McCracken, F. I. 1981. Genetic variation in resistance to canker disease of young American sycamore. *Plant Dis.* **65**(2):140–142.

Conkle, M. T. 1979. "Amount and Distribution of Isozyme Variation in Various Conifer Species. 17th Meet. Can. For. Tree Assoc., Gander, Newfoundland, pp. 109–117.

Connola, D., and Belskafuer, K. 1976. "Large Outdoor Cage Tests with eastern white pine being tested in field plots for white pine weevil resistance." Proc. 23rd Northeast. For. Tree Impr. Conf., State College, Pa., pp 56–64.

Cowling, E. B. 1979. Effects of acid precipitation and atmospheric deposition on terrestrial vegetation. *Environ. Prof.* **1**:293–301.

Cowling, E. B., and Davey, C. B. 1981. Acid precipitation: Basic principles and ecological consequences. *Pulp Pap.* August:182–185.

Day, P. R. 1972. "The Genetics of Rust Fungi. Biol. of Rust Resis. in For. Trees." NATO-IUFRO Adv. Study Inst. on Gen. Impr. for Dis. and Insect Res. of For. Trees, pp. 3–17.

DeBell, D. S. 1971. Phytotoxic effects of cherrybark oak. *For. Sci.* **17**(2):180–185.

Dietrichson, J. 1968. Provenance and Resistance to *Scleroderris lagerbergii* (*Crumenula abietina*). The International Scots Pine Prov. Expt. of 1938 at Matrand. Rep. *Norw. For. Res. Inst.*, No. 92 **25**(6):398–410. (Meddelelser fra Det Norske Skogforsøksvesen nr 92 **25**(6):398–410.)

Dimitri, V. L., and Frohlich, H. J. 1971. Some questions for resistance breeding with red rot of spruce caused by *Fomes annosus*. *Sil. Gen.* **20**(5–6):184–191.

Dinus, R. J. 1969. "Testing Slash Pine for Rust Resistance in Artificial and Natural Conditions." Proc. 10th South. For. Tree Impr. Conf., Houston, Tex., pp. 98–106.

Dinus, R. J., Snow, G. A., Kais, A. G., and Walkinshaw, C. H. 1975. "Variabiliity of *Cronartium fusiforme* Affects Resistance Breeding Strategies." Proc. 13th South. For. Tree Imp. Conf., Raleigh, N.C., pp. 193–196.

Drew, L. K., and Wylie, F. R. 1980. "Tree Injection with Systemic Insecticide to Control Leaf-Eating and Sap Sucking insects." Advisory Leaflet No. 13, Dept. For., Queensland, Australia.

Einspahr, D. W., Wyckoff, G. W., and Harder, M. L. 1979." *Hypoxylon* Resistance in Aspen and Aspen Hybrids," Proc. 1st North Cent. Tree Imp. Conf., Madison, Wis., pp. 114–122.

Elgersma, D. M. 1980. "Resistance Mechanisms of Elms to Dutch Elm Disease." Workshop Gen. Host–Parasite Interactions in For. Wageningen, Holland.

Ferrell, W. K., and Woodard, E. S. 1966. Effects of seed origin on drought resistance of Douglas-fir (*Pseudotsuga menziesii*). *Ecology* **43**(3):499–502.

Filer, T. H., and Randall, W. K. 1978. Resistance of twenty-one sweetgum families to *Botryosphaeria ribis*. *Plant Dis. Rptr.* **62**(1):38–39.

Fisher, R. F. 1980. Allelopathy: A potential cause of regeneration failure. *Jour. For.* **78**(6):346–348.

Fitzgerald, C. H., Schultz, R. C., Forston, J. C., and Terrell, S. 1977. Effects of *Seymeria cassioides* infestation on pine seedling and sapling growth. *South. Jour. App. For.* **1**(4):26–30.

Ford-Robertson, F. C. 1971. *Terminology of Forest Science Technology Practice and Products*. The Multilingual Forestry Terminology Series No. 1, Soc. Amer. For., Washington, D.C.

Forrest, G. I. 1980. "Preliminary Work on the Relation between Resistance to *Fomes annosus* and the Monoterpene Composition of Sitka Spruce Resin." Workshop Gen. of Host–Parasite Interaction in For., Wageningen, Holland.

Frochot, H., Pitsch, M., and Wharlen, L. 1978. "Susceptibility Differences of Mistletoe (*Viscum album*) to Some Poplar Clones (*Populus* sp.)." Congrès des Sociétés Savantes, Nancy, France, pp. 371–380.

Gabriel, W. J. 1975. Allelopathic effects of black walnut on white birches. *Jour. For.* **73**(4):234–237.

Garrett, P. W. 1970. "Early Evidence of Weevil Resistance in Some Clones and Hybrids of White Pine." U.S. Forest Service Research Note NE-117.

Gerhold, H. D. 1962. "Testing White Pines for Weevil Resistance." 9th Northeast. For. Tree Impr. Conf., Syracuse, N.Y., pp. 44–53.

Gerhold, H. D., Schreiner, E. J., Dermott, R. E., and Winieski, J. A. 1966. *Breeding Pest-Resistant Trees.* Pergamon Press, Oxford, England.

Grijpma, P. 1976. Resistance of *Meliaceae* against the shoot borer *Hypsipyla* with particular reference to *Toona ciliata* var. *australis. Tropical Trees*, No. 2:69–77.

Hare, R. C. 1966. Physiology of resistance to fungal diseases in plants. *Bot. Rev.* **32**(2):95–137.

Harris, M. K. 1980. "Genes for Resistance to Insects, Emphasizinig Host–Parasite Interactions." Workshop Gen. of Host–Parasite Inter. in For., Wageningen, Holland.

Hattemer, H. H. 1972. Persistence of rust resistance. In *Biol. of Rust Res. in For. Trees.,* NATO–IUFRO Advanced Study Institute, pp. 561–569.

Hawksworth, F. G., and Wiens, D. 1966. Observations on witches-broom formation, autoparasitism, and new hosts in *Phoradendron. Madrono* **18**(7):218–224.

Henson, W. R., O'Neil, L. C., and Mergen, F. 1970. "Natural Variation in Susceptibility of *Pinus* to *Neodiprion* Sawflies as a Basis for Development of a Breeding Program for Resistant trees." Yale Univ. Bull. No. 78.

Heimburger, C. 1962. Breeding for disease resistance in forest trees. *For. Chron.* **38**(3):356–362.

Heimburger, C. C. 1963. "The Breeding of White Pine for Resistance to Weevil." 1st World Cons. For. Gen. and Tree Impr., Stockholm, Sweden.

Heimburger, C. C., and Sullivan, C. R. 1972. Screening of *Haploxylon* pines for resistance to the white pine weevil. II. *Pinus strobus* and other species and hybrids grafted on white pine. *Sil. Gen.* **21**(6):210–215.

Hertel, G. D., and Benjamin, D. M. 1975. "Tip Moth and Webworm Attacks in Southern Pine Seed Source Plantations." U.S. Forest Service Research Note SE-221.

Heybroek, H. M. 1980. "Monoculture versus Mixture: Interactions Between Susceptible and Resistant Trees in a Mixed Stand." Workshop Gen. of Host–Parasite Inter. in For., Wageningen, Holland.

Holst, M. 1963. "Breeding Resistance in Pines to *Rhyaciona* Moths." 1st World Cons. For. Gen. and Tree Impr., Stockholm, Sweden.

Hood, J. W., and Libby, W. J. 1980. A clonal study of intraspecific variability in radiata pine. I. Cold and animal damage. *Aust. For. Res.* **10**:9–20.

Hornbeck, J. W. 1981. Acid rain—facts and fallacies. *Jour. For.* **79**(7):438–443.

Horsfall, J. G., and Cowling, E. B. 1980. *Plant Disease.* Academic Press, New York.

Hunt, R., and Zobel, B. 1978. Frost hardy eucalypts grow well in the southeast. *South. Jour. Appl. For.* **2**(1):6–10.

Kerr, E. 1972. Trees that resist hurricanes. *Prog. Farmer* (March 1972):628.

King, J. P., and Nienstaedt, H. 1965. Variation in needle cast susceptibility among 29 jack pine seed sources. *Sil. Gen.* **14**(6):194–198.

Kinloch, B. B., and Kelman, A. 1965. Relative susceptibility to fusiform rust of progeny lines from rust-infected and noninfested loblolly pines. *Plant Dis. Reptr.* **49**(10):872–874.

Kinloch, B. B., and M. H. Zoerb. 1971. "Genetic Variation in Resistance to Fusiform Rust Among Selected Parent Clones of Loblolly Pine and Their Offspring." Proc. 11th South. For. Tree Impr. Conf., Atlanta, Ga., pp. 76–80.

Kormanik, P. P., and McAlpine, R. G. 1971. The Response of Three Random Clones of Yellow-Poplar to Simulated Drought and Flooding." 11th Conf. on South. For. Tree Impr., Atlanta, Ga., pp. 18–19.

Kriebel, H. B. 1963. "Selection for Drought Resistance in Sugar Maple." 1st World Cons. on For Gen. and Tree Impr., Stockholm, Sweden.

Kuhlman, E. G. 1972. "Susceptibility of Loblolly and Slash Pine Progeny to *Fomes annosus.*" U.S. Forest Service Research Note SE 176.

Ladeitschikova, E. I. 1980. "Biochemical Aspects of Resistance to Root Rot in Scots Pine." Workshop Gen. of Host–Parasite Inter. in For., Wageningen, Holland.

LaFarge, T., and Kraus, J. F. 1980. A progeny test of (shortleaf × loblolly) × loblolly hybrids to produce rapid-growing hybrids resistant to fusiform rust. *Sil. Gen.* **29**(5–6):197–200.

Lee, J. J., and Weber, D. E. 1979. The effect of simulated acid rain on seedling emergence and growth of eleven woody species. *For. Sci.* **25**(3):393–398.

MacDonald, R. D., Thor, E., and Andes, J. O. 1962. "American Chestnut Breeding Program at the University of Tennessee." 53rd Ann. Meet. North. Nut Growers Assoc., Purdue, Ind.

Marien, J. N. 1980. Juvenile selection of frost resistant *Eucalyptus. AFOCEL,* pp. 225–253.

Marquis, D. A. 1974. "The Impact of Deer Browsing on Allegheny Hardwood Regeneration." U.S. Forest Service Research Paper NE-308.

Martinsson, O. 1979. "Breeding Strategy in Relation to Disease Resistance in Introduced Forest Trees. Sveriges Lantbruksuniversitet, Inter. Rep. NR3.

Martinsson, O. 1980. Stem rusts in lodgepole pine provenance trials. *Sil. Gen.* **29**(1):23–26.

McNabb, H. S., Hall, R. B., and Ostry, M. 1980. "Biological and Physical Modifications of the Environment in Short Rotation Tree Crops and the Resulting Effect upon the Host–Parasite Interactions." Workshop Gen. of Host–Parasite Inter. in For., Wageningen, Holland.

Meinartowicz, L. E., and Szmidt, A. 1978. Investigations into the resistance of Douglas fir (*Pseudotsuga menziesii*) populations to the Douglas fir woolly aphid (*Gilletteella cooleyi*). *Sil. Gen.* **27**(2):59–62.

Miller, T. 1977. "Fusiform Rust Management Strategy in Concept: Site Preparation." Symp. Management of Fusiform Rust in Southern Pines, South. For. Dis. and Insect Res. Coun., Gainesville, Fla., pp. 110–115.

Miller, T., Cowling, E. B., Powers, H. R., and Blalock, T. E. 1976. Types of resistance and compatibility in slash pine seedlings infected by *Cronartium fusiforme*. *Phytopathology*. **66**1229–1235.

Monk, R. W., and Wiebe, H. H. 1961. Salt tolerance and protoplasmic salt hardiness of various woody and herbaceous ornamental plants. *Plant Physiol.* **36**(4):478–482.

Musselman, L. J., Harris, C. S., and Mann, W. F. *Agalinis purpurea*: A parasitic weed on sycamore, sweetgum and loblolly pine. *Tree Plant. Notes*, Fall edition, 1978, pp. 24–25.

Nelson, R. R. 1980. "Host–parasite Interactions and Genetics on the Individual Plant Level. Strategy of Breeding for Disease Resistance." Workshop Gen. of Host–Parasite Inter. in For., Wageningen, Holland.

Nicholls, T. H. 1979. "Dangers of Red Pine Monoculture." Proc. 1st North-Central Tree Imp. Conf., Madison, Wis., pp. 104–108.

Ohba, K. 1980. "Breeding of Pines for Resistance to Wood Nematodes (*Bursaphelenchus lignicolus*)." Workshop Gen. of Host–Parasite Inter. in For., Wageningen, Holland.

Patton, R. L. 1981. "Effects of Ozone and Sulfur Dioxide on Height and Stem Specific Gravity of *Populus* Hybrids." U.S. Forest Service Research Paper NE 471.

Patton, R. F., and Spear, R. N. 1980. "Stomatal Influences on White Pine Blister Rust Infection." Proc. IUFRO Work. Group, Rusts of Hard Pines, Florence, Italy, pp. 1–7.

Phelps, W. R. 1977. Screening center for fusiform rust. *For. Farmer* **36**(3):11–14.

Pisek, A., and Larcher, W. 1954. Zusammenhang zwischen Austrocknungsresistenz und Frost-härte bei Immergrünen [Relationship between drought resistance and frost hardiness in evergreens]. *Protoplasma* **44**(1):30–46.

Porterfield, R. L. 1973. "Predicted and Potential Gains from Tree Improvement

Programs—A Goal-Programming Analysis of Program Efficiency." Ph.D. thesis, Yale University, New Haven, Conn.

Powers, H. R. 1975. Relative susceptibility of five southern pines to *Cronartium fusiforme. Plant Dis. Rep.* **59**(4):312–314.

Powers, H. R., and Duncan, H. J. 1976. Increasing fusiform resistance by intraspecific hybridization. *For. Sci.* **22**(3):267–268.

Powers, H. R., Jr., Matthews, F. R., and Dwinell, L. D. 1978. The potential for increased virulence of *Cronartium fusiforme* on resistant loblolly pine. *Phytopathology* **68**:808–810.

Powers, H. R., and Zobel, B. J. 1978. Progeny of specific loblolly pine clones vary in fusiform rust resistance according to seed orchard of origin. *For. Sci.* **24**(2):227–230.

Radwan, M. A. 1972. Differences between Douglas-fir genotypes in relation to browsing preference by black-tailed deer. *Can. Jour. For. Res.* **2**(3): 250–255.

Randall, W. K. 1971. "Differences Among Willows in Susceptibility to Cottonwood Leaf Beetle." 11th Conf. South. For. Tree Impr., Atlanta, Ga.

Riker, A. J., and Patton, R. F. 1961. Breeding trees for disease resistance. *Recent Adv. Bot.* **2**(14):1687–1691.

Rockwood, D. L., and Goddard, R. E. 1973. "Predicted Gains for Fusiform Rust Resistance in Slash Pine." Proc. 12th South. For. Tree Impr. Conf., Baton Rouge, La., pp. 31–37.

Roth, L. R. 1978. "Genetic Control of Dwarf Mistletoe." Symp. Mistletoe Control Through Forest Management, Pacific Southwest Forest and Range Experiment Station General Technical Report PSW-31.1, pp. 69–72.

Sartor, G. F. and Neel, W. W. 1971. "Variable Susceptibility to *Dioryctria amatella* Among Pines in Clonal Seed Orchards." Proc. 11th Conf. South. For. Tree Imp., Atlanta, Ga., pp. 91–94.

Schmidt, R. A. 1978. "Diseases in Forest Ecosystems: The Importance of Functional Diversity." In *Plant Disease—An Advanced Treatise,* Vol. II, *How Disease develops in Populations,* pp. 287–315.

Schönbach, H., Bellman, E., and Schumann, W. 1966. Die Jugendwuchsleistung, Dürre—und Frostresistenz verschiedener Provenienzen der japanischen Lärche (*Larix leptolepis*) [Early growth and resistance to drought and frost in provenances of Japanese larch]. *Sil. Gen.* **15**(5/6):141–147.

Schreiner, E. J. 1959. "Rating poplars for *Melampsora* Leaf Rust Infection." U.S. Forest Service, Northeastern Forest Experiment Station, Forest Research Note No. 90.

Schreiner, E. J. 1963. "Improvement of Disease Resistance in *Populus.*" 1st World Con. on For. Gen. and Tree Impr., Stockholm, Sweden.

Schumann, W., and Hoffman, K. 1968. Routine testing of frost resistance of 1-year-old spruce seedlings. *Arch. Forstw.* **16**(6/9):701–705.

Shigo, A. L. 1980. "Trees Resistant to Spread of Decay Associated with Wounds." Workshop Gen. Host–Parasite Inter. in For., Wageningen, Holland.

Shirley, H. L. 1937. The relation of drought and cold resistance to source of seed stock. *Minn. Hortic., pp. 1–2.*

Shriner, D. S. 1976. "Effects of Simulated Rain Acidified with Sulfur Acid on Host–Parasite Interactions." Proc. 1st Symp. on Acid Precipitation and the Forest Ecosystem., U.S. Forest Service General Technical Report NE-23, pp. 919–925.

Smith, R. H. 1966. "Resin Quality as a Factor in the Resistance of Pines to Bark Beetles." *Breeding Resistant Trees.* NATO and NSF Advanced Study Institute, Pennsylvania State University, pp. 189–196.

Snow, G. A., and Griggs, M. M. 1980. "Relative Virulence of *Cronartium quercuum f. sp. fusiforme* from Seven Resistant Families of Slash Pine." Proc. IUFRO Work. Group, Rusts of Hard Pines, Florence, Italy, pp. 13–16.

Söegaard, B. 1964. Breeding for resistance to insect attack in forest trees. *Unasylva* **18**(2–3):82–88.

Söegaard, B. 1969."Resistance Studies in *Thuja." Soertryk Det forstlige Forsoegsvoesen Danmark beretning* **245**(31).

Squillace, A. E., and Wilhite, L. P. 1977. "Influence of Oak Abundance and Distribution on Fusiform Rust." Symp. Management of Fusiform Rust in the Southern Pines, Gainesville, Fla., pp. 59–70.

Stephan, B. R. 1973. Über Anfalligkeit und Resistenz von Douglasien Herkunften gegenüber *Rhadbdocline pseudotsugae* [Susceptibility and resistance of Douglas-fir provenances to *Rhabdocline pseudotsugae*]. *Sil. Gen.* **22**(5–6):149–153.

Stern, K. 1972. "The Theoretical Basis of Rust Resistance Testing—Concept of Genetic Gain in Breeding Resistant Trees." Biol. of Rust Res. in For. Trees, Proc. NATO–IUFRO Advanced Study Institute, pp. 299–311.

Stock, M. W., Pitman, G. B., and Guenther, J. D. 1979. Genetic differences between Douglas-fir beetles (*Dendroctonus pseudotsugae*) from Idaho and Coastal Oregon. *Ann. Entomolog. Soc. Amer., pp. 394–397.*

Szöny, L., and Nagy, I. 1968. Klimaresistenz Photsynthese und Stoff Production [Frost resistance and growth of Douglas fir]. *Sonderdruck Tagungsberichte,* No. 100:65–67.

Tamm, C. O., and Cowling, E. B. 1977. Acid precipitation and forest vegetation. *Water, Air Soil Pollu.* **7**:503–511.

Thielges, B. A., and Adams, J. C. 1975. Genetic variation and heritability of Melampsora leaf rust resistance in eastern cottonwood. *For. Sci.* **21**(3):278–282.

Trial, H. 1980. A cartographic history of the spruce budworm in Quebec, Maine and New Brunswick. *Maine For. Rev.* **13**:1–52.

Toole, E. R., and Hepting, G. H. 1949. Selection and propagation of *Albizzia* for resistance to *Fusarium* wilt. *Phytopathology* **39**(1):63–70.

Ulrich, B., Mayer, B., and Khanna, P. K. 1980. Chemical changes due to acid precipitation in a loess derived soil in central Europe. *Soil Sci.* **130**:193–199.

van Buijtenen, J. P., and Isbell, R. 1970. "Differential Response of Loblolly Pine Families to a Series of Nutrient Levels. 1st North Amer. For. Biol. Workshop, Michigan State University, East Lansing.

van Buijtenen, J. P., Bilan, V., and Zimmerman, R. H. 1976. Morpho-physiological characteristics related to drought resistance in *Pinus taeda*. In *Tree Physiology and Yield Improvement*, pp. 349–359. Academic Press, New York.

van Buijtenen, J. P. 1981. Insecticides for seed orchards—a case study in applied research. *South. Jour. Appl. For.* **5**(1):33–37.

Walkinshaw, C. H., Dell, T. R., and Hubbard, S. D. 1980. "Predicting Field Performance of Slash Pine Families from Inoculated Greenhouse Seedlings." U.S. Forest Service, Southern Forest Experiment Station Research Paper SO-160.

Walkinshaw, C., and Bey, C. 1981. Reaction of field resistant slash pines to selected isolates of *Cronartium quercuum* f. sp. *fusiforme. Phytopathology.* **71**:1090–1092.

Waring, R. H., and Pitman, G. B. 1980. "A Simple Model of Host Resistance to Bark Beetles." For. Res. Lab. Research Note 65, Oregon State University, School of Forestry.

Waters, W. E., and Cowling, E. B. 1976. Integrated forest pest management. A silvicultural necessity. *Integ. Pest Mgt.,* pp. 149–177.

Webster, N., and McKechnie, J. L. 1980. *Webster's New Twentieth Century Dictionary—Unabridged Second Edition.* William Collins Publishers, Inc.

Wells, O. O., and Switzer, G. L. 1975. "Selecting Populations of Loblolly Pine for Rust Resistance and Fast Growth." 13th South. For. Tree Imp. Conf., Raleigh, North Carolina, pp. 37–44.

Williston, H. L. 1974. Managing pines in the ice-storm belt. *Jour. For.* **72**:580–582.

Witwer, S. H., and Buckovac, M. J. 1969. The uptake of nutrients through leaf surfaces. In *Handhuch der Pflanzenernährung und Dungung.,* pp. 235–261. Springer-Verlag, New York.

Wood, T., and Bormann, F. H. 1974. The effects of an artificial acid mist upon the growth of *Betula alleghaniensis. Environ Pollut.* **7**:259–268.

Zobel, B. J. 1980a. "Developing Fusiform-Resistant Trees in the Southeastern United States." Workshop Gen. of Host–Parasite Inter. in For., Wageningen, Holland.

Zobel, B. J. 1980b. "The World's Need for Pest-Resistant Forest Trees." Workshop Gen. of Host–Parasite Inter. in For., Wageningen, Holland.

Zobel, B., Blair, R., and Zoerb, M. 1971. Using research data—disease resistance. *Jour. For.* **69**(8):486–489.

Zobel, B. J., and Zoerb, M. 1977 "Reducing Fusiform Rust in Plantations Through Control of the Seed Source." Symp. Man. of Fusiform Rust in Southern Pines, South For. Dis. and Insect. Res. Coun., Gainesville, Fla., pp. 98–109.

Zsuffa, L. 1975. "Some Problems and Aspects of Breeding for Pest Resistance." 2nd World Consul. on For. Dis. and Insects, India, special paper.

Chapter 10

Vegetative Propagation

The use of vegetative propagation is rapidly increasing and is of vital importance to tree improvement. It always has been widely used for the preservation of genotypes in clone banks and for clonal seed orchard establishment. Currently there is an explosion of interest in using vegetative propagation in operational planting programs.

Vegetative propagation has been used successfully for several centuries by horticulturists, and much can be learned from them. The older horticultural practices as well as the new methodology are being increasingly applied in tree improvement programs (Toda, 1974; Rauter and Hood, 1980; Zobel, 1981). Actually, vegetative propagation has been employed in forestry for more than 100 years. There are records in the literature of using rooted cuttings of *Cryptomeria japonica* for planting during the nineteenth and twentieth centuries (Ono, 1882; Kanoo, 1919). Methods of rooting were developed much before that time, and commercial planting of cuttings has been standard practice for this species for many years. However, aside from a few genera like *Populus, Salix,* and *Cryptomeria,* vegetative propagation has not been used extensively in operational forest-planting programs.

This chapter will emphasize the status, value, and use of vegetative propagation in seed production and gene preservation, along with its use and potential in operational forest regeneration programs. Propagation methodology as such will not be covered in detail because it is now being rapidly developed. Much of this development has occurred during the past 5 years. Operational use of vegetative propagation is so new that there still are many questions about how best to employ it, but good progress is being made with southern pines (van Buijtenen et al., 1975), spruce (Birot and Nepveu, 1979; Rauter, 1979), radiata pine (Thulin and Faulds, 1968), *Eucalyptus* (Campinhos and Ikemori, 1980; Destremau et al., 1980) as well as with several other species. Some of the new studies and techniques have not yet been reported. Two helpful references are the series of papers that dealt with various aspects of vegetative propagation published by the Institute for Forest Improvement in Uppsala, Sweden (Anonymous, 1977) and *Micropropagation d'Arbres Forestiers* (Anonymous, 1979).

USES OF VEGETATIVE PROPAGATION

Vegetative propagation has many uses in forestry. These can be summarized as follows: (1) preservation of genotypes through use of clone banks; (2) multiplication of desired genotypes for special uses such as in seed orchards or breeding orchards; (3) evaluation of genotypes and their interaction with the environment through clonal testing; and (4) capture of maximum genetic gains when used for regeneration in operational planting programs.

Some persons prefer to separate the uses of vegetative propagation into research and operational (production) phases. These may be outlined as follows:

A. Research uses for vegetative propagation

   1. Genetic evaluation of plant material, including genotype × environment interaction studies and estimating environmental and genetic correlations, such as juvenile and mature manifestations of the same characteristic.

   2. Determine the magnitude and control of common environmental or C effects that are prevalent in some species.

   3. Preserve genotypes and gene complexes in clone banks and arboreta for scientific purposes and for possible later use in operational programs.

   4. Bring valuable plants to a centralized area, such as to a laboratory or greenhouse for intensive study and breeding.

   5. Speed up the reproductive cycle for accelerated breeding and testing.

   6. For nongenetic studies, to reduce genetic variability (or to obtain the information to handle it statistically) in experiments that will reduce "error variation."

B. Production (operational) uses of vegetative propagation

   1. Develop seed orchards for operational seed production.

   2. Use vegetative propagules directly in operational plantings.

The use of vegetative propagation in forestry will be increasing; it has become one of the most important tools of the tree improvement forester.

Except for a few genera, it is easier to apply regeneration through seed rather than to develop vegetative propagules. Yet, the effort toward vegetative propagation is being strongly sponsored in tree improvement (Fielding, 1963; Thulin, 1969; Libby, 1977; Campinhos and Ikemori, 1980). At present, tests about the comparative performance of vegetative propagules and seedlings are generally inadequate, although some studies list similarities and differences between rooted cuttings and seedlings in growth rate and form (Fielding, 1970; Sweet, 1972; Sweet and Wells, 1974; Roulund, 1978b; Jiang, 1982).

A complete and technical explanation about why vegetative propagation is desired for operational planting would be long, detailed, and complex. Simplified, the advantages of vegetative propagation are the following: (1) the potential to capture greater genetic gain; (2) the potential to obtain greater uniformity of the tree crop than is possible through seed regeneration; and (3) under some situations, the opportunity to speed up results from tree improvement activities.

Genetic variation is partitioned broadly into additive and nonadditive components. When seed regeneration is used, only the additive portion of the genetic variation can be manipulated by the tree improver, unless special efforts such as control pollinations to mass-produce desired seedlots or two-clone orchards are employed. For some characteristics, gains using seed regeneration will be large, but for others that contain significant amounts of nonadditive variance, such as certain growth characteristics, gains through seed production will only be a small portion of the potential that would be possible when vegetative propagation is used (Fielding, 1970). In general terms, the use of vegetative propagation makes it possible to *capture and transfer to the new tree* all of the genetic potential from the donor tree (Figure 10.1). For characteristics such as volume growth that have low narrow-sense heritabilities, it appears posible to more than double short-term genetic gain in many species by using vegetative propagules rather than seed regeneration.

Another advantage of vegetative propagation is the rapidity with which the desired genetic qualities of selected trees can be utilized. It is not necessary to wait for seed production before producing propagules for operational planting. As soon as tests of a tree have proven it to be a good genotype, it can be used directly in operational reforestation by employing vegetative propagation. This is especially true for the easy-to-root species like some in the genus *Populus* in which cuttings from older trees can be readily rooted. However, cuttings from physiologically mature trees of many species are difficult or impossible to root, as will be described later. In sprouting species, such as the eucalypts, the stump sprouts are physiologically juvenile; therefore, they root as juvenile material. However, it takes time to develop a "sprout nursery" that is necessary to produce the number

FIGURE 10.1

Shown are grafts of three greatly differing limb types of radiata pine in Zimbabwe. Any of the types desired could be used operationally when vegetative propagation methods are perfected. These grafts are used to illustrate how well different characteristics will be transferred to the new tree.

of cuttings needed for operational planting and, even under the best of conditions, it takes several years to develop the stock plants needed to supply the cuttings. For more difficult rooters, it is sometimes necessary to undertake expensive and involved procedures to enhance rooting ability. In many species, rooting ability can be maintained to keep the trees in a juvenile stage through methods such as hedging (Libby et al., 1972; van Buijtenen et al., 1975).

If vegetative propagules that grow well with good form at a reasonable cost can be produced, genetic gains and uniformity of growth and wood properties will be greatly enhanced. Vegetative propagation should produce forests with the greatest possible uniformity in size, quality, and wood properties. Variability among trees is a major problem in forestry; use of vegetative reproduction will greatly help to overcome this difficulty.

METHODS OF VEGETATIVE PROPAGATION

There are many types of vegetative propagation (Hartman and Kester, 1983). However, this book on tree improvement will not cover them in detail. Several publications, such as those of Dormling et al. (1976) and Garner (1979), summarize the methodology. Several vegetative propagation methods have been developed especially for use in forestry.

The current emphasis on operational plantings has been in the use of rooted cuttings. Grafting is usually employed to preserve trees in clone banks or for seed orchards in which the objective is large-scale seed production. The newest vegetative propagation method that is receiving a lot of attention and publicity is tissue culture. Although considerble development is still necessary to make tissue culture operational (Durzan and Campbell, 1974; Zobel, 1977), it has considerable potential (McKeand, 1981).

It is important to have a broad understanding of the use, value, and problems of the different methods of vegetative propagation that are being used. Arguments always arise about which are best for regeneration programs. The only answer is to make comparative studies of them under similar, controlled conditions. Comparisons of vegetative propagules with seedlings for growth and form characteristics have been made by Copes (1977), Roulund (1978a, 1978b), and Birot and Nepveu (1979). Sometimes, grafts have more rapid initial growth than rooted cuttings or seedlings. Even when rooted cuttings and seedlings grow at the same rate, form can be quite different, with the cuttings usually having less taper, less butt swell, smaller limbs, and thinner bark (Libby and Hood, 1976). Generally, cuttings grow more slowly than seedlings. However, much depends on the age of the donor tree, how complete the root system of the rooted cutting is, and how the two were handled prior to outplanting.

There are a few terms used in relation to the different methods of vegetative propagation that are now used regularly in forestry and must be understood. The donor tree, the one from which the vegetative propagules have been taken, is called the *ortet*. Individual propagules from an ortet, or from other propagules from the ortet, are each called a *ramet*. The sum of the propagules arising from one ortet is referred to as a group as a *clone*. In forestry, these terms are being used loosely. As an example, it is common to refer to a grafted tree in a seed orchard as being clone ×. In fact, the grafted tree is a ramet of clone ×, which was originally obtained from ortet ×.

Grafts

Grafting has been used from the earliest times and is still used on a large scale to preserve and multiply desired genotypes (Dimpflmeier, 1954; Bouvarel, 1960). It is a basic tool for the horticulturist and has been used widely in forestry for clone preservation and seed orchard establishment. Methods of grafting are numerous; these are covered in many texts, among which are Hartmann and Kester (1983), Dorman (1976), Garner (1979), and documents such as that by Struve (1981). It is sometimes immaterial which method of grafting is used, although special methods have been developed for the very difficult conditions in field grafting in forestry where the environment cannot be controlled (Hoffmann, 1957; Webb, 1961). Some species, especially certain hardwoods, do not graft as easily as most conifers, and adjustments to the usual methods must be made to obtain a reasonable degree of sucess (Hatmaker and Taft, 1966). This is especially true for some oak species (*Quercus*), although seed orchards of oaks have been established (Enkova and Lylov, 1960; Farmer, 1981). Grafting in walnut has been widely developed, as has been explained by Beineke and Todhunter (1980).

Often, it is not poor grafting technique that results in failure, but rather the poor care given the scion or rootstock before or during grafting or in release following grafting. (The *scion* is the piece grafted that has been obtained from the ortet; the *rootstock* is the plant on which the graft is made.) Many more grafts are killed by poor management than by incorrect grafting per se. Methods worked out for the southern pines, where grafting success has climbed from "mediocre" to above 90%, have been described by Dorman (1976).

A major problem with grafting is incompatibility between the stock and the scion (Hong, 1975). Because incompatibility is strongly clonal, it has a major effect on tree improvement programs through loss of clones, especially in such species as Douglas fir (Duffield and Wheat, 1964). Nearly all species show incompatibility to some extent (see Burgess, 1973, for eucalypts). For most species it is an inconvenience, but one can work around it. For example, about 20% of the grafts made in loblolly pine seed orchards show some degree of incompatibility. Loss of whole clones can be a serious result when the lost clone happens to be one of the best genotypes. A series of studies has been made about the cause and possible control of incompatibility (Corte, 1968; Copes, 1969, 1970; Lantz, 1973; Slee and Spidy, 1970; McKinley, 1975). Good results have been obtained in overcoming or avoiding graft incompatibility in species like Douglas fir (Copes, 1981). Graft incompatibility is a problem that must be circumvented as much as possible, or avoided entirely by use of other methods of vegetative propagation. Only rarely is graft incompatibility so serious that it makes a program inoperative.

There are several different types of incompatibility, each with differing symptoms, that can be recognized with experience and careful observation. The most common incompatibility symptom is a swelling above the point of grafting and a *scion overgrowth* of the rootstock caused by blockage of the phloem; this is commonly referred to as a *saddle overgrowth* (Figures 10.2 and 10.3). Saddle overgrowth sometimes appears during the first or second year following grafting in pines, but often it does not become evident until the fourth or fifth year, or even later. Foliage abnormalities are usually evident before the actual overgrowth can be seen, with the needles or leaves, including those of the current year, usually being small with brown tips. In most conifers, abundant resin exudation is evident

FIGURE 10.2

Although graft incompatibility is best known for the pines and Douglas fir, it occurs in most species. Shown is an incompatible *Eucalyptus* in Minas Gerais, Brazil, with the dramatic swelling, or saddle overgrowth, above the graft. Incompatibility is so severe in a few species that it precludes an efficient use of grafts for seed orchard use.

FIGURE 10.3

Typical incompatibility is shown for loblolly pine. There is a saddle overgrowth above the graft union (*a*). Often the union at point of graft is poor as is shown by the peeled specimen (*b*). It should be noted that the roots of the graft were below the knot.

on the foliage. The needles in pines often group themselves parallel to the branch and become sparse, giving the tree a distinctive abnormal appearance. Often there is an accompanying general yellowing or graying of the foliage. One almost certain sign of incompatibility, which is accompanied by the abnormal foliage, is the presence of excessively heavy flower production. In loblolly and slash pines, a 1- or 2-year-old graft that produces abundant male flowers is usually incompatible, although this is not necessarily true for early flowering species such as Virginia pine. The saddle-type incompatibility is strongly clonal, and it usually results in the death of the grafted tree within a few years. Among individual clones, from very few to as many as 100% of the ramets may show incompatibility.

Death of large grafts with *no overgrowth* appears to result from a different type of incompatibility. This form is less common and can appear on all ramets of a single clone throughout an orchard at about the same time. Affected trees die rapidly, particularly in association with extreme environmental conditions such as heat and drought. Like the overgrowth type, damage is quite clonal, and it is not unusual for nearly every graft of the affected clone to die.

In some pines an unusual cambial growth exists (usually restricted to the understock) that results in incompatibility with bark ridges and dead zones, a condition that is usually referred to as *fluted bark*. The cause of this abnormality is not known. A pathological condition has been suspected, as have been deficiencies or excesses in nutrients, but these have never been confirmed. Adverse soil conditions seemed to be a causative factor because damage appears to be most

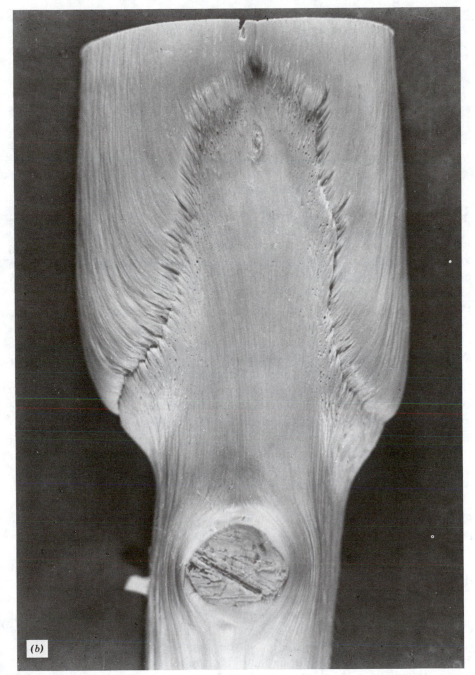

(b)

FIGURE 10.3 (*continued*)

prevalent on heavy clay soils, showing up most frequently following severe droughts. Deep, drastic subsoiling and heavy nitrogen fertilizaton has helped to correct, or at least arrest, the fluted bark condition in several pine orchards. Fluted bark has a clonal tendency that is related to the scion used (even though the abnormality is in the understock), but the tendency is not so strong clonally as are the most standard incompatibilities. Infected trees have leader dieback, with sparse, graying, or yellow foliage that sheds early; the needles of the pines are strongly appressed to the branches. Such trees may live for many years and, while alive, produce exceedingly heavy seed crops. Seeds produced on incompatible grafts often are weak or are not viable.

Rooted Cuttings

The method of vegetative propagation that is currently being developed most rapidly is *rooted cuttings*. This technique has been in use for some species for a long time (Ozawa, 1904). Methods and techniques are many, and information about them can be obtained from the books previously mentioned in the section on grafting and from numerous articles such as the following: Yim (1962), Wheat (1964), Hare (1970), Brix and Barker (1971), Hinesley and Blazich (1981), and Pousujja (1981). The rapid developments in just the past few years in the eucalypts (Franclet, 1963; Hartney, 1980; Laplace and Quillet, 1980) show what can be done to develop and use rooted cuttings in a short time (Figure 10.4). Progress with the eucalypts may well indicate what might happen in many other species. It is only a matter of time before several of the principal conifers (Cameron, 1968) and some hardwoods will be planted operationally as rooted cuttings, but much technological development is still needed for most species. As an example, great progress is being made on spruce (Rauter, 1979).

A major deterrent to using rooted cuttings is their dependency on age. Young trees will often root readily, but the same trees may be almost impossible to root when they become older. This is particularly frustrating to the tree improver who works with proven desirable genotypes. When the trees are left to grow long enough to prove their genetic worth, it is then often too late to root them. As mentioned previously, there are indications that trees that are vegetatively propagated from older individuals will grow more slowly than those taken from younger trees (Talbert et al., 1982).

Another major restriction to the use of rooted cuttings from older trees is that the propagules sometimes do not grow into a normal tree form; this concept is covered in a later section in this chapter on *abnormal growth*.

A deterrent to making maximum gains using rooted cuttings in operational planting is the very large clonal variability in rooting ability; this is especially strong in older individuals. Variation in rooting ability often dictates what trees are available to the research or applied program (Hyun, 1967; Shelbourne and Thulin, 1974; Kleinschmit and Schmidt, 1977). Clonal variation in rooting occurs, no matter what method is employed. The percentage of clones that propagate satisfactorily is so vital and sometimes so low that it is difficult to keep a sufficiently broad genetic base, and in some species so few parent trees respond well enough to rooting that an initially broad genetic potential may be reduced to an alarming degree. If one selects or develops 100 outstanding trees but only 10 of these root well enough to use operationally, the effectiveness of the tree improvement program will be greatly limited. Improved techniques will help to some extent, but losses of large numbers of otherwise excellent genotypes are one of the most serious obstacles to the operational use of rooted cuttings. Satisfactory rooting percentages will vary with species and the needs of the organization involved. As an example, for *Eucalyptus* in Aracruz, Brazil, a 75% rooting is considered minimal for use in the planting program (*Campinhos and Ikemori, 1980*).

There is no question that the use of rooted cuttings will become operational on a large scale in forestry; the actual time will depend on how quickly methodology is perfected. The key to success is the development of methods to bypass age, growth, and form problems. The hope is that researchers will learn to treat proven older trees so that they revert to the juvenile condition. In some species, mostly confined to the hardwoods, the sprouts formed following cutting or injuring the tree are juvenile; therefore, the age problem is bypassed. The ability to sprout and thus root as juvenile material is used on a large scale in some eucalypts.

Rooting Needle Fascicles

The method of taking needle fascicles of pine and rooting them has been used for a long time (Zak and McAlpine, 1957; Hoffman and Kummerav, 1966). A moderate amount of work has been done in the intervening years, as has been summarized by Girovard (1971), and interest has grown recently (Struve, 1980). Initially, rooting of needle fascicles was sometimes successful, but shoots often did not develop normally. With more knowledge and better use of chemicals and hormones, this problem has become less severe, and balanced plants can be produced.

The original idea that appealed so greatly to tree improvement workers was the opportunity to get a large number of propagules from an individual tree. Prospects appear good now, but it is too early to predict how successful the use of needle fascicles will be on an operational scale. The objective of increasing numbers is still of importance, but current interest has centered on being able to produce vegetative propagules from needle fascicles with juvenile characteristics from older trees that have already proven their genetic worth.

Air Layers

The process of air layering is one in which roots are generated on an intact branch by girdling, which is usually accompanied by hormone application (Kadambi and Dabral, 1954; Hoekstra, 1957; Chonard and Parrot, 1958). A similar process that occurs naturally in some species is called *layering*; this is when roots are formed on branches that touch or become buried in the soil (Cooper, 1911).

Air layering has several uses. One is to produce propagules directly that are needed to establish seed orchards and thus avoiding graft incompatibility (Barnes,

FIGURE 10.4

The stages in rooting cuttings of *Eucalyptus* in Espirito Santo, Brazil, are shown. Sprouts from a stump that will be used for cuttings (*a*). A rooted cutting ready for field planting (*b*). A commercial planting of rooted cuttings that have been in the field for several months (*c*). (Photos courtesy Aracruz Florestal, Espirito Santo, Brazil.)

1969). It is quite satisfactory for some species in which needs for rooted propagules are small. It is also sometimes used as an intermediate method to obtain roots for species in which rooted cutting success is marginal. This is done to *P. caribaea* in Venezuela. The air layers are detached after callus formation and rooted in conventional beds with a good degree of success.

Tissue and Organ Culture

Tissue culture is the newest and currently the most publicized of the vegetative propagation methods. Because the method is developing so very rapidly, any discussion will soon be out of date. Tissue culture has great potential, but it must

FIGURE 10.4 (*Continued*)

be viewed realistically (Rediske, 1976; Zobel, 1977; Sommer and Brown, 1979; Bonga, 1980). For example, the uniformity of individuals within clones of identical genetic makeup are sometimes quite dissimilar, often showing as much variability as seedlings from individual seeds from a given tree. Also, it is difficult to move the plantlets from the environment in which they are formed so they will grow normally under the uncontrolled forest conditions. Using current methods, tissue culture plantlets are expensive because of the multiple handling that is now required. With time, however, new technology should reduce this problem.

A number of books and articles have been written about tissue culture production (see Mott, 1981, for a review and references). The major concerns with tissue culture are now being vigorously studied, and it is clear that much more research and development are needed before propagation by tissue culture will be satisfactory for operational use (Kelly, 1978; Leach, 1979; Amerson et al., 1981). It is especially important to produce plantlets that are essentially uniform within a clone before tissue culture will be of value for operational plantings. This is the case because plant uniformity is one of the major attractions of vegetative propagation.

The best immediate use for tissue culture will be as a research tool and as a rapid method of utilizing improved genetic stock (McKeand, 1981). Whether tissue culture plantlets will be important for operational planting depends on the "unknowns" listed previously. After the methodology is better developed, including transfer to field conditions, then uniformity will undoubtedly increase. Once that happens, the major job will be to make the system cost-effective. Utility of tissue culture plantlets will depend on their relative growth and form compared to other kinds of vegetative propagules and seedlings. To be of most value, it will be necessary to produce plantlets from older, previously tested, and proven trees.

ABNORMAL GROWTH

Plagiotropic and Orthotropic Response

A major problem that occurs when working with vegetative propagules is that genetically identical propagules often grow differently, depending on their origin on the original plant and the age of the donor when the propagule is taken. *Plagiotropic growth* refers to the situation in which a vegetative propagule *does not assume tree form* but continues to grow like a branch. *Orthotropic* means the propagule assumes an upright or normal tree form. Causes for propagules growing differently, even though they are "genetically identical," are many and will only be mentioned here briefly. They involve age, location on the tree, and other environmental effects. The differential in growth and development when plagiotropism is present is very frustrating. The common idea that cuttings from a tree are genetically identical and therefore they should grow alike is misleading. The cuttings *are* identical in the sense that they have the same genotype. Obviously, some genes are more effective than others, or they may be "turned on" or "turned off" by the environment, age, and position within the parent plant or by outside treatment; this will in turn affect the physiology of the tree. As a result, propagules sometimes do not grow in the same pattern, or they may not have the same tree form, even though they are genetically identical. Plagiotropic growth is common in such genera as *Abies, Picea, Araucaria,* and *Sequoia,* and it is found to a lesser extent in *Pseudotsuga.* Plagiotropism is not common in *Pinus* or in most hardwoods.

The Effect of Age and Location Where Propagules Are Obtained

The major problems in operational (and research) uses of vegetative propagules relate to the effect of the age (Ducci and Locci, 1978) and location of the propagule from the parent plant and its ability to grow as a tree.

Age differences are highly important. As Franclet (1979) has written, "There is a progressive loss with age of the aptitude for vegetative propagation." Physiologically mature tissue has a lower rooting percentage, takes longer to initiate roots, and develops fewer roots than does physiologically juvenile material. In addition, plagiotropic growth often occurs.

The term *cyclophysis* is sometimes applied to age effects and *topophysis* to location or origin effects. These are serious problems in some species (Jang et al., 1980), but studies indicate that solutions may become available. Although many physiologists consider them to be related (often using the term *topophysis*), Corriveau (1974) and Olesen (1978) consider cyclophsis and topophysis as two separate processes. Olesen defines them as follows: (1) *cyclophysis* is the process of maturation of the apical meristem; and (2) *topophysis* is the phenomenon that occurs when scions, buddings, and rooted cuttings maintain for some time the branchlike growth habit (plagiotropic growth) they had as shoots on the ortet. A third cause of variation is *periphysis,* which refers to locations in different environments, such as shade and sun shoots on an individual tree.

Although the concepts of cyclophysis and topophysis are widely recognized (Naes-Schmidt and Soegaard, 1960; Wright, 1976; Land, 1977), their effects are not always understood. They include not only growth variation but also unseen physiological and morphological changes. It is clear that problems with topophysis can be reduced and the use of vegetative propagation increased, if juvenile material can be used or if rejuvenation from the mature to the juvenile stage can be accomplished. Juvenile material tends to assume orthotropic growth habits much more readily than mature material.

Methods of developing juvenility (Franclet, 1979) or maintaining juvenility by methods such as hedging (Libby et al., 1972; Libby and Hood, 1976; Brix and van Driessche, 1977) are essential to further developments for operational planting of vegetative propagules (Figure 10.5). The utility of root sprouts or sprouts from adventitious buds in developing juvenility is known, but only a few studies have been done on pines that sprout (Santamour, 1965). This may have a major value if these sprouts have juvenile characteristics like the eucalypts, because several of the most important tropical pine species do sprout rather profusely. In some

FIGURE 10.5

To overcome or bypass the poor rooting and often poor form on
cuttings from old trees, the method hedging is being developed. It
consists of trimming the tree and keeping it low. This is shown for
radiata pine in New Zealand. Cuttings from the hedged plants tend to
maintain their young physiological age. (Photo courtesy of Mike
Wilcox, Forest Research Institute, New Zealand.)

species, such as red maple (*Acer rubrum*), rooted cuttings obtained from grafts
respond as though they are juvenile, even though the original graft came from an
old tree.[1] This ability of cuttings taken from grafts to root readily is a great
advantage in some hardwood species, especially those that do not sprout from the
stump.

Rejuvenation of mature clones sometimes occurs during the tissue culture
process. This phenomenon could become of great value once it is better understood and more reliable. Cell culture, leading to embryoids that can be coated and
treated like seeds, though not yet well developed or operational, could be a most

---

[1]Personal communication from Dr. Dan Struve, Horticulture Department, Ohio State University,
Columbus.

important contribution to the use of vegetative propagation in forestry. With current technology, systems involved with the mass production of rooted cuttings of loblolly pine would be easier to devise if rejuvenation of mature genotypes were possible (Foster et al., 1981). An example is that needle fascicles of loblolly pine cultured *in vitro* often form adventitious buds in the basal region of the fascicles at the juncture of the needles (Mehra-Palta et al., 1977). It is highly probable that these are, in fact, juvenile buds and can be cultured, using the techniques of Mott and Amerson (1981) to produce shoots and roots. As pointed out by Lyrene (1981), reversion from mature to juvenile characteristics may be a general phenomenon of tissue culture.

Although loss of rooting potential and sexual maturity tend to occur at about the same age, they do not appear to be strongly related to each other. Sexual maturity is usually retained following vegetative propagation, whether or not a juvenile state for rooting occurs. It is sometimes possible, especially in hardwoods, to induce good rooting from heavily flowering, vegetatively produced trees.

In many trees, especially the hardwoods, there is a "cone of juvenility" near the base of the tree that retains juvenility (de Muckadell, 1954). This can be observed on *Quercus* and *Fagus* at the time of leaf fall, when the lower leaves in the juvenile area are retained longer than are the mature leaves higher up. Cuttings taken from this area are much easier to root than those taken from the branches in the mature area of the tree.

VEGETATIVE PROPAGATION AND TREE IMPROVEMENT

Vegetative propagation is used for species in the genera *Populus, Salix, Eucalyptus, Cryptomeria, Sequoia, Picea,* and a few others. Methods for the first four genera are known and fully operational, while that for *Picea* is becoming used more widely (Figure 10.6). The major criticism of current programs is not in methodology of vegetative propagation per se, but in the lack of ongoing breeding programs to produce better trees for the vegetative regeneration program. Although some organizations have genetic improvement programs combined with their vegetative regeneration, most do not. Many merely select within natural stands or plantations or produce F_1 hybrids from which to choose superior genotypes. Hybrids are usually little better than the parents used, and a genetics program to improve the parents before new hybrids are made is essential for long-term gain if hybrids are to be utilized in tree improvement programs using vegetative propagation.

With the exception of *Cryptomeria* in Japan, the vegetative propagation programs with most conifers are just getting started. Good initial gains will be possible by using outstanding genotypes from current stands, but this is not enough, and new and improved trees need to be developed. It is not recommended that a lot of time and money be spent on developing sophisticated

FIGURE 10.6

Great improvements in wood quality, growth, and uniformity are possible with rooted cuttings. Shown are 3-year-old plantations of *Eucalyptus* at Aracruz Florestal in Brazil. Trees are about 65 ft tall (20.2 m) and uniform in height by clone (*a*). Note how uniform in diameter each 3-year-old tree from 1 clone is (*b*).

vegetative propagation techniques unless there is a parallel intensive genetic improvement program.

The use of vegetative propagules has a special place in forestry; it should be used for high-value products or on sites where there is a special need. For example, if a certain kind of wood is desired, it can often be supplied by vegetative propagules, even though it may be necessary to restrict the genetic base to fill this special need. Many eucalypts have interlocked grain or wood that is under internal stresses that cause splitting when the trees are felled. Occasional trees are straight grained without internal stress and make fine high-quality plywood or furniture. The few clones with suitable wood can be used to supply the special need for *Eucalyptus* with high-quality wood. Often, disease-free trees produce disease-free rooted cuttings; a prime example is *Diaporthe cubensis* on *Eucalyptus* in Brazil. The same special usage of vegetative propagules could be made to produce trees with other special, uniform, or otherwise desirable qualities (see Figure 10.6).

Although not generally operational, studies on rooting cuttings for some quality hardwoods in the southeastern United States have been done on sweet-gum (*Liquidambar styraciflua*) by Brown and McAlpine (1964); on black walnut (*Juglans nigra*) by Carpenter (1975); on black cherry (*Prunus serotina*) by Farmer and Basemann (1975); on sugar maple (*Acer saccharum*) by Gabriel et al. (1961); and on yellow poplar (*Liriodendron tulipifera*) by McAlpine and Kormanik (1972). Several of the authors mentioned feel that vegetative propagation in hardwoods can be developed operationally, and there is no doubt of its value for the species with high-quality woods, such as in some of the tropical hardwoods (Lee and Rao, 1980).

OPERATIONAL USE OF VEGETATIVE PROPAGATION

All kinds of genetic advantages as well as problems could be listed that are relative to the operational use of vegetative propagation. If one brings all considerations to a common denominator, they add up to "gain versus risk;" that is, how much extra gain can be achieved through vegetative propagation while still retaining an acceptable level of risk? This basic question is widely argued, but conclusions are rare because of differing emphases on the relative risks by the various investigators and sometimes because of a lack of knowledge about achievable gains. It is not important to come to a concensus. What is important is to be aware of the gains and risks and to make a conscious decision about their relative importance before an operational vegetative-propagation program is undertaken. The decision is not simply one of biology but includes operations and policy. Thus, the silviculturist, forest manager, and administrator must all share in any decision that is made.

The first concern that is always raised is that of the danger of planting large acreages with the same or similar genotypes. Many persons decry planting large acreages of trees of the same species, which to them represents a dangerous monoculture, no matter how variable the genotypes are that are planted (Anderson, 1975). This very real problem is often blown out of perspective when it is being argued. On the other hand, some persons cite agriculture and its widespread use of crops with very narrow genetic bases with outstanding importance to society as reasons why there should be little objection. (Monocultures are covered in the following section.)

A common mistake made by some foresters is to assume that members of a clone will be adapted only to a narrow range of environmental conditions. Although clones are usually less broadly adapted than mixtures of full- or half-sibs, and even though each member of a clone has the same genotype, the individual genotype can possess a considerable ability for adaptation to differing pests or to adverse environments. It is possible to select clones that possess greater adaptability than that possessed by the average seedling. A forest tree needs great adaptability merely to survive and to reproduce. The danger from planting large areas with a single clone arises when the adaptability of the clonal genotype is exceeded by adverse conditions. The result will then be that *all* trees of the clone will be subject to injury in the adverse environment. Trees seem to be less well buffered to attacks by pests than to weather extremes, especially to pests that have come from outside the natural range of the tree species.

How Many Clones Should Be Used?

The problem of what is a suitable clone number for extensive field planting is always of importance, but it is especially true when vegetative propagules are used. There is no absolute answer, although a good estimate can be made that is based on experience and knowledge of the variation in the species used. The question again relates to "gain versus risk," that is, how can one achieve the greatest improvements by selecting only the best clones while staying within acceptable limits of the danger of destructive losses in the forest? Usually, biologists are timid and they sacrifice too much gain to be certain they take little risk, whereas financially-oriented people often emphasize gain to the point of real danger from destruction by pests or adverse environments. Realistically, some risk is always involved in any breeding program, but when one knows the species well, the risk can be kept low.

The important question relating to how many clones are necessary to assure reasonable safety and to make optimal gain is usually answered as, "It all depends on rotation age, on intensity of forest management, on genetic variability of the species and clones involved, the likely risks and the acceptable loss levels" (Libby, 1981). But this answer needs to be better defined. It is unlikely that hundreds of clones would be required, as the more cautious advise, and there is no question that a single clone, as has been suggested by a few specialists, will not be safe. As a generalized guide, Libby's (1981) recommendation that 7 to 30 clones be used would appear to be a safe and reasonable number.

In an area in South America in which Zobel is involved, about 600 outstanding clones of *Eucalyptus* have been selected. The trees are fast growing and variable with a 6- to 8-year rotation. It was found (Zobel et al., 1982) that the best 100 clones could nearly double the gain, compared to the situation if all 600 clones had been used. The area has six different districts, some of which have quite different environments. The best 15 clones to be used in each district are being determined by testing. The same clones were not best in all environments, and it was found that about a total of 50 clones was needed. For the eucalypts, with their variability and short rotations of 5 to 6 years, there is no unusual danger by using the best 15 clones in each environment.

Generally, the more severe the stress a given species or provenance may face (i.e., they are poorly adapted to the environment), the greater the number of clones that should be used. There are special conditions, such as severe insect or disease areas or very severe environments, in which very few clones may be justified. This is because only a few that have the needed qualities are available. This short-term and rather risky strategy becomes necessary when only those few that will survive such things as freezing weather or disease must be used if the forest is to be productive.

Before vegetatively propagated clones are used for operational regeneration, they need to be tested in all the major environments of the proposed planting area. Ultimately, all the potentially good clones need to be tested on each different site. The best clones for each soil site area should then be used to reforest the area. Not all clones will be suitable for all sites. Because of site × genotype interaction, different clones will often be used on different planting sites. Overlooking this refinement, as is too often done, can lead to significant losses or even to disaster.

How Should the Clones Be Deployed?

Closely related to the number of clones that should be used is their deployment; that is, should they be planted randomly and intermixed, or should they be planted in small blocks of pure clones? This was argued as early as 1918 by Hirasiro, who felt mixtures of clones were the best.

The immediate reaction of most specialists is that clones should be planted in mixture. They argue that mixtures will be better safeguards against pest attack and spread, and thus they will provide some measure of insurance for success of the plantings. This is true for root diseases, but it is not necessarily so for airborne diseases and many insects. Leaf diseases or canker diseases carried by windborne spores seem to spread about equally rapidly in pure or mixed-species stands. Some persons argue against the superiority of multiclonal plantings. For example, the analysis of Libby (1981) indicates that mosaics of monoclonal plantings may be the best strategy. What follows are some arguments that have been given in favor of block planting.

1. Each clone tends to have a different growth curve and developmental pattern. Some clones will never be able to develop properly in mixture and will be severely suppressed by competition from other faster-starting clones.

At the very least there will be differences in size and quality of trees in the intermixed plantings, reducing one of the greatest advantages from vegetative propagation, that is, greater uniformity.

2. Planting and "nursery" operations are much simplified when planting is done by blocks.

3. Wood uniformity among trees is maximum within a block of trees from the same clone. The time will probably come when trees of different blocks will be used for special products, such as plywood or sawtimber.

4. If a really serious problem develops within a given clone, a whole block can be harvested and replaced to keep the forest in maximum productivity. Even if it was possible to salvage an individual clone in mixtures, it could not be replaced, and the result would be a forest with low stocking. Generally, salvage from mixtures is not economically feasible, and often the salvage operation causes more damge to the residuals than the net return from the salvage operation.

Although opinions about clonal distribution have changed over the years, most of the published literature indicates a preference for multiclonal mixtures. This controversy must be settled through trials rather than by the expression of opinions. However, some large operations are currently following the approach of maximizing gains by block plantings. For species with reasonable variability and short-rotation ages, the recommendations have been for 10 to 20 ha monoclonal blocks of vegetatively propagated material. Many persons feel these are too large; however, anything much smaller than 10 ha becomes inefficient to operate as a unit. With more experience, this recommendation may well change, but with what is now known it appears that 10 to 20 ha blocks are about right.

Special Considerations in the Use of Vegetative Propagation

A great danger in using vegetative propagation is to assess the worth of the "parent" tree at too young an age, or to assess the value of the cuttings or plantlets before they have had a chance to show their worth. This is especially true for growth characteristics. The time frame for testing vegetative propagules should be little different from that of seedling genetic tests. Time can be shortened, however, if the clones are from families that have already been progeny tested. Although there are occasional reports of good juvenile–mature correlations for volume growth, most of the literature for most species indicates that a reliable estimate of mature performance of individual genotypes cannot be obtained until about one-half rotation age (Wakeley, 1971; Franklin, 1979). Lambeth (1979) feels that for a 30-year rotation, an assessment at 6 to 8 years of age is the most economical time to select. This is not the place to argue this most important concept, but those who have had widespread experience with a number of species *over a long period of time* are disturbed about bad decisions that are currently being made from too early assessments regarding the genetic superiority of trees used in vegetative propagation programs. From the production standpoint, there is great advantage in having physiologically young material, but foresters too often assume that if the tree is superior when very young it will still be superior at rotation age. This enables them to vegetatively propagate young material. A too-early assessment probably is the most serious error being made when vegetative propagation is used operationally. The error is not easily observed, no matter what the rotation age, although it is more evident under long-rotation conditions.

Although it is evident to biologists that a good phenotype may or may not produce a good plant when it is vegetatively propagated, many persons assume that a good-looking tree will automatically produce good cuttings or other propagules. Many organizations do not even make tests, and they assume that propagules from a good-looking parent will produce good trees.

Rarely has any consideration been given in the clonal use of forest trees to "within-clone" variation that may show up after several generations of propagation. This is common in horticulture, showing up as decreased rooting ability, form and growth degradation, or reduction in flowering. These can result from such things as virus infections, unexplainable internal physiological changes, or simply by root deformation (Figure 10.7). The horticulturists have observed such

FIGURE 10.7

Sometimes lack of uniformity within clones is caused by root deformation that is a result from the handling of the propagules in containers. Results of such deformation are shown on the roots of a grafted tree.

clonal degradation to be common; they establish "foundation stocks" where true-to-type, pest-free plants are maintained. Foresters must be on the alert for such late-appearing within-clone variabilities, as well as for occasional somatic mutations that do occur. These can be especially dangerous in seed orchards in which many ramets are produced from a given graft.

Another general concept that must be considered with respect to vegetative propagation is the cost of vegetative propagules versus seedlings. Usually, vegetative propagules are much more expensive to produce and are more difficult to establish than are seedlings. As methods are developed and experience is gained, costs of vegetative propagules can be reduced greatly (Kleinschmit and Schmidt, 1977) and even be no greater than seedlings (Campinhos and Ikemori, 1980). Cost comparisons alone are really not useful because one must weigh the added gains against the costs. Often, a considerable additional cost per planted tree becomes insignificant when assessed on a cost-per-acre and gain-per-acre basis. For example, just a couple of percentage points of improvement resulting from using the better cuttings can often more than justify a doubling or tripling of cost of the propagules established in the field.

RESTRICTING THE GENETIC BASE AND MONOCULTURE

Although the subject of the breadth of the gene base has been touched on in many places in this book, perhaps the most suitable time to discuss it is in relation to vegetative propagation, because genetic restriction can be at its most extreme in this area.

What Is Monoculture?

There is much misunderstanding about the meaning of monoculture and its potentials and problems. There is no universally accepted understanding of what monoculture means. Basically, the very broad definition in Webster,[2] "cultivation of a single crop or product without using the land for other purposes," must be accepted. This definition has been altered by some biologists as follows: "growing extensive areas of plants that are closely related genetically." The important question related to the biological definition is, How close is closely related? For example, Glasgow (1975) interprets monoculture as raising even-aged stands of one or a very limited number of species in blocks large enough to have a significant ecological impact. On the other hand, Feret (1975) feels that planting a single species does not necessarily predicate monoculture; it all depends on how uniform the species is. The understanding of monoculture by many tree improvers is as follows: "extensive plantings of similar genotypes of forest trees that are

---

[2]*Webster's New Twentieth Century Dictionary.* Unabridged 2nd ed., 1980.

homogeneous enough so that the dangers from pests or environmental extremes become too great a risk."

Using Webster's definition of monoculture as being a single crop or product carries no connotation of *good* or *bad* and is not closely tied to genetic uniformity. Therefore, growing sawtimber on a given area of land would constitute a monoculture, even if more than one species was involved. In fact, most of the biologists' or foresters' special definitions that relate to a restriction of the gene base really refer to the intensity of silviculture on land management being used, not to monoculture as it was originally defined. According to Webster's definition, nearly every forest tree plantation would be a monoculture, neither inherently good nor bad.

Whether or not the term *monoculture* is used, the tree improver must be concerned about the increase in risk resulting from reducing the genetic base in forest plantations. The variability of the base population from which the trees have been obtained is highly important. For example, plantings from most seed orchards in which selections have been restricted to one per stand, have a broad gene base for adaptability and are not as restricted genetically as collections from a few individual trees from a single stand. The most extreme form of gene base reduction is the establishment of large monoclonal plantings that has both great advantages and dangers in forestry. In truth, arguments about what constitutes a monoculture are useless. It is the status of the gene base that is important (Zobel, 1972). According to Webster, even-aged management can be called a monoculture, but whether or not it is a risk depends on the heterogeneity of the gene base involved.

A question that is often raised is whether true monocultures occur in nature. Using Webster's definition, they are usual, whereas in the biological sense they are rare. Monoculture-like populations may be prevalent in some species like *P. resinosa* that seem to have a very restricted gene base (Fowler, 1961). Restricted gene bases certainly occur in some of the sprouting species, such as the aspens (*Populus* sp.), where rather extensive areas can be populated with all the trees being the same clone, growing on the same root system (Baker, 1934).

Restricting the Genetic Base

The idea of growing forest trees with relatively similar genotypes over large areas is a cause of considerable concern among foresters (Tucker, 1975). It is interesting how many of the same people will look over an agricultural field with millions of plants of nearly identical genotypes and not feel much concern. Part of the reason for this difference is the relative ease with which environments can be manipulated and controlled in the two situations. Compared with forestry, environmental manipulation is much easier, more flexible, and economically justifiable in agriculture, making it possible to control those factors that can cause catastrophic losses of similar genotypes. As forest management is intensified, the forester will move in the same general direction of greater environmental control. But no matter how well people try to manage the environment in forestry, nature will usually prevail, and stressful conditions will always be encountered for the tree crop. When the planted trees have similar genotypes, as when only a few clones are used, a catastrophic loss can occur, resulting in many years of cumulative productivity loss if the trees are not yet old enough to salvage. When a catastrophe occurs in agriculture and the crop is damaged or lost, adjustments can be made and the farmer can start again the next season. The agriculturist usually grows a crop for only a short period during the year in which the environment is optimal for crop development. But in forestry the plants must survive for many years throughout all seasons, giving a much greater chance for an unusual happenstance to damage or destroy the tree crop. Also, it takes much longer for a forester to locate, develop, and use new genetic material.

If the genetic base in forestry is narrowed, the possibility for losses will become greater. Although yields are enhanced, the forester must come to grips with the following question: How much risk of death or loss of yield or quality can be tolerated for a given amount of additional product uniformity, volume, or quality? This complex question will be answered differently by different people and organizations; a major point to be stressed is that such a decision will be easier if proper attention has been paid to keeping the genetic makeup of the forest trees used in regeneration broad enough to avoid or reduce losses.

An example of restricting the gene base and its ramifications in an operational forestry program were mentioned in the discussion related to vegetative propagation. But the problem also arises with the use of seeds. For example, one large company in the southeastern United States plants seedlings by mother tree blocks. This company collects seeds by mother tree, grows seedlings in the nursery by mother tree, and outplants in 20-ha blocks by mother tree for all of its approximately 12,000 ha per year planting program. The program started with about 40 seed orchard parents; when in short supply, seed was collected from the best 30 clones. As seed became more abundant, the total annual acreage was planted with seed from fewer clones until now they use the seven best open-pollinated families. Such planting of seed from seed orchards by clone (mother tree) appears to be most successful after 6 years and has shown some obvious advantages. Many arguments have been offered both in favor and against this degree of genetic restriction. It is a major problem that will be faced by all programs as imporved seed become plentiful. The question is: How much risk is involved to obtain the additional gain resulting from restricting the number of clones in the seed orchard in order to use only the best to obtain maximum gain?

Dangers of a Restricted Genetic Base

Two categories of problems can result from an interaction of an adverse environmental condition or pest attack and a restricted genetic base. The first is death or severe damage; this is easily understood and recognized. Indeed, most persons equate the adverse effects of gene restriction only in such grossly observable terms when death results. The second kind of adverse effect results in a reduction in growth and vigor; this often goes unnoticed, except by the keen observer, but can be of considerably more importance than outright kill. Assessment of this kind of damage is exceptionally difficult. Often, the loss is ignored or even accepted with the general comment that "I'd rather have unhealthy Species X than healthy Species Y because X still produces more volume." The insidious aspect of planting trees with poor adaptability is that the blame is often placed on a secondary causative factor, such as insects or fungal infection, rather than on the primary cause, which is poor adaptability of the plant to the environment.

Extreme care must be taken not to equate problems arising from using the wrong species or the wrong geographic source of a species to losses from genetic restriction related to monoculture. Use of the wrong species or geographic source can be avoided with proper understanding and care. Problems such as those resulting from using Monterey pine in the southeastern United States or in central Brazil were caused by a lack of species adaptability, not because of the common belief that the trees were planted as a "monoculture." Likewise, poor growth or even death of an eastern source of douglas fir planted in the western Cascade Mountain range of Washington should fall under the category of *carelessness* but not be blamed on the adverse effects of monoculture. The trees would have grown no better if they had been mixed with other seed sources. *The concern with so-called monoculture* by the tree-improver must be *related to the dangers from reducing the genetic potential of an otherwise well-adapted species that may result from a tree-breeding program and development of improved strains.*

The danger from growing restricted genotypes of forest trees over large acreages hardly needs to be documented. Disasters from using certain *Cryptomeria* or *Populus* clones have been both of the dramatic and insidious types. Kills of large stands have occasionally occurred, whereas decline of stands with resultant slowdown in growth caused by changing environments, or attack by pests, have been quite common. Examples of troubles in sexually propagated stands have been less commonly documented, probably because most are of the insidious type; however, losses are more widespread than is usually recognized. The difficult job is to assess accurately whether the death or decline is simply a result of offsite planting or has been caused by a narrowed genetic base.

How Much Can the Genetic Base Be Reduced?

The most severe restriction of the genetic base occurs with vegetative propagation, a situation in which one or a few superior genotypes is planted over large acreages. It is not unusual to find members of one clone destroyed or damaged by diseases, insects, cold, or drought, while another apparently similar clone remains healthy. Few foresters will argue about the danger of using a single clone, and warnings against such a practice have been sounded for many years.

Reduction of the gene base may also be a problem with sexually propagated trees. The real problem facing the forester working with sexually propagated trees is to determine what constitutes a "too restricted genetic base" that can become "too great a risk." Is it necessary to leave mediocre or "less than the best" parents in a seed orchard to achieve a broader genetic base? For example, if 12 of 20 pine clones are really outstanding, and the rest produce only moderately good progeny, is it safe to remove the inferior 8? Should 2-clone orchards be used for special problem areas? Answers to these questions are not yet available, and decisions must be based on a knowledge of the variation within the species and how this will affect the response of a plantation to adverse environments or pests.

There is another complication in defining the size of the genetic base that is needed. In many species, the response to pests or environmental extremes often changes with tree age, and the response by young trees can be very misleading with respect to the response of the same trees with advancing age. A tree must survive and grow through a complete life cycle to be a successful member of a population. There is danger in inadvertently breeding a strain of trees that are quite adaptable for most of the strain's life span but not for one specific life stage. Wide experience indicates this can be serious in the juvenile period, but we have less information on the potential loss of growth or death at a later period in life. Just because a planting may grow well for the first few years is no guarantee that it will continue to do so.

Based on observation and extensive progeny tests over many years with many species, both in the temperate and tropical areas, it is evident that fears of disaster due to the narrowing of the genetic base and planting forest trees in monocultures have been overemphasized—at least for the first two cycles of breeding. Tree improvers cannot quickly reduce the genetic base for complex, multigene characters that are common to forest trees. But in advanced generations and/or use of vegetative propagation, the possible adverse effect of restricted genotypic plantings will increase.

Although it will vary by species, we feel that first-generation production seed orchards of *Pinus taeda* can be reduced to as low as six cross-pollinating clones with only minimal danger from monocultural problems. This differs dramatically from the commonly expressed idea that a dangerous monoculture automatically results from plantations using a single species.

The job of the tree breeder is to keep the adaptive base broad enough so that catastrophic damage will not occur. This requires planning several generations ahead and, in certain circumstances, sacrificing some immediate gain for long-term adaptability. Production seed orchards should be looked at essentially as being "dead end," that is, the genetic base may be reduced so that it is still safe for field planting but does not retain enough variability for future generations of an ongoing breeding program. It is essential that production orchards be paired with the developmental orchards or research clone banks in which many genotypes are held and crossed to serve as a broad base from which future production orchards will be drawn. Development and maintenance of proper research clone banks require a lot of money, effort and real skill.

Great care needs to be taken in invoking the horrors of a too-restricted genetic base, but it can be a horror if ignored; the tree improver must always be cognizant of the potential gains and dangers from restricting the genetic base.

LITERATURE CITED

Amerson, H. V., McKeand, S. E., and Mott, R. L. 1981. "Tissue Culture and Greenhouse Practices for the Production of Loblolly Pine Plantlets." 16th South. For. Tree Impr. Conf., Blacksburg, Va., pp. 168–175.

Anderson, W. C. 1975. "Some Economics of Monoculture." Pine Monoculture in the South. Workshop of the South. Forest Economics Workers, Biloxi, Miss.

Anonymous. 1977. Proc. Symp. *Vegetative propagation of forest trees— physiology and practice.* The Inst. for For. Impr. and the Dept. of For. Gen., College of Forestry, the Swedish University of Agricultural Sciences, Uppsala, Sweden.

Anonymous. 1979. *Micropropagation d'Arbres Forestiers. AFOCEL,* No. 12, Assoc. Foret-Cellulose, Paris, France.

Baker, F. S. 1934. *Theory and Practice of Silviculture.* McGraw-Hill, New York.

Barnes, R. D. 1969. A method of air-layering for overcoming graft incompatibility problems in pine breeding programs. *Rhodesia Sci. News* 3(4):102–107.

Beineke, W. F., and Todhunter, M. N. 1980. "Grafting Black Walnut," For. and Nat. Res. FNR-105, Purdue University, West Lafayette, Ind.

Birot, Y., and Nepveu, G. 1979. Clonal variability and ortet–ramet correlations in a population of spruce. *Sil. Gen.* 28(2–3):34–47.

Bonga, J. M. 1980. Plant propagation through tissue culture, emphasizing woody species. In *Plant Cell Cultures: Results and Pespectives*, pp. 253–264. Elsevier /North-Holland, Amsterdam.

Bouvarel, P. 1960. Les vieux pino laricio greffes de la forêt de Fontainebleau [The old Corsican pine grafts in the forest of Fontainebleau]. *Sil. Gen.* 9(2):43–44.

Brix, H., and Barker, H. 1971. Trials in rooting Douglas fir cuttings by a paired cutting technique. *Can. Jour. For. Res.* 1(2):120–125.

Brix, H., and van Driessche, R. 1977. Use of rooted cuttings in reforestation. Brit. Col. For. Ser. and Can. For. Ser. Joint Report.

Brown, C. L., and McAlpine, R. G. 1964. "Propagation of Sweetgum from Root Cuttings." Ga. For. Res. Paper 24, Ga. For. Res. Coun., Macon, Ga.

Burgess, I. P. 1973. Vegetative propagation of *Eucalyptus grandis. N.Z. Jour. For. Sci.* 4(2):181–184.

Cameron, R. J. 1968. The propagation of *Pinus radiata* by cuttings. *N.Z. Jour. For. Sci.* 13(1):78–89.

Campinhos, E., and Ikemori, Y. K. 1980. "Mass Production of *Eucalyptus* spp. by Rooting Cuttings," IUFRO Symposium: Genetic Improvement and Productivity of Fast-Growing Trees, São Pedro, São Paulo, Brazil.

Carpenter, S. C. 1975. Rooting black walnut cuttings with Ethephon. *Tree Planters' Notes* **25**(2):31–34.

Copes, D. L. 1969. "External Detection of Incompatible Douglas-fir grafts." Proc. Inter. Plant Propagators Soc. Annual Meeting, pp. 97–102.

Copes, D. L. 1970. Double grafts in Douglas fir. *For. Sci.* **16**(2):249.

Copes, D. L. 1977. Comparative leader growth of Douglas fir grafts, cuttings and seedlings. *Tree Planters' Notes* **27**(3):13–16.

Copes, D. L. 1981. "Selection and propagation of highly Graft-Compatible Douglas fir rootstocks—A Case History." U.S. Forest Service Research Note PNW-376. Pacific Northwest Forest and Range Experiment Station, Portland, Ore.

Corriveau, A. G. 1974. "The Clonal Performance of Loblolly and Virginia Pines: A Reflection of Their Breeding Value." Ph.D. thesis, North Carolina State University, Raleigh.

Corte, R. 1968. Note sur l'incompatibilité de greffe chez les coniferes [Note on graft incompatibility in conifers]. *Sil. Gen.* **17**(4):121–130.

de Muckadell, M. S. 1954. Juvenile stages in woody plants. *Phys. Plant.* **7**:782–796.

Destremau, D. X., Marien, J. N., and Boulay, M. 1980. "Sélection et multiplication végétative d'hybride d'eucalyptus résistant au froid." IUFRO Symposium: Genetic Improvement and Productivity of Fast-Growing Trees, São Pedro, São Paulo, Brazil.

Dimpflmeier, R. 1954. Propfungen an Waldbäumen durch Prof. Dr. Hendrick Mayr vor 55 bis 60 jahren [Grafting of forest trees by Dr. Mayr 55 to 60 years ago]. *Z. Forstgen. Forstpflanz.* **3**:122–125.

Dorman, K. W. 1976. *The Genetics of Breeding Southern Pines.* USDA, U.S. Forest Service, Agricultural Handbook No. 471, Washington, D.C.

Dormling, I. C., Ehrenberg, D., and Lundgren, D. 1976. "Vegetative Propagation and Tissue Culture [Vegetativ Förökning och Vävnadskultur]." 16th IUFRO For. Cong., Oslo, Norway, No. 22.

Ducci, F., and Locci, A. 1978. Prove de radicazione di tales provenienti da piante mature di *Pseudotsuga menziesii* [Research on rooting scions from old trees of *Pseudotsuga menziesii*]. *Ann. Ist. Sper. Silvicol.* **IX**:37–70.

Duffield, J. W., and Wheat, J. G. 1964. Graft failure in douglas fir. *Jour. For.* **62**(3):185–186.

Durzan, D. J., and Campbell, R. A. 1974. Prospects for the mass production of improved stock of forest trees by cell and tissue culture. *Can. Jour. For. Res.* **42**(2):151–174.

Enkova, E. I., and Lylov, G. I. 1960. The use of grafting in the creation of oak seed orchards. *Lesn. Hoz.* **12**(6):34–36.

Farmer, R. E. 1981. Variation in seed yield of white oak. *For. Sci.* **27**(2):377–380.

Farmer, R. E., and Besemann, K. D. 1975. Rooting cuttings from physiologically mature black cherry. *Sil. Gen.* **23**(4):104–105.

Feret, P. P. 1975. "Biological Significance and Problems of Pine Monoculture." Pine Monoculture in the South. Workshop of the Southern Forest Economics Workers, Biloxi, Miss.

Fielding, J. M. 1963. "The Possibility of Using Cuttings for the Establishment of Commercial Plantations of Monterey Pine." 1st World Consul. on For. Gen. and Tree Impr., Stockholm, Sweden.

Fielding, J. M. 1970. Trees grown from cuttings compared with trees grown from seed (*P. radiata*). *Sil. Gen.* **19**(2–3):54–63.

Foster, G. S., Bridgwater, F. E., and McKeand, S. E. 1981. "Mass Vegetative Propagation of Loblolly Pine—A Reevaluation of Direction." Proc. 16th SFTIC, pp. 311–319.

Fowler, D. P. 1961. Initial studies indicate *Pinus resinosa* little affected by selfing. 9th NE For. Tree Impr. Conf., Syracuse, N.Y., pp. 3–8.

Fowler, D. P. 1964. Effects of inbreeding in red pine, *Pinus resinosa. Sil. Gen.* **13**(6):170–177.

Franclet, A. 1963. "Improvement of *Eucalyptus* Reforestation Areas by Vegetative Multiplication." 1st World Consul. on For. Gen. and Tree Impr., Stockholm, Sweden.

Franclet, A. 1979. "Micropropagation of Forest Trees—Rejuvenation of Mature Trees in Vegetative Propagation." *AFOCEL*, No. 12, pp. 3–18. Assoc. Foret-Cellulose, France.

Franklin, E. C. 1979. Model relating level of genetic variance to stand development of four North American conifers. *Sil. Gen.* **28**(5–6):207–212.

Gabriel, W. J., Marvin, J. W., and Taylor, F. H. 1961. "Rooting Greenwood Cuttings of Sugar Maple—Effect of Clone and Medium." Sta. Pap. No. 144, Northeast Forestry Experimental Station, U.S. Forest Service, Upper Darby, Pa.

Garner, R. J. 1979. *The Grafter's Handbook.* Oxford University Press, New York.

Girovard, R. M. 1971. "Vegetative Propagation of Pines by Means of Needle Fascicles—a Literature Review." Centre de Recherche Forestière des Laurentides, Quebec Region, Quebec. Info. Report No. Q-X-23, Canadian Forest Service.

Glasgow, L. L. 1975. "Public Attitudes towards Monoculture." Pine Monoculture in the South, Workshop of the Southern Forest Economics Workers, Biloxi, Miss.

Hare, R. C. 1970. "Factors Promoting Rooting of Pine Cuttings." 1st North American For. Biol. Workshop, Michigan State University, East Lansing.

Hartman, H. T., and Kester, D. E. 1983. *Plant Propagation: Principles and Practices*. Second Edition Prentice-Hall, Englewood Cliffs, N.J.

Hartney, V. J. 1980. Vegetative propagation in the eucalypts. *Aust. For. Res.* **10**(3):191–211.

Hatmaker, J. F., and Taft, K. A. 1966. Successful hardwood grafting. *Tree Planters' Notes* **79**:14–18.

Hinesley, L. E., and Blazich, F. A. 1981. Influence of postseverance treatments on the rooting capacity of Fraser fir stem cuttings. *Can. Jour. For. Res.* **11**(2):316–323.

Hirasiro, M. 1918. A proposal to grow seedlings and cuttings of *Cryptomeria* in thorough mixture. In *Abstracts of Japanese Literature in Forest Genetics and Related Fields* (R. Toda, ed.), Vol. I, Part A, p. 163. Noorin Syuppan Co., Ltd., Tokyo, Japan.

Hoekstra, P. E. 1957. Air layering of slash pine. *For. Sci.* **3**(4):344–349.

Hoffman, C. A., and Kummerow, J. 1966. Anatomische Beobachtungen zur Bewurzelung der Kurztriebe von *Pinus radiata* [Anatomic observations on the rooting of short shoots of *Pinus radiata*]. *Sil. Gen.* **15**(2):35–38.

Hoffman, K. 1957. Freilandsommerpfropfungen mit Nadelhölzern [Summer field grafting of conifers]. *Forst Jagd* **7**(5):202–203.

Hong, S. O. 1975. "Vegetative Propagation of Plant Material for Seed Orchards with Special reference to Graft-Incompatibility Problems." For. Comm. Bull. 54. Britain, HMS:38–48.

Hyun, S. K. 1967. "Physiological Differences among Trees with Respect to Rooting." IUFRO Congress (Munich), **III**(22):168–190.

Jang, S. S., Kim, J. H., and Lee, K. J. 1980. "Rooting Characteristics of Cuttings and Effect of Topophysis on the Rooting of *Pinus thunbergii*." Res. Rept. Inst. For. Gen. No. 16, Korea, pp. 41–48.

Jiang, I. B. 1982. "Growth and Form of Seedlings and Juvenile Rooted Cuttings of *Sequoia sempervirens* and *Sequoiadendron giganteum*." M.S. thesis, School of Forestry, University of California, Berkeley.

Kadambi, K., and Dabral, S. 1954. Air layering in forestry practice. *Indian Forester* **80**:721–724.

Kanoo, K. 1919. An outline of forestation by *Cryptomeria* cuttings in Obi, Miyazaki prefecture. In *Abstracts of Japanese Literature in Forest Genetics and Related Fields* (R. Toda, ed.), Vol. I, Part A, p. 113. Noorin Syoppin Co., Ltd., Tokyo, Japan.

Kelly, C. W. 1978. "Intra-clonal Variation in Tissue Cultured Loblolly Pine (*P. taeda*)." M.S. thesis, School of Forestry Research, North Carolina State University, Raleigh.

Kleinschmit, J., and Schmidt, J. 1977. "Experience with *Picea abies* Cuttings in Germany and Problems Connected with Large-Scale Application." Symposium: vegetative Propagation of Forest Trees—Physiology and Practice, Uppsala, Sweden, pp. 65–86.

Lambeth, C. C. 1979. "Interaction of Douglas-Fir Full Sib Families with Field and Phytotron Environments." Ph.D. thesis, North Carolina State University, School of Forest Resources, Raleigh.

Land, S. B. 1977. "Vegetative Propagation of Mature Sycamore." Proc. 14th South. For. Tree Impr. Conf., Gainesville, Fla., pp. 186–193.

Lantz, C. 1973. "Survey of Graft Incompatibility in Loblolly Pine." Proc. 12th South. For. Tree Impr. Conf., pp. 79–85.

Laplace, Y., and Quillet, G. 1980. "Sylviculture intensive pour matériel végétal amélioré [Intensive silviculture for rooted cuttings]." IUFRO Symposium: Genetic Improvement and Productivity of Fast-Growing Trees, São Pedro, São Paulo, Brazil, pp. 1–8.

Leach, G. N. 1979. Growth in soil of plantlets produced by tissue culture. *Tappi* **62**(4):59–61.

Lee, S. K., and Rao, A. N. 1980. Tissue culture of certain tropical trees. In *Plant Cell Cultures: Results and Perspectives*, pp. 305–311. Elsevier/North-Holland, Amsterdam.

Libby, W. J. 1977. "Rooted Cuttings in Production Forests." 14th South. For. Tree Impr. Conf., Gainesville, Fla., pp. 13–27.

Libby, W. J. 1981. "What Is a Safe Number of Clones per Plantation?" Proc. IUFRO Meeting on Genetics of Host–Pest Interaction, Wageningen.

Libby, W. J., and Hood, J. V. 1976. Juvenility in hedged radiata pine. *Acta Hortic.* **56**:91–98.

Libby, W. J., Brown, A. G., and Fielding, J. M. 1972. Effects of hedging radiata pine on production, rooting and early growth of cuttings. *N.Z. Jour. For. Sci.* **2**(2):263–283.

Lyrene, P. M. 1981. Juvenility and production of fast-rooting cuttings from blueberry shoot cultures. *J. Am. Soc. Hort. Sci.* **106**:396–398.

McAlpine, R. G., and Kormanik, P. P. 1972. "Rooting Yellow-Poplar Cuttings from Girdled Trees," USDA. U.S. Forest Service Research Note SE-180.

McKinley, C. R. 1975. "Growth of Loblolly Scion Material on Rootstocks of Known Genetic Origin. 13th South. For. Tree Impr. Conf., Raleigh, N.C., pp. 230–233.

McKeand, S. E. 1981. "Loblolly Pine Tissue Culture: Present and Future Uses in Southern Forestry," Tech. Report No. 64, North Carolina State University, School of Forest Resources.

Mehra-Palta, A., Smeltzer, R. H., and Mott, R. L. 1977. "Hormonal Control of Included Organogensis from Excised Plant Parts of Loblolly Pine (*Pinus taeda* L.)." Tappi For. Bio. Wood Chem. Conf., Chicago, Ill., pp. 15–20.

Mott, R. L. 1981. Trees. In *Cloning of Agricultural Plants via in Vitro Techniques* (B. V. Conger, ed.), pp. 217–254. CRC Press, Boca Raton, Fla.

Mott, R. L., and Amerson, H. V. 1981. "A Tissue Culture Process for the Clonal Production of Loblolly Pine Plantlets," North Carolina Agr. Res. Service Tech. Bull. No. 271.

Naes-Schmidt, K., and Soegaard, B. 1960. Podehojdens indflydelse paa podekvistens vaékstrytme og form [The influence of the grafting height on the development of the scion]. *Soertryk Det forstlige Forsogsvaesen Danmark* **26**:315–324.

Olesen, P. O. 1978. On cyclophysis and topophysis. *Sil. Gen.* **27**(5):173–178.

Ono, M. 1882. Which grow faster, seedlings or cuttings in the case of *Cryptomeria* or hinoki cypress? In *Abstracts of Japanese Literature in Forest Genetics and Related Fields*, (R. Toda, ed), Vol. I, Part A, p. 22. Noorin Syuppin Co., Ltd., Tokyo, Japan.

Pousujja, R. 1981. "Potentials and Gains of a Breeding Program, Using Vegetative Propagation Compared to Conventional Systems." Ph.D. thesis, North Carolina State University, Raleigh.

Rauter, M. 1977. "Collection of Spruce Cuttings for Vegetative Propagation," For. Res. Note No. 11, Ministry of Natural Resources, Ontario, Canada.

Rauter, M. 1979. "Spruce Cutting Propagation in Canada. Breeding Norway Spruce, Norway Spruce Provenances." IUFRO, Div. 2, Bucharest, Rumania.

Rauter, R. M., and Hood, J. V. 1980. "Uses for Rooted Cuttings in Tree Improvement Programs." 18th Can. Tree Imp. Conf., Duncan, British Columbia.

Rediske, J. H. 1976. "Tissue Culture and Forestry." Proc. 4th Amer. For. Biol. Workshop, pp. 165–178.

Roulund, H. 1973. "The Effect of Cyclophysis and Topophysis on the Rooting Ability of Norway Spruce Cuttings." Forest Tree Improvement No. 5, Arboretet, Hørsholm, Akademisk Forlag.

Roulund, H. 1978a. "Stem Form of Cuttings Related to Age and Position of Scions (*Picea abies*)." Forest Tree Improvement No. 13, Arboretet Hørsholm, Akademisk Forlag.

Roulund, H. 1978b. A comparison of seedlings and clonal cuttings of Sitka spruce (*Picea sitchensis*). *Sil. Gen.* **27**(3–4):104–108.

Santamour, F. S. 1965. Rooting of pitch pine stump sprouts. *Tree Planters' Notes*, No. 70, pp. 7–8.

Shelbourne, C. J. A., and Thulin, I. J. 1974. Early results from a clonal selection and testing program with radiata pine. *N.Z. Jour. For. Sci.* **4**(2):387–398.

Slee, M. U., and Spidy, T. 1970. The incidence of graft incompatibility with related stock in *P. caribaea hondurensis*. *Sil. Gen.* **19**(5–6):184–187.

Sommer, H. E., and Brown, C. L. 1979. Application of tissue culture to forest tree improvement. In *Plant Cell and Tissue Culture: Principles and Applica-*

tions W. R. Sharp, P. O. Larson, E. F. Paddock, and V. Raghaven, (eds.), pp. 461–491. Ohio State Univ. Press, Columbus.

Struve, D. K. 1980. "Vegetative Propagation of *Pinus strobus* by Needle Fascicles and Stem Cuttings." Ph.D. thesis, North Carolina State University, Raleigh.

Struve, D. K. 1981. *Conifer Grafting Manual* (Mimeo). Depatment of Horticulture, Ohio State University, Columbus.

Sweet, G. B. 1972. "Effect of maturation on the Growth and Form of Vegetative Propagules." 2nd North American Biol. Workshop, Corvallis, Ore., Society of American Forestry.

Sweet, G. B., and Wells, L. G. 1974. Comparison of the growth of vegetative propagules and seedlings of *Pinus radiata. N.Z. Jour. For. Sci.* **4**(2):399–409.

Talbert, J. T., Wilson, R. A., and Weir, R. J. 1982. "Utility of First Generation Pollen Parents in Young Second Generation Loblolly Pine Seed Orchards." 7th North American For. Biol. Workshop, Lexington, Ky.

Thulin, I. J. 1969. "Breeding of *Pinus radiata* through Seed Improvement and Clonal Afforestation." 2nd World Consul. on For. Tree Breeding, Washington, D.C.

Thulin, I. J., and Faulds, T. 1968. The use of cuttings in breeding and afforestation of *Pinus radiata. N.Z. Jour. For.* **13**(1):66–77.

Toda, R. 1974. Vegetative propagation in relation to Japanese forest tree improvement. *N.Z. Jour. For. Sci.* **4**(2):410–417.

Tucker, R. E. 1975. "Monoculture in Industrial Southern Forestry." Pine Monoculture in the South, Workshop of the Southern Forest Economics Workers, Biloxi, Miss.

van Buijtenen, J. P., Toliver, J., Bower, R., and Wendel, M. 1975. Mass production of loblolly and slash pine cuttings. *Tree Planters' Notes* **25**(2):4, 26.

Wakeley, P. C. 1971. Relation of thirtieth year to earlier dimensions of southern pines. *For. Sci.* **17**(2):200–209.

Webb, C. D. 1961. "Field Grafting Loblolly Pine," Tech. Rept. No. 10, School of Forest Resources, North Carolina State University, Raleigh.

Wheat, J. G. 1964. Rooting of cuttings from mature Douglas fir. *For. Sci.* **10**(3):319–320.

Wright, J. W. 1976. *Introduction to Forest Genetics.* Academic Press, New York.

Yim, K. B. 1962. "Physiological Studies on Rooting of Pitch Pine (*Pinus rigida*) Cuttings," Res. Report No. 2., Inst. For. Gen., Forest Experimental Station, Suwon, Korea.

Zak, B., and McAlpine, R. G. 1957. "Rooting of Shortleaf and Slash Pine Needle Bundles." Research Note No. 112, Southeastern Forest Experimental Station.

Zobel, B. J. 1972. "Genetic Implications of Monoculture." 2nd North Amer. For. Biol. Workshop, Oregon State University, Corvallis.

Zobel, B. J. 1977. Tissue culture (as seen by a plant breeder). *Tappi* **61**(1):13–15.

Zobel, B. J. 1981. "Vegetative Propagation in Forest Management Operations." Proc. 16th South. For. Tree Imp. Comm. Meet., Blacksburg, Va., pp. 149–159.

Zobel, B., Campinhos, E., and Ikemori, Y. 1982. "Selecting and Breeding for Wood Uniformity," Proc. TAPPI Research and Development Meeting, Ashville, N.C., pp. 159–168.

CHAPTER 11

Hybrids in Tree Improvement

There has been controversy in the tree improvement area of study between those researchers who favor the hybrid approach to developing better trees and others who feel that tree improvement efforts could be better expended in other areas. Hybrids have the distinct advantage in that the tree breeder can create something different that may not occur in nature. Hybridiztion can produce new combinations of genes (Piatnitsky, 1960), but Wagner (1969) has emphasized that the parental species of the hybrids establish the poles of diversity, and hybrids only fill in the gap in variation that lies between the two poles. He has stated that nothing "new" is created and that hybridization only counteracts the extremes in characteristics produced by the evolutionary forces of mutation, selection, and genetic drift. By developing hybrids, however, the breeder can bring together desirable characteristics of the parents and "tailor make" trees that are not otherwise available in nature. The potential of hybridization was early recognized by such persons as Stockwell and Righter (1947), Righter (1955), and Duffield and Snyder (1958).

There has been a series of reviews on hybrid forest trees; for example, those for the genus *Pinus* by Critchfield (1975) and Little and Righter (1965), for *Quercus* by Palmer (1948), for *Picea* by Wright (1964), and for *Juniperus* by Hall (1952).

Despite the advantages and potential value of hybrids that are recognized by most tree improvers, they generally have been used only sparingly in operational forest regeneration. The key to success in the future use of most hybrids will be the degree to which vegetative propagation can be used operationally. Obtaining seeds of hybrids is usually difficult and expensive; therefore, their mass production using seed has been successful in only a few instances. One noticeable success has been the pitch × loblolly pine hybrid in Korea (Hyun, 1976). Hybrids have been used most extensively in species relatively easy to reproduce vegetatively, such as the poplars and willows (Stout et al., 1927; Hunziker, 1958; Schreiner, 1965). A number of scientists have urged a broader use of hybrids in tree improvement (Lester, 1973; Schmitt, 1973; Fowler, 1978). Their use is rapidly expanding as better techniques are developed, and hybrid programs, such as the one for *Eucalyptus* in Brazil (Campinhos, 1980), have shown great promise (see Figure 10.6). Although production of hybrids among forest trees is considered by some to be a new activity, hybrid forest trees were recognized a long time ago (Tanaka, 1882).

WHAT IS A HYBRID?

Two of the most popular and often misused terms in forestry are *hybrid* and *hybrid vigor*. Definitions are not exact, and the word *hybrid* is used to describe several different types of crosses. To most foresters a *hybrid* is a cross between two species. To some tree improvers, the word *hybrid* also includes crosses between different geographic races within a species (Scamoni, 1950; Nilsson, 1963; Wright, 1964; Howcroft, 1974). Many botanists and agricultural crop breeders carry the definition of hybrid further to include crosses between any two unlike genotypes;

this is similar to the definition by Snyder (1972) who wrote that "a hybrid is the offspring of genetically different parents." If this definition is used, nearly every cross made between forest trees would produce a hybrid. In this book the word *hybrid* generally refers to crosses between species. However, it is not incorrect to apply "hybrid" to intraspecific crosses, such as between races, or even to crosses among two different genotypes in the same population.

Hybrids are sometimes given a "mystical value" with respect to growth and quality by people who do not understand the genetics of hybrid formation. For example, hybrids are frequently made and tested using the most convenient trees from the parental species; sometimes the major criterion for use as a parent is the ease with which a tree can be climbed. Hybrids inherit the characteristics of their parents. Therefore, if hybridization is to successfully produce improved growth, form, or pest resistance, the individual parents must be carefully chosen to provide the desired characteristics. Some persons believe that only the best and desirable characteristics of the parental species will be evident in the hybrid, but the worst characteristics of each parent can also appear in the hybrid progeny. Differences in hybrids have been reported to depend upon the particular parental combinations that happened to be used (Eifler, 1960). Ways to assure good parental combinations have been discussed by Conkle (1969) and Chaperon (1979). Often, the first crosses between species have been made on garden, park, or arboretum trees either by accidental open pollination or by controlled pollination. In either instance, the number and variability of the parental trees used in the crosses are small.

Usually, the hybrid has characteristics intermediate between its parents (Figure 11.1). Occasionally, however, the hybrid will strongly carry a desired characteristic of one parent and not show intermediacy. For example, the Coulter × Jeffrey pine hybrid (*Pinus coulteri* × *Pinus jeffreyi*) carries resistance to the pine reproduction weevil that the Coulter pine parent possesses (Miller,

FIGURE 11.1

Shown are mature cones of *P. coulteri, P. jeffreyi,* and the natural hybrids between the two. Note the intermediacy in size and conformation of cone scales of the mature cones, with Jeffrey pine at the left, Coulter pine at the right, and the hybrid in the center.

FIGURE 11.2

Sometimes hybrids are outstanding. Shown is a natural hybrid on the lands of Aracruz Florestal in Brazil with outstanding growth and form. This tree, 7 years old, can be used in the vegetative propagation program.

1950), although the hybrid is generally intermediate otherwise (Zobel, 1951a). Similarly, the loblolly × shortleaf pine hybrid is nearly resistant to fusiform rust, which is similar to the shortleaf parent, even though the cross is otherwise intermediate.

Sometimes hybrid crosses result in rare individuals that have characteristics outside the range of the parental species. Such individuals can be quite remarkable and have great potential value, especially when vegetative propagation can exploit the unusual and desirable genotypes (Figure 11.2).

HYBRID VIGOR (HETEROSIS)

The appeal of hybrids in forestry includes the potential to bring together unlike genotypes to create genetic combinations different from the parental species that may have some special value for growth or desired product. There is also the possibility of the presence of what is called *hybrid vigor*. Some persons assume all hybrids will display *hybrid vigor*, and this phenomenon is often used as the major reason why hybrids should be included in tree improvement programs. However, hybrid vigor may or may not occur in crosses between unlike forest tree genotypes.

What is *hybrid vigor*, or as it is often called, *heterosis*? Like many terms, there is no single definition that is acceptable to all. Many people define it as a growth superiority in which the hybrid exceeds that of both parents (Györfey, 1960), but others define hybrid vigor as being an increase over the mean of the parents (Snyder, 1972). Heterosis usually relates to traits of size (most often height), but it also includes yield or general thriftiness. Some persons apply the term *hybrid vigor* when the hybrid excels in *any* measurable characteristic such as wood qualities, cold hardiness or the ability to withstand pests, and not necessarily for size alone. A commonly observed result of hybridization is precocity in flowering of the cross relative to the parents; this also can be called a kind of hybrid vigor (Venkatesh and Sharma, 1976). In this book, the term *hybrid vigor* will normally refer to a size superiority over both parents, but it is essential to understand that the term may be properly used for characteristics other than size. It is beyond the scope of this book to cover the causes or probable causes of heterosis. This is a very complex subject, but a good discussion can be found in the literature, for example, the study by Jinks and Jones (1958).

In conifers, hybrid vigor is not evident from the seeds. In his article in 1945, Buchholz made it clear that embryos of pine hybrids can grow faster, but the seeds cannot be larger than those of the female parent. He states that "the mega-gametophyte is formed before fertilization of the egg. The seed of a pine is fully grown and has reached its ultimate size at time of fertilization—there can be no enlarging effect on seed size due to the activity of the contained embryo."

There is considerable argument about the extent and importance of hybrid vigor in forestry. The general feeling has been that hybrid vigor usually is not present in meaningful amounts in forest trees (van Buijtenen, 1969). Fowler (1978) stated that "heterosis or hybrid vigor in species and provenance hybrids is the exception rather than the rule." He explained this as partially being the result of crossing highly adapted parental species that may produce hybrids that are less well adapted to the habitats where the parents are growing. For pines, Little and Somes (1951) reported little exceptional vigor in hybrids involving pitch pine, which is similar to hybrids in many other species. However, there have been a number of reports of hybrid vigor in the poplars and willows (Stout et al., 1927; Pauley, 1956; Schreiner, 1965; Chiba, 1968). More recently, significant superior vigor in hybrids has been reported in *Eucalyptus* (Chaperon, 1976; Venkatesh and Vakshasya, 1977; Camphinos, 1980), in larch (Keiding, 1962; Miller and Thulin, 1967; Wang, 1971), in spruce (Hoffmann and Kleinschmitt, 1979), and in larch intraspecific hybrids (Nilsson, 1963). With time, more complete crossing and study, significant and useful hybrid vigor in forest trees may well be found, but as of now, it must be considered only moderately important except in a few genera or for a few specific crosses. Hybrids have historically had more value as a source of new combinations of genes than for extra vigor, but this could well change as more becomes known.

One must be very careful in the assessment of hybrid vigor because time and location are most important. Hybrids can express vigor at one stage in their development while showing none at other stages, or under certain environmental conditions and not under others (Johnson, 1955). This sometimes becomes very complex such as is the case for *Pinus sondereggeri,* the cross between loblolly and longleaf pine (Chapman, 1922; Namkoong, 1963). In the nursery bed the hybrid is often considerably taller and huskier than either parent. After outplanting, the loblolly usually outgrows the hybrid and longleaf pine until perhaps the fortieth year when the longleaf may become the largest. With such a pattern can one then say that the hybrid *P. sondereggeri* has hybrid vigor?

Another problem in the definition of hybrid vigor occurs when the hybrid is planted in a habitat in which only it and one parent can survive and grow but in which the other parent may not grow well or may even die. Such a situation occurs with the pitch × loblolly pine hybrid (Little and Trew, 1976). When grown on cold, poor sites the hybrid pine does well and often surpasses the pitch pine, whereas the loblolly parent is killed or is severely stunted by cold. The question has been raised whether the pitch × loblolly pine does have hybrid vigor under this condition. According to the definition followed in this book it does, because it is outperforming both parents.

NATURAL HYBRIDS

Many hybrid combinations occur naturally. Quite frequently these are over-looked and are considered to be variants of one of the parental species. Such hybrids can be very difficult to determine, especially when the parents are quite similar. Sometimes hybrids are easy to distinguish because of their intermediacy between two greatly differing species; a prime example of this is the *P. coulteri* × *P. jeffreyi* hybrid (Zobel, 1951a). Hybrids quite often appear when species are grown as exotics (Marien and Thibout, 1978), and especially when they are grown in arboreta where species that never occur together in nature are brought together so that gene exchange is possible. A classic example of this type of situation involves *Eucalyptus* in the arboretum at Rio Claro, Brazil, from which numerous hybrids have arisen (Brune and Zobel, 1981).

The more intensively one studies natural populations of forest trees, the more hybrids are found. But care needs to be taken; just because a tree differs considerably from its "type" species does not mean that it is necessarily a hybrid. The most difficult job in working with hybrids is to delineate between them and normal variants within a species (Figure 11.3). Often the F_1 hybrids (first cross or first filial generation as it is called) are not too difficult to distinguish, but as one encounters F_2 (crosses between F_1 hybrids), F_3, and other hybrid combinations, the job of distinguishing between hybrids and variants within a species becomes most difficult. To add to the confusion, often the hybrids cross back to the parental species (called *backcrosses* and designated as B_1, B_2, etc.), creating a group of intergrading individuals, and it becomes nearly impossible to determine accurately whether the individuals are hybrids, backcrosses, or species variants.

When two species that are growing together hybridize freely and backcross, a population known as a *hybrid swarm* is produced in which many combinations are

FIGURE 11.3

Natural hybrids are rather common in some species, and the more intensively species are studied, the more hybrids that are found. They must be considered in tree improvement programs; some are excellent. The *P. taeda* × *P. serotina* hybrid graft illustrated was selected as *P. taeda* and put into the seed orchard. It flowered out of synchronization with the other clones and had to be removed, even though its progeny were excellent.

formed with few apparent barriers being evident. Hybrid swarms have been studied under natural conditions by several investigators. For example, Namkoong (1963) did this for *Pinus sondereggeri*, (the loblolly × longleaf cross), Hardin (1957) for *Aesculus*, Kirkpatrick (1971) and Clifford (1954) for *Eucalyptus*, and Hall (1952) for *Juniperus*. The most intensive studies of natural hybridization have been for the oaks (*Quercus*). A few of the published articles are the following: McMinn et al. (1949), Muller (1952), Chisman (1955), Tucker (1959),

Bray (1960), and Burk (1962). Hybridization in the oaks has occurred so frequently that some persons feel that for some groups of related species there is an almost total intergradation and that species categorization means little. Intensive studies have also been made within the genus *Eucalpytus* by such persons as Pryor (1951, 1976) and Clifford (1954). Certain species in this genus hybridize freely, and the number of species varies greatly, depending upon whether one is a taxonomic "splitter" or "lumper." The pines in Mexico and Central America are a major group of species in which hybridization in natural stands is of key importance (Figure 11.4). It is not uncommon to find one species at the bottom of a mountain that gradually changes and intergrades as one goes higher up the mountain until it is called another species. There are no distinct breaks in variation between the two species. Some workers prefer to break such groups of interrelated and intergrading species into *complexes* that contain one to several species that frequently hybridize.

Introgression (also called *introgressive hybridization*) may be defined as the limited spread of genetic material from one species into another species as the result of hybridization that is followed by repeated backcrossing of the hybrid and its progeny to one or both of the parental species. Thus, the result of introgression

FIGURE 11.4

An area known for extensive hybridization and introgression is in Mexico and Central America. The area shown (*a*) in Nuevo León, Mexico, has very diverse environments, with trees growing on isolated mountain peaks. Pine stands, such as the one shown in Guatemala (*b,*) often are essentially hybrid swarms with all degrees of speciation, hybridization, and backcrossing.

is that in some cases of genes one species are eventually incorporated into the gene pool of another. Through introgression, an occasional individual or population of a desirable species may be found that possesses certain good characteristics from a generally less desirable species. For example, the rust resistance of the western loblolly pine populations could well be the result of introgression with shortleaf pine A similar result is obtained from an intensive breeding program in which the desired character(s) of one species are incorporated into another by hybridization and subsequent backcrossing. The general feeling among tree improvers is that introgression has a major effect on population structure and thus upon selection and breeding efforts. The subject of *introgression* has been well covered by Anderson (1968) in his book entitled *Introgressive Hybridization* and by Mettler and Gregg (1969).

Regardless of the long-term effects on evolution and speciation, hybridization

(b)

FIGURE 11.4 (*Continued*)

has created an excellent pool of genetic variability that the plant breeder needs to recognize, assess, and use. Too often this source of variability in forest trees has been overlooked.

Where Natural Hybrids Occur

Hybrids are most often found in geographically disturbed areas, such as mountainous or volcanic regions (see Figure 11.4). Such disturbed regions are particularly prevalent in Central America, where hybridization in both conifers and hardwoods is common. Large changes in the environment occur within short distances, thus affording the opportunity for species mixing and for hybrid establishment. Similar results occur when the habitat has been disturbed through the activities of people.

More hybrids are produced in nature than are ever found growing in natural settings because their seeds do not germinate and grow to maturity. Usually the species found growing on an area are well adapted to their environments so any hybrids produced have little or no selective advantage and, in many instances, would be at a disadvantage for survival, growth, and reproduction. But when the habitat is very diverse and has been changed naturally or altered by human beings, an "environmental niche" is often available that is ideally suited to the hybrid and in which it will flourish. This has been referred to as "hybridization of the habitat" by Anderson (1948) and usually is a prerequisite for the large-scale establishment of hybrids or their derivatives. All kinds of environmental upsets, such as frequent or very severe burning that affects the soil, or logging with heavy equipment that drastically disturbs the soil, will result in a disturbed and potential hybrid habitat. Slash-and-burn farming, which is associated with shifting agriculture in the tropics, has undoubtedly produced many "hybrid habitats." Zobel has observed a small valley in the Sierra Nevada of California that formerly contained predominantly ponderosa pine (*P. ponderosa*) but that now has a considerable proportion of ponderosa × Jeffrey pine hybrids following logging and several severe fires. Sometimes the impact on the habitat is so extensive that it enables the establishment of a hybrid swarm (Chapman, 1922; Namkoong, 1963). Whole counties in Louisiana and North Carolina contain hybrid swarms of *P. sondereggeri* (*P. palustris* × *P. taeda*) following severe and continued disturbance of the sites through farming, grazing, and burning. Some trees have developed into the so-called "Erambert's hybrid" that is made up of several species, although Schmidtling and Scarborough (1968) feel it is primarily longleaf × loblolly.

Serious mistakes have been made in assessing whether two species can hybridize in natural populations. Frequently, hybrids may be formed, but they do not have a chance to become established because the environment is more suitable to one or both parental species. The only real way to tell if hybrids are being produced is to collect seeds from trees in the suspected area of hybridization and then grow them under controlled conditions. Such an assessment must be done on a large scale because the number of hybrids formed can vary from a few (less than 1%) to a relatively large percentage. For example, in areas where loblolly pine and longleaf pine grow sympatrically (i.e., occur together), individual *P. palustris* trees can be found whose seed contain no longleaf × loblolly (*P. sondereggeri*) hybrids, whereas other trees may produce well over 75% hybrids. Based on seedling production in large nurseries, where the parental species are sympatric, loblolly × longleaf hybrid seed are formed at a 5 to 15% rate even though usually only a very few hybrids grow to maturity in naturally regenerated forests.

If one hopes to find or study hybrid trees growing in wild populations, sampling should be done where there are a fair number of both parental species growing intermixed in the stand. If one species is predominant, seed collections should be made from the rare species. One should also look for forest sites that have been severely disturbed and that will enable the hybrids to become established. In undisturbed stands, one often observes what Pryor (1978) found for *Eucalyptus*, where the occurrence of interspecific hybrids was common. Hybrids were growing at stand junctions, but their progenies were limited, and they did not invade the parental stands. It is amazing how often natural hybrids will be found if searched for in disturbed or intermediate habitats. A basic necessity for working with natural hybrids is to have a complete knowledge of the characteristics of both parental species so that a hybrid can be recognized.

What are the barriers to the formation of natural hybrids when the parental species are not separated geographically and can produce hybrids artificially? The most common barrier is the time of flowering. Many otherwise compatible crosses do not occur in nature because the flowering times of the parents do not overlap. A prime example of this is *P. taeda* × *P. echinata*. Receptivity of conelets and pollen release are often 3 weeks apart between these two species, and normally receptivity of the female strobili of loblolly pine is completed long before shortleaf pollen is shed. Yet, despite this apparent isolating time barrier, hybrids do occur (Zobel, 1953). In certain years the receptive period and time of pollen shed of the two species may overlap somewhat as the result of unusual weather, which is usually a cold, wet spring. Sometimes insects affect the male strobili of shortleaf trees, causing early pollen maturity with pollen shed 2 or 3 weeks before what is normal. Similarly, insect attack may kill the resting bud of loblolly and cause a new bud to arise in the spring from which the female strobilus will emerge. These late females are sometimes receptive in synchronization with shortleaf pollen shed. Similar happenings also occur in other species as have been reported for aspen (*Populus*) by Einspahr and Joranson (1960).

Determination of Natural Hybrids

Before one can determine whether a tree is a hybrid, it is essential to know the limits of the variation of the parental species. This sounds easy but is often most difficult and requires an extensive assessment of "pure" stands of the species. It is the word *pure* where the problem lies. How is a pure species to be defined? Even in cases not affected by crossing with other species, most species have geographic races that have developed through differential selection under different environments. It is not possible to travel to dissimilar environments to study a species and assume that the same characteristics will be found when the species is growing in a different geographic area. Therefore, the limits of variation of the parental species must be determined in the area near to where the hybrid grows.

Even more confusing and hard to handle in a determination of whether a species is pure or not, is to ascertain whether it is carrying genes from another species as a result of past hybridization and introgression. For example, *P. taeda* hybridizes with *P. echinata, P. palustris, P. elliottii, P. rigida, P. serotina,* and perhaps with other species. Because of this, it is not possible to define what pure *P. taeda* really is; everywhere throughout its range, this "promiscuous" species hybridizes and carries genes in it from one to several other species. Under conditions in which parental species are not pure, hybrid determination becomes most difficult, and very sophisticated methodology is required (Smouse and Saylor, 1973; Saylor and Kang, 1973). On the other hand, the study of the *P. coulteri* × *P. jeffreyi* hybrid was relatively simple in that it was not difficult to find "pure" parental populations, and the hybrid was intermediate between two quite different species (Zobel, 1951a).

Many methods have been developed to assess natural hybrids, but details are beyond the scope of this volume. They vary from the simple system of Anderson's "hybrid index" (Anderson, 1953) to very complex methods of canonical analyses (Smouse and Saylor, 1973) and other sophisticated statistical methods, such as the one developed by Namkoong (1963). The more different and discrete the species are, the easier it is to determine hybridity and the more suitable are the simple methods, such as Anderson's "hybrid index." When species are similar and the characteristics of the hybrids strongly overlap within the range of variation of the pure species, the more complex statistical methodology is required.

In the past, most hybrid determinations in forest trees have been made using morphological and anatomical characteristics, usually based on foliage and seeds or other reproductive structures (Mergen, 1959; Schütt and Hattemer, 1959). Occasionally wood differences have been employed, but usually wood anatomy has not been too useful. In recent years, certain chemical and enzymatic methods have been developed to determine hybridity more precisely. In the pines, characteristics of the resins have been used effectively for many years (Zobel, 1951b; Mirov, 1967). Some of the other methods reported are vapor-phase chromatography to study resins (Bannister et al., 1959), paper chromatography, and flavonoid compounds (Riemenschneider and Mohn, 1975). Feret (1972) used peroxidase isoenzymes to study elm (*Ulmus*) hybrids, and isoenzymes have become a favored tool for taxonomic studies. Electrophoretic methods have been greatly improved, enabling better studies of hybridization, and positive results have been reported by many researchers. In *Fraxinus*, Santamour (1981) used flavonoids and counarius to identify hybrids. Serological techniques were used by Moritz (1957).

Recent work that combines anatomical, morphological, and chemical methods as well as known crossability patterns is making the determination of hybridity in natural stands a much more exact undertaking. The best tool of all is to have some certified artificial hybrids between the species involved that can be used as a standard against which the putative hybrids can be judged. Unfortunately, such known and certified hybrids are often not available, but when they are, the job of determining natural hybridity becomes much simplified.

The Value of Natural Hybrids to the Tree Improver

Other than in a few genera like *Eucalyptus*, *Salix*, and *Populus*, in which vegetative propagation can be readily used, natural hybrids have not been widely planted on an operational scale. Hybrids do have great value for the tree improver as a source of variation for breeding programs. As vegetative methods are improved, hybrids will become more widely used..

It is extremely important for the forest geneticist and tree improver to recognize that hybridization frequently does occur naturally in forest trees and that although the crossing patterns may pose real problems from a taxonomic standpoint, the results of hybridization are invaluable for tree-breeding programs because a large number of greatly differing genotypes have been created from which the breeder can select. Skill and intensive testing are required to sort out the most useful gene combinations available from natural hybridization.

The importance of hybridization in the processes of speciation and evolution has been widely discussed and argued by such persons as Anderson (1953) and Stebbins (1959, 1969). Smouse and Saylor (1973) made an in-depth analysis of the pitch pine (*P. rigida*), pond pine (*P. serotina*), and loblolly pine (*P. taeda*) hybridization patterns and the effect of these on speciation.

It would serve no purpose to make a long list of natural hybrids that occur, but they are found in many genera in addition to those listed previously. For example, yellow birch (*Betula alleghaniensis*) and paper birch (*Betula papyrifera*) hybridize (Barnes et al., 1974), as do the aspens (Pauley, 1956), and a number of other hardwoods and conifers.

ARTIFICIAL HYBRIDS

Natural hybrids can be used in applied tree improvement programs, but usually this is rather a hit-or-miss approach. If hybrids are to be considered seriously in a breeding program, artificial hybrids should be developed from suitably good individuals from the parental species. What is of great importance is that artificial hybridization can bring together species and individuals that would not otherwise have an opportunity to cross. It enables the production of genotypes that incorporate desired characteristics from two species into a single individual or group of individuals (Diller and Clapper, 1969).

Making Artificial Hybrids

Making artificial hybrids is often not easy. Frequently, the techniques used must be developed, and even after they have been perfected, a common result is that low yields of viable seeds are obtained. When crosses are made between distantly related species, viable seeds are sometimes never obtained, or if a few sound seed are obtained, they will often not germinate satisfactorily. Occasionally, use of special methods, such as excision of the embryo or removing the seed coat, makes the production of viable plants possible. This method was used to produce viable *P. lambertiana* × *P. armandii* seedlings (Stone and Duffield, 1950). In addition to having erratic germination, seeds of artificial hybrids are often small and sometimes lack vigor when they germinate.

The difficulties of obtaining seed of hybrids vary greatly by species and genus. For example, in genera such as *Quercus* (oaks), only one seed per pollinated flower can be obtained, whereas for other species (*Pinus, Platanus, Populus*), dozens or hundreds of viable seeds may be obtained from the pollination bag. The most desirable and economical method of propagation to mass produce hybrid plants will depend on the ease with which viable hybrid seed may be obtained or the ease of vegetative propagation. For example, in the oaks, vegetative propagation of some form would be necessary to mass-produce hybrid plants economically. In those species where large quantities of viable hybrid seed may be obtained more easily, mass pollination techniques may eventually be developed. For hybrids with special high values, standard control hand pollination may be possible to obtain plantable amounts of seed.

The major obstacle to the widespread use of hybrids in operational forestry has been the inability to mass-produce them easily and economically. In some instances, this has been partially overcome where labor is cheap, such as in Korea where control pollinations to produce the hybrid *P. taeda* × *P. rigida* have been done by hand on a mass scale (Hyun, 1976). Attempts have been made to produce hybrids using supplemental mass-pollination techniques for a number of species (Hadders, 1977; Bridgwater and Trew, 1981). This method distributes large quantities of pollen from one of the desired species to the receptive female structures of the other. Many different kinds of hand- and mass-pollination methods have been tried, sometimes sucessfully, sometimes with marginal success. For example, in tests employing mass pollination to produce the longleaf- × loblolly pine hybrid, success by individual mother tree has ranged from essentially no hybrids being produced to as many as 85% hybrids. Attempts to isolate large groups of flowers in tentlike structures and then to mass pollinate within this isolation zone have not been successful.

A method used successfully to produce hybrid *Eucalyptus* in Florida and in Brazil consists of establishing an orchard of one species within which individuals of another species are planted and from which hybrid seed will be collected (Campinhos, 1980; Dvorak, 1981). The individuals from which seeds are collected must be relatively or completely self-incompatible. The method also works well for dioecious species where interplanting species that flower in synchrony and are compatible, has resulted in an easy production of large amounts of hybrid seed.

As mentioned several times before, problems in mass-producing hybrids can be overcome for a number of genera through the use of vegetative propagation. After hybrids are produced, tested, and selected, the most outstanding trees can then be propagated vegetatively on a large scale for production planting. This method has been used for decades for poplar and willow, and more recently it has been used operationally for *Eucalyptus* hybrids (Chaperon, 1976; Campinhos, 1980). Many millions of hybrid trees are now being planted using vegetative propagation.

Often, hybrids are difficult to produce because methods of crossing have not been sufficiently well developed. Reproductive structures of forest trees are very fragile, and careless methods can result in failure or very low seed yields. Before successful hybridization can be accomplished, intensive study of the reproductive system of the species involved is required. An excellent example is the very exhaustsive compilation of information about pines in the *Pollen Management Handbook*.[1] When such summaries of the methods of pollination and fertilization are obtained for different species, much better crossing success will result.

What Hybrids Can Be Made?   Everyone knows how difficult it is to produce hybrids from distantly related species. However, closely related species sometimes will not cross easily, even though this has been tried many times (Eklund, 1943). There are many reasons given for incompatibilities resulting in failure of seed production (McWilliam, 1958; Saylor and Smith, 1966; Krugman, 1970). These reasons sometimes are morphological or anatomical (Kriebel, 1972); this is especially true for some of the tropical hardwoods with very specialized reproductive systems. Barriers may also be chemical or physiological. Unexplained results are common; for example, two individuals from two different species may hybridize freely, whereas two other individuals from the same species may not produce viable seed. It is common to have problems with reciprocal crosses in which the hybrid can be made easily in one direction but with difficulty or not at all for the reciprocal cross with the same parents.

Hybrids quite often can be made in one environment and not in another. The literature contains numerous statements about the lack of crossability between two species based on a few trials, sometimes with only one tree of a species from an arboretum or greenhouse with an environment that is different from that where either parent grows naturally. Conclusions based on such restricted crossing are not necessarily valid, because if a greater number of trees of the two species are tried or if the crosses are made under more normal environments, efforts are often then successful. Lists of crosses that cannot be made must be carefully assessed with knowledge of how many parents have been used or where the crosses have been tried or the conditions under which germination of the seed obtained was made. It is not unusual to obtain sound seed that appear normal but still will not germinate.

It would require a book to list all the successes and failures in producing hybrids, and such a list would soon be out of date. More combinations are constantly being found (Fowler, 1978), and there is the need to intensify and in particular to systematize the crossing data. This is emphasized strongly by Nikles

---

[1]U.S. Department of Agriculture Handbook No. 587. (C. Franklin, ed.). U.S. Forest Service, Washington, D.C.

FIGURE 11.5

A most useful hybrid is the one used in Queensland, Australia, to plant on excessively wet sites. The cross between *P. caribea* v. *hondurensis* and *P. elliottii* is much superior to either parent tree on the wet problem sites. The hybrid was produced by the Queensland Forest Service.

(1981) who has successfully produced a *P. caribaea* × *P. elliottii* hybrid that has grown well on wet sites that were not suited to either parent (Figure 11.5). The tendency is to emphasize hybridization in temperate species, but much good work has been done among some tropical species such as mahogany (*Swietenia*) (Whitmore and Hinojosa, 1977).

Wide Crosses   Usually, the more distant the relationship, the more unlikely it is that a cross will produce viable seeds. However, sometimes unusual and seemingly unlikely wide crosses can be made. In the pines, no known successful crosses have been made between the two major categories of the *Haploxylon* (soft) and the *Diploxylon* (hard) pines. The genus *Pinus* is divided into different subdivisions by several authors (e.g., Martínez, 1948; Duffield, 1952; Little and Critchfield, 1969), and usually crosses cannot be made between subdivisions. There are exceptions, however, such as the *P. elliottii* × *P. clausa* crosses made by Saylor and Koenig (1967). The more trials that are made, the more intergroup crosses there are that seem to be possible. Other genera are also divided into major categories, as in *Eucalyptus*. Extensive crossing trials have been made, and some wide crosses have been successful, but generally successful crossing is restricted to groupings within genera. Crosses between species that are widely separated geographically (Santamour, 1972) are often successful; it is surprising how often species that have been separated for thousands of years by hundreds or thousands of kilometers cross with little difficulty.

Sometimes closely related species will not hybridize. In *Pinus*, the western hard pines will not cross with the eastern hard pines. For example, *P. taeda* and *P. ponderosa* have not been successfully crossed. *Pinus resinosa* (red or Norway pine) is noted for its inability to cross with other closely related members of the genus. In an attempt to see just what might be possible, a whole group of closely and more distantly related species of pine was crossed using *P. taeda* as the female parent. Seed and seedlings were obtained from some rather unexpected species combinations, and the putative hybrids are being tested in the field (Williford et al., 1977). These wide crosses have not yet been confirmed at older ages.

Hybrids sometimes can be made between very distantly related species and in a few instances even between genera. The most well known of these is the Leyland cypress, which is a cross between *Cupressus macrocarpa* and *Chamaecyparis nootkatensis* (Jackson and Dallimore, 1926). This hybrid grows well, and there have been suggestions that it would have value for operational forestry uses. In an attempt to reinterpret existing taxa as being of hybrid origin, crosses have been hypothesized between *Tsuga* and *Picea* (Campo-Duplan and Gaussen, 1949; Vabre-Durrieu, 1954). Undoubtedly, other very wide crosses within and even between genera will be made in the years ahead as technology improves and greater efforts are made. Obviously, the intergeneric nature of hybrids depends on the definition of the genera involved. For example, in the *Rosaceae,* where the taxonomic status keeps changing, many intergeneric hybrids have been recognized.

Crossing Hybrids and Hybrid Breakdown

Some hybrid individuals are spectacular, with the result that there is a strong desire to use them. The best and simplest method of obtaining large numbers of good hybrid propagules is to locate good parental combinations, multiply the parents by grafting or rooting, and use these to create large quantities of proven quality F_1 trees. But the drawback of such a method is the difficulty of making the crosses and producing the vegetative propagules (Figure 11.6). Too often hybrid programs are based only on F_1 genotypes from the original wild or unimproved populations. Continued progress requires producing hybrids from advanced-generation parent trees.

A major question relates to how occasional outstanding F_1 hybrid individuals can be used in advanced-generation breeding. One suggested method is to use the hybrids ($F_1 \times F_1$) to produce F_2 seed; seed orchards have been established using such good F_1 parents. This is an exceedingly dangerous practice because the F_2 progeny will vary across the whole range from one parent through the intermediate hybrids to the other parent. This phenomenon is the result of genetic

FIGURE 11.6

A difficulty with working with hybrids is making the controlled crosses. (Shown are pine yearlings of a pine cross.) It is difficult to mass-produce hybrid seed by controlled crossing; the best use of hybrids occurs when vegetative propagation can be used to mass-produce the desirable hybrids.

recombination that creates parental genotypes at some gene loci. A very common and basic result is that nonuniformity will be found in the F_2 progeny and later (F_3, F_4, etc.) generations. Despite this biological fact of segregation, F_1 seed orchards have been recommended by persons such as Hyun (1976) and Nikles (1981) who reported they did not find the expected segregation in the F_2. Certainly, hybrid seed orchards should never be made using F_1 individuals *until* tests have proven that the trees produced have the desired uniformity. There is an ongoing argument by some geneticists who say that since more trees are planted per unit area in forestry than can reach maturity and since thinning may be planned, it therefore is not serious if some inferior individuals are planted. This may be true for the final harvest of crop trees, but the forester certainly wants to get the most from a thinning operation. The general trend is to widen spacing as more uniform, improved trees become available. However, this requires that a high proportion of desirable trees be planted.

An outstanding example of hybrid breakdown, combined with the adverse effects of related matings, is the so-called Brazilian source of *Eucalyptus grandis* (Brune and Zobel, 1981). A number of species were planted in the arboretum at Rio Claro in São Paulo, Brazil, and these hybridized freely. Seeds were collected and operationally planted from the best *E. grandis* trees in the arboretum. The progeny grew very well, and seed were collected from these plantations for additional planting. After using this method to obtain seed for a number of generations, the quality of trees degenerated, with unwanted species, hybrids, and dwarfs appearing. In some plantations, as many as 5 to 15% of the trees never grew into a satisfactory tree form. Stands from these seeds are very nonuniform and have low productivity. In these same stands there is an occasional superoutstanding individual with remarkable form and growth (see Figure 11.2). Some appear to be hybrids and some are similar to the parent species. Because of the unknown parentage and relatedness, these can only be used directly in operational planting through vegetative propagation. They can also be used by letting them cross, testing them, and then using the very best of the progeny through vegetative propagation. If a hybrid program is to be efficient and successful, it is essential to avoid coming to the point of hybrid breakdown and relatedness that was attained in the Brazilian source of *E. grandis*.

One of the best methods for the use of hybrids is to cross them back to outstanding individuals of one of the parental species. When done correctly, backcrossing will enable the transference of the desired characteristic into the best parental species. For example, the *P. coulteri* × *P. jeffreyi* cross is quite resistant to the pine reproduction weevil. Jeffrey pine is relatively slow growing, of good form, but very sensitive to the weevil. Coulter pine is a fast grower but of relatively poor form. Backcrossing the relatively fast-growing and insect-resistant hybrid back to the best Jeffrey pine parents and making the proper selections will result in a faster-growing, better-formed tree than the F_1 hybrid. The backcross hybrid will look more like Jeffrey pine and will also be more weevil resistant.

Future Hybrid Research

The potentials and the research left to do in forest tree hybridization are indicated in Table 11.1 (from Critchfield, 1975), which shows the number of species and the number of hybrid combinations that have been made in the genus *Pinus*. When the number of potential combinations of pines possible (4500) is compared to those successfully made (95), it is evident how much is left to do in a genus on which so much effort has already been expended. Critchfield (1975) states that "because of the economic importance of many pines, species hybridization has been more fully explored in this genus than in most other plant genera." It is true that much has already been learned. For example, some pines, like *P. pinea*, have not been sucessfully crossed with other species; others, like *P. resinosa*, have crossed with difficulty, and still others, such as *P. taeda*, cross relatively easily with a large number of species. For the spruces, another genus on which considerable hybridization work has been done, Kleinschmitt (1979) states that only 156 species crosses have been made of the 1260 possible. He found some spruce hybrids to be quite outstanding. It is sufficient to say that the possibilities for making hybrids among forest trees have only been touched.

TABLE 11.1
Successful Interspecific Hybridizations within and Between Subsections of *Pinus*[a]

Subsections(s)[b]	Number of Species	Estimated Number of Hybrid Combinations
Cembrae (white pines)	5	1
Cembrae × *Strobi*		2
Strobi (white pines)	14–15	18
Cembroides (pinyon pines)	8	6
Balfourianae (foxtail pines)	2–3	2
Sylvestres (Eurasian hard pines)	19	19
Australes (southern and Caribbean hard pines)	11	15
Australes × *Contortae*		2
Contortae (small-cone pines)	4	2
Sabinianae (big-cone pines)	3	2
Sabinianae × *Ponderosa*		2
Ponderosae (western and Mexican hard pines)	13–15	16
Ponderosae × *Oocarpae*		2
Oocarpae (closed-cone pines)	7	6

[a]From Critchfield, (1975).

[b]The eight species in the other five subsections have no verified interspecific hybrids.

HYBRID NOMENCLATURE

In the past, a number of hybrids were given names and specific rank, especially in the genus *Populus* in which a whole new and sometimes confusing nomenclature has been developed. This has also been done for some pines, such as for *Pinus sondereggeri* (loblolly × longleaf pines) or *Pinus attenuradiata* (Monterey × knob-cone pine) or the so-called Erambert hybrid (a cross possibly involving several southern pines), whose parents have not been accurately defined. It is evident that the practice of giving specific rank or new names to every hybrid forest tree would lead to chaos. The acceptable methods of naming hybrids are outlined in the *International Code of Botanical Nomenclature* adopted by the Twelfth International Botanical Congress in Leningrad, USSR (Stafleu et al., 1978).

One method is to use a "collective" name preceded by an X sign, such as *P. X sondereggeri* or *P. X attenuradiata*. Publication of these must follow the international code for validity. This is an option generally used for hybrids of natural origin. Another is to utilize a formula such as *P. attenuata* X *radiata*, using the names of the parental species. The names may be alphabetically arranged, or if parentage is known, the female parent should be listed first. The third method, which is not totally satisfactory, is to use cultivar names. These cannot be Latin or Latin combinations but may be combinations of the common names. They must be published with a description (in English), and it is recommended that a type be designated and placed in an herbarium. The cultivar name begins with a capital letter and is either placed in single quotation marks or is preceded by "cv." The name could be something like the following: *P.* (*coulteri* X *jeffreyi*) = 'Jeffcoult' or P. cv. Jeffcoult. Obviously, this method could lead to hybrid names that have little relation to the names of the parents and could result in confusion unless extreme care is taken.

A compromise for control crosses may be to use code numbers, always keeping a record of the parentage of the trees used to make the cross. The most important criterion is to keep accurate records in such a manner that one can quickly determine the makeup of the cross rather than be saddled with learning a lot of meaningless new specific names (or numbers), many of which are most difficult to pronounce or spell and which give no clue to the origin of the hybrid involved. Use of code numbers becomes more efficient when computer facilities are available, but code numbers are not descriptive except for those who made the hybrids.

J. W. Duffield[2] suggests that natural hybrids need not be named; after a hybrid has been produced artificially, it becomes a cultivated plant, subject to the rules used for cultivated plants. As Pryor (1965) says, when discussing this major problem in the eucalypts, "some binomials have been based on hybrid type specimens and these, since they purport to describe species, must be discarded because hybrids cannot be equated with species. Thus in the literature there are some fifty or so binomials which cannot be names of valid species." Natural hybrids will be handled differently than man-produced hybrids (Little, 1960). This is typified by a recent letter from J. Perry[3] who raises the question about when is it suitable to call a tree a hybrid: Is a study of the phenotype sufficient, or does one have to rear and test offspring? The area of hybrid nomenclature in forest trees is difficult and needs to be clarified.

THE FUTURE OF HYBRIDS IN APPLIED TREE IMPROVEMENT

As mentioned previously, the chief restrictions to the use of hybrids have been the following: (1) the inability to mass produce them from seed; (2) the lack of suitable methods to produce hybrids vegetatively on a mass scale for operational planting; and (3) the inability to carry on advanced breeding programs with F_1 plants because they do not produce F_2's that are uniform enough for operational planting. Because of these restrictions, hybrids in tree improvement have generally been considered as oddities, toys, or something to develop that might someday have value. Fortunately, some of the objections have been overcome or will soon be overcome (Denison and Franklin, 1975; Matthews and Bramlett, 1981), resulting in the rapid operational use of hybrids like what was done many years ago for the poplars (Stout et al., 1927). Hybridization of forest trees is just now arriving on the tree improvement scene as an important improvement

---

[2]Personal communication.
[3]Personal communication.

method. As techniques are better developed and understood, there will be an explosive use of hybrids in operational forestry.

Before the use of hybrids can reach its optimum, a general realization of the biological fact must come about to the effect that the *goodness of a hybrid depends on the goodness of its parents*. This means that a hybridization program must include a breeding program for the parental species to improve them to be used later as parents of new and improved hybrids. This need has been largely overlooked in current hybridization programs in which all the effort has gone into crossing species without a parallel program to produce better parents to make new crosses. A few programs have followed this pattern; for example, the rapidly moving *Eucalyptus* program referred to by Campinhos (1980) will develop several generations of improved parental trees to be used for future crossing for hybrid production. Lack of parental improvement will make a number of hybrid programs less effective than they should be.

Why do tree improvers have such enthusiasm about the future importance of hybrids? This attitude generally is due to the new methodology now being developed that will enable the vegetative reproduction of desired hybrids. Most workers are not too optimistic about an economic mass production of hybrid seeds, although such methods as mass pollination, which was described previously, are making this somewhat more feasible. But when vegetative propagation becomes feasible, a whole new world is opened up. For example, if $F_1 \times F_1$ do not produce uniform F_2 progeny, the occasional chance combination "super F_2" can be used. Further progress of the program will then depend both on improvement of parental species, which is followed by hybridization along with crossing among hybrids and choosing the occasional outstanding offspring. Even if seed yields are very low, which is a problem with some hybrids, enough seed will be produced to grow progeny from which to select for vegetative propagation.

Most people look for improvement in morphological characteristics in hybrids. But the greatest gains may well be in physiological characteristics that are related to adaptability. For example, Eriksson et al. (1978) have stated that "by selection of suitable parents interprovenance crosses can be used to produce hybrids with desired photoperiodic characteristics and temperature requirements." Improvements in cold hardiness, drought hardiness, and the ability to withstand nutrient deficiencies are possible through hybrid improvement efforts. One of the most lucrative areas is in the field of pest resistance, in which the hybrid or its derivative may carry the desired resistance (Miller, 1950; Sluder, 1970; LaFarge and Kraus, 1980, and many others). The advantage of hybridization is that it is possible to develop something that is entirely different from what exists in nature by the proper choice of parents.

As a hybrid breeding program progresses, it becomes possible through backcrossing to move a desired trait from an otherwise less desired species into a more desirable one. This has been done by LaFarge and Kraus (1980) who were able, by backcrossing the loblolly × shortleaf pine hybrid to the loblolly parent, to obtain trees that were equal to or better in growth than the parent loblolly pine but that carried considerably more resistance to fusiform rust than loblolly pine.

The potential for the use of hybrids in forest tree improvement is only now becoming recognized and usable. In the future, programs must operate in this area more seriously than they have in the past.

LITERATURE CITED

Anderson, E. 1948. Hybridization of the habitat. *Evolution* **2**:1–9.

Anderson, E. 1953. Introgressive hybridization. *Biol. Rev.* **28**:280–307.

Anderson, E. 1968. *Introgressive Hybridization.* Hafner, Co. New York.

Bannister, M. H., Brewerton, H. V., and McDonald, I. R. 1959. Vapour-phase chromatography in a study of hybridism in Pinus. *Svensk Papp. Tidn.* **62**(16):567–573.

Barnes, B. V., Dancik, B. B., and Sharik, T. L. 1974. Natural hybridization of yellow birch and paper birch. *For. Sci.* **20**(3):215–221.

Bray, J. R. 1960. A note on hybridization between *Quercus macrocarpa* and *Quercus bicolor* in Wisconsin. *Can. Jour. Bot.* **38**(5):701–704.

Bridgwater, F. E., and Trew, I. F. 1981. Supplemental mass pollination. In *Pollen Management Handbook,* (Carlyle Franklin, ed.) pp. 52–57. U.S. Forest Service Agricultural Handbook No. 587, Washington, D.C.

Brune, A., and Zobel, B. J. 1981. Genetic base populations, gene pools and breeding populations for Eucalyptus in Brazil. *Sil. Gen.* **30**(4–5):146–149.

Buchholz, J. T. 1945. Embryological aspects of hybrid vigor in pines. *Science* **102**(2641):135–142.

Burk, C. J. 1962. An evaluation of three hybrid-containing oak populations on the North Carolina outer banks. *Jour. Elisha Mitchell Soc.* **78**(1):18–21.

Campinhos, E. 1980. More wood of better quality through intensive silviculture with rapid growth improved Brazilian *Eucalyptus. Tappi* **63**(11):145–147.

Campo-Duplan, M., and Gaussen, H. 1949. Sur quatre hybrids de genres chez les Abietineae. *Bull. Soc. Hist. Nat. Toulouse* **84**(1/2):105–109.

Chaperon, H. 1976. "Amélioration génétique des Eucalyptus hybrides au Congo Brazzaville [Production of hybrid eucalypts in Brazzaville, Congo]." 3rd World Cons. For. Tree Breed., Canberra, Australia.

Chaperon, H. 1979. Nouvelles perspectives D'Amélioration génétique induites par le bouturage du pin maritime [New prospects of genetic breeding through vegetative propagation of *Pinus pinaster*]. In Assoc. Foret-Cellulose (AFOCEL) Annual Report, 1979, Paris, France, pp. 31–53.

Chapman, H. H. 1922. A new hybrid pine. *Jour. For.* **20**(7):729–734.

Chiba, S. 1968. "Heterosis in Forest Tree Breeding." 9th Symp. Jap. Soc. of Breed. Tech. Note No. 69.

Chisman, H. H. 1955. The natural hybrid oaks of Pennsylvania. *Res. Pap. Pa. State For. Sch.* **22**:1–4.

Clifford, H. T. 1954. Analysis of suscepted hybrid swarms in Eucalyptus. *Heredity* **8**(3):259–269.

Conkle, M. T. 1969. Hybrid population development and statistics. *Sil. Gen.* **18**(5–6):197.

Critchfield, W. B. 1975. "Interspecific Hybridization in *Pinus*.: A Summary Review." Proc. 14th Mtg. Can. Tree Impr. Assoc., pp. 99–105.

Denison, H. P. and Franklin, E. C. 1975. **Pollen management.** I *Seed Orchards* (A. Faulkner, ed.) Forestry Commission Bulletin No. 54, London, England, pp. 92–100.

Duffield, J. W. 1952. Relationships and species hybridization in the genus *Pinus*. *A. Forstgen. Forstpflanz.* **1**(4):93–99.

Duffield, J. W., and Snyder, E. B. 1958. Benefits from hybridizing American forest tree species. *Jour. For.* **56**(11):809–815.

Dvorak, W. 1981. "*Eucalyptus robusta*: A Case Study of an Advanced Generation Hardwood Breeding Program in Southern Florida." M.S. thesis, School of Forest Research, North Carolina State University, Raleigh.

Eifler, I. 1960. Untersuchungen zur individuellen Bedingtheit des Kreuzungserfolges zwischen *Betula pendula* und *Betula pubescens* [The individual results of crosses between *B. pendula* and *B. pubescens*]. *Sil. Gen.* **9**(6):159–165.

Einspahr, D. W., and Joranson, P. N. 1960. Late flowering in aspen and its relation to naturally occurring hybrids. *For. Sci.* **6**(3):221–224.

Eklund, C. 1943. Species crosses within the genera *Abies, Pseudotsuga, Larix, Pinus* and *Chamaecyparis*, belonging to the family Pinaceae. *Svensk Papp. Tidn.* **46**:55–61, 101–105, 130–133.

Eriksson, G., Ekberg, I., Dormling, I., and Matern, B. 1978. Inheritance of bud flushing in *Picea abies*. *Theor. Appl. Gen.* **52**:3–19.

Feret, P. P. 1972. Peroxidase isoenzyme variation in interspecific elm hybrids. *Can. Jour. For. Res.* **2**(3):254–270.

Fowler, D. P. 1978. Population improvement and hybridization. *Unasylva* **30** (119–120):21–26.

Györfey, B. 1960. Hybrid vigor in forest trees and the genetic explanation of heterosis. *Erdeszeti Kut.* **56**(1/3):327–340.

Hadders, G. 1977. "Experiments with Supplemental Mass Pollination in Seed Orchards of Scots Pine (*Pinus sylvestris*)." 3rd World Cons. on For. Tree Breed., Canberra, Australia.

Hall, M. T. 1952. Variation and hybridization in junipers. *Ann. Missouri Bot. Gar.* **39**(1):1–64.

Hardin, J. W. 1957. Studies in the Hippocastanaceae. IV. Hybridization in *Aesculus*. *Rhodora* **59**(704):185–203.

Hoffman, D., and Kleinschmitt, J. 1979. "An Utilization Program for Spruce Provenance and Species Hybrids." IUFRO Norway Spruce Meeting, Bucharest.

Howcroft, N. H. 1974. "A racial hybrid of *Pinus merkusii*," Trop. For. Res. Note SR28, Papua, New Guinea.

Hunziker, J. H. 1958. Estudios citogeneticos in *Salix humboldtiana* y en sauces hibridos triploides cultivados en la Argentina [Cytogenetic studies of *S. humboldtiana* and triploid hybrid willows cultivated in Argentina]. *Rev. Invest. Agríc.* **12**(2):155–171.

Hyun, S. K. 1976. Interspecific hybridization in pines with special reference to *P. rigida* × *taeda*. *Sil. Gen.* **25**(5–6):188–191.

Jackson, B., and Dallimore, W. 1926. A new hybrid conifer, *Cupressus Leylandi* (between *Cupressus macrocarpa* and *Chamaecyparis nootkatensis*). *Bull. Miscel. Info.* **3**:113–116.

Jinks, J. L., and Jones, R. M. 1958. Estimation of the components of heterosis. *Genetics* **43**(2):223–234.

Johnson, A. G. 1955. "Southern Pine Hybrids, Natural and Artificial." 3rd South. Conf. For. Tree Impr., New Orleans, La., pp. 63–67.

Keiding, H. 1962. Krydsningsfrodighed Hos Laerk [Hybrid vigor in larch]. *Dans. Skovforen. Tidsskr.* **47**:139–157.

Kirkpatrick, J. B. 1971. A probable hybrid swarm in *Eucalyptus*. *Sil. Gen.* **20**(5–6):157–159.

Kleinschmit, J. 1979. "Present Knowledge in Spruce Provenance and Species Hybridization Potential." IUFRO Norway Spruce Meeting, Bucharest.

Kriebel, H. B. 1972. Embryo development and hybridity barriers in the white pines (Section Strobus). *Sil. Gen.* **21**(1–2):39–44.

Krugman, L. 1970. "Incompatibility and Inviability Systems among Some Western North American Pines." Proc. Sex. Repro. Forest Trees, IUFRO, Sect. 22, Finland, Part 2.

LaFarge, T., and Kraus, J. F. 1980. A progeny test of (shortleaf × loblolly) × loblolly hybrids to produce rapid growing hybrids resistant to fusiform rust. *Sil. Gen.* **29**(5–6):197–200.

Lester, D. T. 1973. "The Role of Interspecific Hybridization in Forest Tree Breeding." Proc. 14th Mtg. Can. Tree Impr. Assoc., pp. 85–97.

Little, E., and Critchfield, W. B. 1969. "Subdivisions of the genus *Pinus*." U.S. Forest Service Misc. Pub. 1144.

Little, E. L. 1960. Designating hybrid forest trees. *Taxon* **9**:225–231.

Little, E. L., and Righter, F. I. 1965. Botanical description of forty artificial pine hybrids. *Tech. Bull. No. 1345*.

Little, S., and Somes, H. A. 1951. "No Exceptional Vigor Found in Hybrid Pines Tested." Northeast. For. Expt. Sta. Res. Note No. 10.

Little, S., and Trew, I. F. 1976. "Breeding and Testing Pitch × Loblolly Pine Hybrids for the Northeast." Proc. 23rd Northeast. For. Tree Impr. Conf., pp. 71–85.

Marien, J. N., and Thibout, H. 1978. Hybridization naturelle d'eucalyptus plantés dans le Sud de la France [Natural cross pollination of *Eucalyptus* planted in Southern France]. *AFOCEL* **1**:89–112.

Martínez, M. 1948. *Los Pinos Mexicanos*. Ediciones Botas, Mexico City.

Matthews, F. R., and Bramlett, D. L. 1981. Cyclone pollinator improves loblolly pine seed yields in controlled pollinations. *South. Jour. App. For.* **5**(1):42–46.

McMinn, H. E., Babcock, E. B., and Righter, F. I. 1949. The Chase oak, a new giant hybrid oak from Santa Clara Co., Calif. *Madroño* **10**(2):51–55.

McWilliam, J. R. 1958. "Pollination, Pollen Germination and Interspecific Incompatibility in *Pinus*." Ph.D. thesis, Yale University.

Mergen, F. 1959. Applicability of the distribution of stomata to verify pine hybrids. *Sil. Gen.* **8**(4):107–109.

Mettler, L. E., and Gregg, T. G. 1969. *Population Genetics and Evolution*. Prentice-Hall, Englewood Cliffs, N.J.

Miller, J. M. 1950. "Resistance of Pine Hybrids to the Pine Reproduction Weevil." For. Res. Note No. 68, Calif. For. and Range Experiment Station.

Miller, J. T., and Thulin, I. J. 1967. "Five-year Survival and Height Compared for European, Japanese and Hybrid Larch in New Zealand." Res. Leaflet No. 17, New Zealand Forest Service.

Mirov, N. T. 1967. *The Genus Pinus*. Ronald Press Co, New York.

Moritz, O. 1957. Serologische Differenzierung von Arten als Voraussetzung der Fruhdiagnose des Hybridcharakters [Serological differentiation of species as a hypothesis for early diagnosis of the hybrid character]. *Der Zuchter* **4**:75–76.

Muller, C. H. 1952. Ecological control of hybridization in *Quercus*: A factor in the mechanism of evolution. *Evolution* **612**:147–161.

Namkoong, G. 1963. "Comparative Analyses of Introgression in Two Pine Species." Ph.D. thesis, School of Forestry, North Carolina State University, Raleigh.

Nikles, D. G. 1981. "Some Successful Hybrid Breeds of Forest trees and Need for Further Development in Australia." Proc. 7th Meet. RWG No. 1—Forest Genetics, Traralgon, Victoria.

Nilsson, B. O. 1963. "Intraspecific Hybridization and Heterosis within *Picea abies*." 1st World Cons. on For. Gen. and Tree Impr., Stockholm, Sweden.

Palmer, E. J. 1948. Hybrid oaks of North America. *Jour. Arnold Arbor.* **29**:1–48.

Pauley, S. S. 1956. "Natural Hybridization of the Aspens." Minnesota For. Notes 47.

Piatnitsky, S. S. 1960. "Evolving New Forms of Oaks by Hybridization." 5th World For. Cong., Seattle, Vol. 2, pp. 815–818.

Pryor, L. D. 1951. "A genetic analysis of some *Eucalyptus* species. *Proc. Lin. Soc. N. S. W.* **76**(3–4) :140–147.

Pryor, L. D. 1976. *The Biology of Eucalyptus,* Studies in Biology No. 61. Camelot Press Ltd., Southampton, England.

Pryor, L. D. 1978. Reproductive habits of the eucalypts. *Unasylva* **30**(119–120):42–46.

Riemenschneider, D., and Mohn, C. A. 1975. Chromatographic analysis of an open-pollinated Rosendahl spruce progeny. *Can. Jour. For. Res.* **5**(3):414–418.

Righter, F. I. 1955. "Possibilities and Limitations of Hybridization in *Pinus.*" Proc., 3rd South. Conf. For. Tree Breed., New Orleans, La., pp. 54–63.

Santamour, F. S. 1972. Interspecific hybridization in *Liquidambar. For. Sci.* **18**(1):23–26.

Santamour, F. S. 1981. "Flavenoids and Coumarines in *Fraxinus* and Their Potential Utility in Hybrid Verification." Proc. 27th Northeast. For. Tree Imp. Conf., Burlington, Vt., pp. 63–71.

Saylor, L. C., and Smith, B. W. 1966. Meiotic irregularity in species and interspecific hybrids of *Pinus. Am. J. Bot.* **53**:453–468.

Saylor, L. C., and Kang, K. W. 1973. A study of sympatric populations of *Pinus taeda* and *Pinus serotina* in North Carolina. *Jour. Elisha Mitchell Sci. Soc.* **89**(142):101–110.

Saylor, L. C., and Koenig, R. L. 1967. The slash × sand pine hybrid. *Sil. Gen.* **16**(4):134–138.

Scamoni, A. 1950. Uber die weitere Entwicklung kunstlicher Kiefernkreuzungen in Eberswalde [The further development of artificial pine crosses in Eberswalde]. *Der Zuchter* **20**(1–2):39–42.

Schmidtling, R. C., and Scarborough, N. M. 1968. "Graphic Analysis of Erambert's hybrid." U.S. Forest Service Research Note SO-80.

Schmitt, D. M. 1973. "Interspecific Hybridization in Forest Trees: Potential Not Realized." 14th Mtg. Can. Tree Impr. Assoc., Part 2, pp. 57–66.

Schmitt, D., and Snyder, E. B. 1971. Nanism and fusiform rust resistance in slash × shortleaf pine hybrids. *For. Sci.* **17**(3):276–278.

Schreiner, E. J. 1965. "Maximum Genetic Improvement of Forest Trees through Synthetic Multiclonal Hybrid Varieties." Proc. 13th Northeast. For. Tree Impr. Conf., Albany, N.Y., pp. 7–13.

Schütt, P., and Hattemer, H. H. 1959. Die Eignung von Merkmalen des Nadelquerschnitts für die Kiefern Bastarddiagnose [The suitability of characteristics of the transverse section of needles for the analysis of pine hybrids]. *Sil. Gen.* **8**(3):93–99.

Sluder, E. R. 1970. "Shortleaf × Loblolly Pine Hybrids Do Well in Central Georgia." Ga. For. Res. Coun. Res. Paper No. 64.

Smouse, P. E., and Saylor, L. C. 1973. Studies of the *Pinus rigida—serotina* Complex II. Natural hybridization among the *Pinus rigida—serotina complex, P. taeda* and *P. echinata. Ann. Missouri Botanical Gardens* **60**(2):192–203.

Snyder, E. B. 1972. "Glossary for Forest Tree Improvement Workers." Southern Forest Experiment Station, U.S. Forest Service, New Orleans, La.

Stafleu, F. A., and Committee. 1978. "International Code of Botanical Nomenclature." Assoc. for Plant Tax., Vol. 97.

Stebbins, G. L. 1959. The role of hybridization in evolution. *Proc. Am. Phil. Soc.* **103**(2):232–251.

Stebbins, G. L. 1969. The significance of hybridization for plant taxonomy and evolution. *Taxon* **18**:26–35.

Stockwell, P., and Righter, F. I. 1947. Hybrid forest trees. *Yearb. Agr.* **1943–1947**:465–472.

Stone, E. C., and Duffield, J. W. 1950. Hybrids of sugar pine by embryo culture. *Jour. For.* **48**(3):200–203.

Stout, A. B., McKee, R. H., and Schreiner, E. J. 1927. The breeding of forest trees for pulpwood. *Jour. N.Y. Bot. Gar.* **28**:49–63.

Tanaka, Z. 1882. A variation of pine. *Bull. Jap. For. Assoc.* **7**:32.

Tucker, J. M. 1959. A review of hybridization in North American oaks. *Proc. Int. Bot. Cong.* **II**:404 (abstracts).

Venkatesh, C. S., and Sharma, V. K. 1976. Heterosis in the flowering precocity of *Eucalyptus* hybrids. *Sil. Gen.* **25**(1):28–29.

Venkatesh, C. S., and Vakshasya. 1977. "Effects of Selfing, Crossing and Interspecific Hybridization in *Eucalyptus camaldulensis*." 3rd World Cons. For. Tree Breed., Canberra, Australia.

Vabre-Durrieu, A. 1954. L'Hybride *Tsuga-picea hookeriana* et ses parents. Étude des plantules [The hybrid Tsuga-picea hookeriana and its parents. Study of seedlings]. *Toulouse Soc. Hist. Nat. Bul.* **89**:47–54.

van Buijtenen, J. P. 1969. Applications of interspecific hybridization in forest tree breeding. *Sil. Gen.* **18**(5–6):196–200.

Wagner, W. H. 1969. The role and taxonomic treatment of hybrids. *BioScience* **19**(9):785–788.

Wang, C. W. 1971. "The Early Growth of *Larix occidentalis* × *P. leptolepis* Hybrid." University of Idaho, Stat. Note 17.

Whitmore, J. L., and Hinojosa, G. 1977. "Mahogany (*Swietenia*) Hybrids." Research Paper ITF-23, U.S. Forest Service, Puerto Rico.

Williford, M., Brown R., and Zobel, B. J. 1977. "Wide Crosses in the Southern Pines." Proc. 14th South. For. Tree Impr. Conf., Gainesville, Fla., pp. 53–62.

Wright, J. W. 1964. Hybridization between species and races. *Unasylva* **18**(2–3):73–74.

Zobel, B. J. 1951a. The natural hybrid between Coulter and Jeffrey pines. *Evolution* **5**(4):405–413.

Zobel, B. J. 1951b. Oleoresin composition as a determinant of pine hybridity. *Bot. Gaz. (Chicago)* **113**(2):221–227.

Zobel, B. J. 1953. Are there natural loblolly–shortleaf pine hybrids? *Jour. For.* **51**(7):494–495.

CHAPTER 12

Wood and Tree Improvement

WOOD QUALITY

Specific Gravity and Wood Density

The Importance of Specific Gravity Variations in Forestry
Variation Patterns in Wood

Variation within Trees—Juvenile and Mature Wood
Importance of within-Tree Variation
Variation Patterns among Trees
Wood Variation among Sites and Geographic Areas
Methods of Sampling for Wood Properties

CONTROLLING WOOD QUALITY

Tree Form

Genetics of Wood

Wood Specific Gravity
Fiber and Tracheid Length
Other Wood Properties
Genetic Relationships among Wood Properties
Effect of Growth Rate on Wood Properties

Fertilizers and Wood Qualities

WOOD QUALITIES OF EXOTICS

Exotic Softwoods

Exotic Hardwoods

LITERATURE CITED

Although the primary emphasis of most tree breedings programs is to obtain faster-growing, better-formed, well-adapted, and pest-resistant trees, improved wood properties can also be obtained from the same programs. Research has shown that most wood qualities, as well as tree form and growth characteristics that affect wood, are inherited strongly enough to obtain rapid economically important gains through genetic manipulation.

The differing wood properties have a significant effect on the quality and yield of pulp and paper products and on strength and utility of solid wood products. Publications by Artuz-Siegel et al. (1968), Higgins et al. (1973), Barker (1974), and Foelkel et al. (1975a) are only a very few of the many publications that relate the effect of wood properties to product quality.

There is often considerable reluctance by some persons to include wood as part of a tree improvement program. Deterrents that are given include difficulties in deciding on the desired wood properties, difficulties of prediction about what type wood will be desired in future years, and the realization that no single type of wood is ideal for every product. Another argument put forth in the early years of tree improvement programs was that wood qualities are so strongly affected by growth and environment that genetic manipulation would not be successful. Some investigators felt that the changes in harvesting and utilization (i.e., lowering rotation ages and using small tops and limbs) would result in the use of wood that was so different from that now harvested that any changes caused through genetics would be relatively minor.

No matter what specific type of wood is desired for the future, improvement in wood characteristics will be of value in almost every program. Wood is notably non-homogeneous, both within and among trees of a species, as well as among species and geographic sources. Genetic manipulation of wood in a breeding program can result in a higher proportion of desired wood. This is by far the most useful improvement that can be made for wood qualities. The easier and more consistent conversion of uniform raw materials results in cheaper production of a higher-value final product (Zobel, 1983). In general, tree improvement programs that have wood production as their goal should include knowledge of the manipulation of wood qualities.

WOOD QUALITY

The question always arises about which wood properties should be changed. Even though nearly all wood properties that have been studied respond satisfactorily to breeding, each has different economic value and importance. Because any genetic program will achieve the greatest gains from concentration on a few characteristics, it is necessary to determine the most important wood property and not try to include everything.

There is no question that wood specific gravity, or wood density, is by far the most important within-species wood characteristic for nearly all products (Eins-pahr et al., 1969; Barefoot et al., 1970). In some breeding programs, fiber or tracheid length are also included. Both specific gravity and fiber and tracheid length have strong inheritance patterns, and changes in them can have a significant effect on the final product. For certain special products, other wood properties can take on great importance, especially for quality hardwoods in which grain or color may be key characteristics.

The woods of softwoods and hardwoods are varied and different. Breeders attempting to improve wood must be aware of its complexity in the hardwoods and the relative simplicity of the wood of the softwoods. There are numerous cell-type differences within hardwoods that sometimes respond quite differently to genetic and silvicultural manipulation than do the simpler coniferous woods. One obvious example is the difference between ring-porous and diffuse-porous hard-wood species. In ring-porous trees, vessels formed early in the growing season are much larger in diameter than those formed later in the year and greatly influence wood properties. Examples of ring-porous species are the oaks and ashes. In diffuse-porous species such as the poplars, birches, and eucalypts, the vessels are smaller in diameter and are of essentially the same size regardless of the time of formation. The general principles related to wood and breeding strategies are the same for softwoods and hardwoods, but details can be grossly different.

Specific Gravity and Wood Density

It is important to understand the meaning of the terms *specific gravity* and *wood density* thoroughly. *Specific gravity* is the ratio of the weight of a given volume of wood to the weight of an equal volume of water. It is unitless. *Wood density* is the weight of wood expressed per cubic volume, for example, 450 kg/m^3 or 28.0 lb/ft^3.

$$\text{Specific gravity} = \frac{\text{weight of a given volume of wood}}{\text{weight of an equal volume of water}}$$

$$\text{Wood density} = \text{weight of wood per unit volume such as kg/m}^3 \text{ or lb/ft}^3$$

Specific gravity and wood density are different ways of expressing how much wood substance is present. Specific gravity will be primarily used in this book. Either specific gravity or density can be calculated by knowing the other. In the metric system, density/1000 = specific gravity. Thus 450 kg/m^3 is equal to a specific gravity of 0.45. In the English system, density (lb/ft^3) is equal to specific gravity $\times$ 62.4 (the weight of 1 lb of water). Thus 62.4 $\times$ 0.45 = approximately 28 lb/ft^3.

Specific gravity *is not a simple wood characteristic but is a combination of characteristics*, each of which has a strong inheritance pattern of its own. Combined, they determine what is called *specific gravity* (van Buijtenen, 1964).

Despite its complexity, specific gravity is considered to be a single property in most breeding programs. Specific gravity is primarily determined by three different wood characteristics, which will be described below and which are illustrated in Figure 12.1

1. **Amount of summerwood** Some trees start formation of thick-walled summerwood (latewood) cells early in the year—others later. Because summerwood has a high specific gravity (see Figure 12.2), this pattern results in an overall high wood specific gravity for the early summerwood producer, no matter where the tree may be growing. Percentages of summerwood can vary by 100% among trees of the same species and age that are growing at the same rate with their roots intertwined.

2. **Cell size** Occasionally, trees that are otherwise similar have essentially the same wall thickness but different-sized cells. When this happens, the tree with small cells will have the highest specific gravity.

3. **Thickness of cell wall** Although summerwood is defined by Mork's definition as *cells that have double wall thickness as great or greater than lumen*

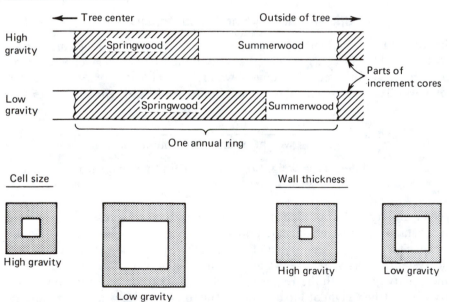

FIGURE 12.1

Specific gravity is not simple and is determined by a number of wood components. Three of these, summerwood percentage, cell size, and wall thickness, are illustrated. Despite its complexity, the combined wood property called specific gravity has a strong inheritance pattern.

diameters (Mork, 1928), the actual wall thickness of summerwood cells varies greatly among trees. Some trees have such thick-walled summerwood that the lumen is very small, resulting in a "rodlike" cell that gives a black or dark summerwood in pines. Other trees have summerwood cell walls that barely fit Mork's definition; in this instance, the summerwood is light brown When two trees have cells of essentially the same size, but one has thicker walls than the other, differences in specific gravity will be considerable.

The differences between springwood and summerwood cells in *P. taeda* are shown in Figure 12.2. In conifers, it is generally the characteristics of the summerwood that determine the differences in wood properties between trees, whereas in hardwoods, specific gravity can relate to other things such as the vessel volume or the amount of ray cells (Taylor, 1969). Usually, little importance is given to the effects of springwood (earlywood) cell differences on specific gravity variations among trees, although springwood does have some effect. Springwood cell properties may vary some from tree to tree, but the effect of differing springwood cells is generally small enough so that resulting influences on the overall specific gravity of the tree are not large.

The Importance of Specific Gravity Variations in Forestry   Specific gravity is of key importance to foresters because it has a major effect on both yield and quality of the final product (Barefoot et al., 1970) and because it is strongly inherited (Zobel, 1961; van Buijtenen, 1962; Harris, 1965).

Overall biomass productivity cannot be determined unless wood specific gravity is known. When volume production is assessed by cubic measurements such as a cord, cubic meter, or by green weight, an inaccurate estimate of productivity will result due to the variation in specific gravity and in moisture content (which is negatively related to specific gravity). To illustrate this, green weights are compared to dry weights for a given volume of wood in Table 12.1 for loblolly pine. Note the change in the ratio of green to dry weights as age increases.

Specific gravity determinations permit construction of dry weight yield tables that give a useful prediction of the productivity per unit area of land. Dry weight yields from different age classes permit the determination of the biological productivity potential of stands of trees.

TABLE 12.1
Green and Dry Weight off Wood Varies with Tree Age, as Shown for *P. taeda*

Age of Trees (years)	Green Weight of 100 ft³ (lb)	Dry Weight of 100 ft³ (lb)	Ratio of Green/Dry Weight
18	6230	2696	2.31
30	5880	2759	2.13

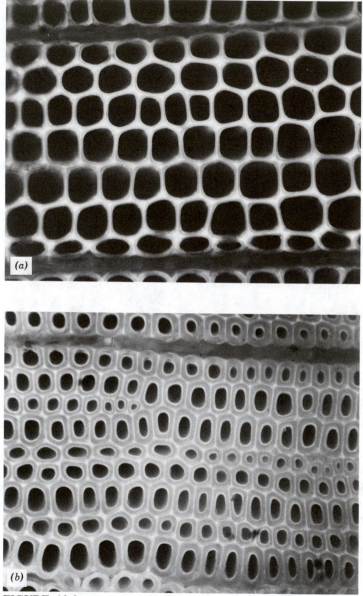

FIGURE 12.2

Within an annual ring springwood (earlywood) and summerwood
(latewood), cells are formed that are very different from one
another. These are illustrated side by side: (*a*) springwood, (*b*)
summerwood. Both were taken from a mature *P. taeda* tree. The
summerwood cells have the major effect on wood and product
properties.

Much has been written about the effect of specific gravity on the quality of pulp and paper; a few publications summarizing these findings are those of Barefoot et al. (1970), Kirk et al. (1972), Bendtsen (1978), and Zobel (1981). It is clear from these and many other summaries that the importance of specific gravity many times overshadows the importance of other wood properties; this is especially true for the key paper characteristic referred to as *tear strength*. It is so important that in most programs which have pulp and paper as final products specific gravity is the only wood characteristic manipulated. Because of its effect on quality and yield and its high heritability, specific gravity has become of major interest in most tree improvement programs—no matter if the objective is to produce fiber or solid-wood products (Zobel et al., 1978).

Variation Patterns in Wood

It is essential to have a good knowledge of wood variability if its quality and yield are to be manipulated. In addition to the well-known and recognized differences among species, variability in wood also occurs as follows: (1) within a given tree; (2) among trees of the same species; (3) sometimes, between populations of a species growing in a single locality; and (4) frequently, between populations of a species growing in different geographic areas (Zobel et al., 1960a). It is the purpose in this chapter to summarize briefly and to simplify the mass of information relative to the various wood variation patterns. Many examples are from loblolly pine (*P. taeda*), the most completely studied forest tree species with respect to wood properties. A full treatment of wood variation would require several hundred pages; the reader is cautioned, therefore, that a simplified résumé must obviously contain numerous generalizations that are designed to reflect the usual trends and that exceptions will sometimes occur.

Variation in wood is caused by many factors, including tree form, genetic differences, growth variations, differing environments, and evolutionary history. For the latter, Wakeley (1969) has emphasized the simplicity of ray and tracheid development of wood of the softwoods through evolutionary processes compared to the much more complex and heterogeneous hardwoods with their fibers, vessels, rays, fiber tracheids, and other cell types (Taylor, 1977). The wood of hardwoods is much more difficult to work with than is that of the softwoods. For example, in the hardwoods, specific gravity is influenced not only by cell wall dimensions but also by the relative amounts of ray and vessel elements (Taylor, 1969).

Variation within Trees—Juvenile and Mature Wood   It has been recognized for a very long time that wood properties may vary greatly within a tree, from the pith (or center) outward or from the stump upward. Near the center of the tree, juvenile wood is formed; this is illustrated in Figure 12.3. The concept, qualities, and problems associated with juvenile wood have been covered in many papers and reviews, including an in-depth treatment in the "TAPPI Short Course on Fast-Grown Plantation Wood" in Melbourne, Australia, in 1978. Early authors,

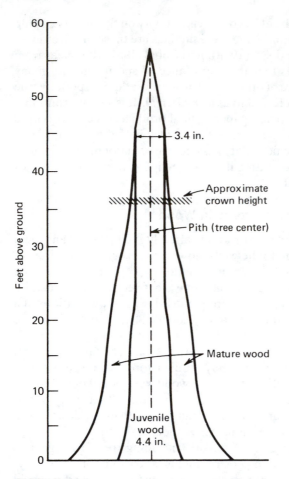

FIGURE 12.3

Juvenile wood exists as a semicylinder within the tree. This is especially evident for the pines as illustrated by the tree cross section representing the average of a number of 17-year-old *P. taeda.*. The trees will produce mostly mature wood at their bases and juvenile wood near their tops. The number of rings from the pith, or tree center, is related to the duration of juvenile wood production.

such as Paul (1957), referred to juvenile wood as "crown-formed" wood. This was explained by Kozlowski (1971) and Larson (1969) as being the result of the relative abundance of growth regulators and carbohydrates in the cambial zone near the crown. The similarity in wood qualities between top wood of older trees and wood near the center of the tree at its base has long been recognized (Marts,

1949). Although juvenile-wood characteristics differ from those of mature wood, it is not necessarily "bad" wood, and it is ideal for certain products. Juvenile-wood characteristics of most pines can be briefly summarized as follows.

1. Juvenile wood is formed near the pith of the tree (Figure 12.4). The number of rings from the tree center during which juvenile wood is formed varies from about 20 in *P. ponderosa* to 10 in *P. taeda*, 7 in *P. elliottii*, and 5 or 6 in *P. caribaea*. The actual age of the tree is not of importance, but the number of rings from the tree center or pith determines whether juvenile wood will be formed. Juvenile- and mature-wood formation apparently are related to the maturity of the cambial cells as influenced by the hormone balance, and juvenile wood is produced near the top of the tree no matter how old the tree is. In some species (*P. elliottii*, *P. caribaea*), the change between juvenile and mature wood can be quite abrupt (Figure 12.5), whereas in others, it is much more gradual (e.g., *P. taeda*, *P. radiata*, *P. oocarpa*) (see Figure 12.4). In no

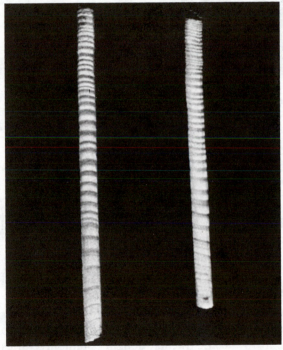

FIGURE 12.4

Juvenile wood has wood qualities that differ greatly from those of mature wood. This can be seen on the two *P. taeda* increment cores. The pith (brown fleck) is at the center of the tree. Note the color and "consistency" of the wood and how it changes out to about the seventh and eighth annual rings where typical mature wood is produced to the bark.

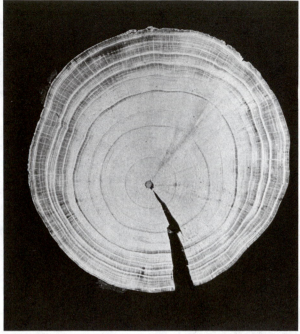

FIGURE 12.5

The transition from juvenile to mature wood is very abrupt in some species. Illustrated is a *P. caribaea* from a tropical area showing the very abrupt transition between the two kinds of wood. Compare it with *P. taeda,* shown in Figure 12.4.

species is there an abrupt shift from juvenile to mature wood during 1 year, and the change from juvenile wood before "typical" mature wood is formed usually is a gradual process occurring over several years (Figure 12.6).

2. Juvenile-wood specific gravity is lower than that of mature wood. The low gravity results primarily from thin cell walls and the low relative amount of summerwood-type cells (see Figure 12.2).

3. Juvenile-wood tracheids are short. Tracheid length is shortest near the center of the tree and increases rapidly toward mature wood, where it stabilizes to some extent. For example, in *P. taeda,*, tracheid length commonly varies from less than 2 mm in the juvenile wood to 3.5 to 5.5 mm in mature wood.

4. Juvenile wood is unstable when dried, because it shrinks longitudinally much more than does mature wood. The instability is the result of relatively flat fibril angles and causes major problems on drying (Meylan, 1968).

5. Juvenile wood produces quite different yields and quality in paper when compared to mature wood (Kirk et al., 1972). In addition to different cell

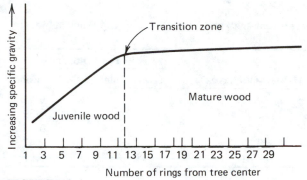

FIGURE 12.6

There usually is a gradual transition in wood qualities from the center of the tree outward. For example, the lowest specific gravity of most softwoods is nearest the tree center; it rises rapidly for a time and then tends to level off. This trend is shown schematically, indicating there is not an abrupt change but a transitional change from juvenile-to-mature wood qualities.

morphology, it has a chemical composition that differs in such things as hemicelluloses, lignins, and resinous constituents. As a result of this and due to poor liquor penetration, it pulps differently than does mature wood.

6. Because of low specific gravity and thin cell walls, juvenile wood is weak when used as a solid-wood product (Pearson and Gilmore, 1980). Nevertheless, it is widely used in the manufacture of structural lumber.

The variation in wood properties associated with height up the tree in pines can perhaps be best visualized if one considers that a core of "juvenile" wood with short tracheids and low specific gravity exists in the form of an inner "cylinder" at the center of the tree bole. This cylinder extends from the base of the bole all the way up to the top of the tree (see Figure 12.3). The result is that top logs consist mainly of juvenile wood, whereas the butt log of the same tree has more mature wood with a higher specific gravity (Stern, 1963; Zobel et al., 1972). Thus, when total tree chips or plywood cores are used, the proportion of juvenile wood is high. Topwood and bolewood differences in several wood properties are shown for 30-year-old loblolly pine trees in Table 12.2. Also shown are values for 11-year-old trees. Compare the topwood values of the 30-year-old trees with those of the 11-year-old trees.

The differences shown in Table 12.2 are of real economic significance, for example, 100 ft^3 (2.83 m^3) of wood from the basal logs of loblolly pine will yield about 500 lb (227 kg) more dry fiber [equivalent to about 200 to 300 lb (91 to 136 kg) of kraft pulp] than the same volume of the upper logs from the same trees. The

TABLE 12.2
Wood Qualities of Young *P. taeda* (11-Year-Old) and Lower Bole and Top Wood of Mature (30-Year-Old) Trees

		30-Year-Old Trees	
Wood Property	**11-Year-Old Trees**	**Lower Bole**	**Topwood**
Specific gravity	0.42	0.48	0.41
Tracheid length (mm)	2.98	4.28	3.59
Cell wall thickness (μm)	3.88	8.04	6.72
Lumen size (μm)	42.25	32.78	32.47
Cell diameter (μm)	40.01	48.86	45.91

fresh green weights will not be greatly different because differences in specific gravity of the wood within a species are masked by the higher moisture content of lower-gravity wood. Within a species there is a strong negative relationship between specific gravity and the moisture content of green wood, that is, low-gravity green wood has high moisture. Representative values for moisture content, resin content, and density are shown in Table 12.3 for juvenile and mature loblolly pine wood at several locations along the length of the stem.

TABLE 12.3
Wood Properties of 30-Year-Old Loblolly Pines Reported by 5-ft Bolts from Base to Top of Tree, by Juvenile and Mature Trees[a]

	Wood Density[b] After Resin Extraction (lb/ft³)		Percentage Resin Content		Percentage Moisture Content[b]	
5-ft Bolt No.	**Juvenile Wood**	**Mature Wood**	**Juvenile Wood**	**Mature Wood**	**Juvenile Wood**	**Mature Wood**
Base of tree						
1	27.5	34.3	3.0	2.2	110	74
2	26.8	33.7	2.9	2.0	122	85
3	25.0	31.8	2.8	2.2	133	97
4	25.0	31.2	2.6	2.1	137	102
5	25.0	29.3	2.5	2.4	139	110
6	25.0	29.3	2.5	2.0	145	117
7	24.3	—	2.6	—	151	—
8	24.3	—	2.5	—	153	—
9	23.1	—	2.7	—	163	—
Merchantable top[c]						

[a]Data are based on an average of 63 trees that were felled, divided into 5-ft bolts to a 4-in. top and into juvenile (first 10 rings from tree center) and mature wood.

[b]Expressed as a percentage of dry weight. For example, 163% moisture content means there are 1.63 lb of water for every pound of dry wood.

[c]4 in. (10.1 cm).

Importance of Within-Tree Variation Compared with material from older stands, wood from young pine plantations will have low-cellulose and high-hemicellulose yields when pulped. As an example, in loblolly pine, yields of paper were as much as 3% lower per unit of dry weight of wood from 12-year-old trees, compared with 30-year-old trees (Kirk et al., 1972). The thinner-walled cells associated with a high proportion of low-gravity juvenile wood produce kraft paper with low tear strength. Because the thinner-walled cells collapse more during manufacture, paper produced from juvenile wood has greater tensile strength, and generally the burst strength is good.

Chemical requirements and overall costs of pulping are greater for wood from young plantations. A 5% increase has been estimated in chemical costs for pulping 12-year-old versus 30-year-old loblolly pine (Kirk et al., 1972). In one study, low-specific-gravity wood (0.37) produced only 90% as much pulp as did wood of normal specifiic gravity (0.44) (Table 12.4).

A knowledge of within-tree variation can help answer questions about whether *volume* or *dry weight* of wood should be optimized in a forestry operation. If decisions are based solely on gross volume but wood weight is desired, major errors may result unless the effects of within-tree variation can be assessed. For example, if one can grow as much volume on an area with two 15-year rotations as is possible with one 30-year rotation, what will be the difference in yield of pulp? In loblolly pine, considerably more weight of usable wood fibers would be obtained from the single 30-year rotation than from the two 15-year rotations, although the volumes produced are the same. The green weight of the young stand would be high but only because of high moisture content (Table 12.5). Older stands continue to add considerable wood weight, even as growth modifies, because of the higher specific gravity of the greater amount of mature wood produced.

A number of studies have been made showing the importance and effects of tree age in forestry operations. One of the best of these was by the Hammermill Paper Company in Alabama (Kirk et al., 1972). The differences in mill yields and quality of paper between old and young pine trees were large and reflects the differing amounts of juvenile wood. Even more graphic was a study by the Federal Paper Board Company (Semke and Corbi, 1974) that showed the effect of age and within tree location of the wood used. They found that slabs gave the highest specific gravity and longest tracheids; following in order were roundwood, top

TABLE 12.4

Pulp Yields from Three Specific Gravity Categories of Wood from Young Loblolly Pine Trees

Specific Gravity	0.37	0.42	0.48
Pulp yield (% dry wood)	44	46	47
Kg pulp/m^3 green wood	160	191	224
Mill production/% of 30-year-old[a]	76	90	107

[a]Average gravity of 0.44.

TABLE 12.5

Generalized Yields from 100 Solid ft³ (2.83 m³) of Green, Debarked Loblolly Pine Wood of Different Ages

Stand Age (year)	Dry Clear Wood		Water		Resin Extractives		Other[a]	Total
	lb	%	lb	%	lb	%		lb
25	2700	43	3450	55	100	1.5	?	6440
30	2950	47	3200	51	120	1.8	?	6420
35	3030	48	3100	50	140	2.2	?	6440

[a]Include knots, resin associated with knots, "includes" bark, and so forth. It appears this material constitutes between 3 to 7% of woods-run logs.

wood, 15-year-old trees and 10-year-old trees. As one example, they found the average wall thickness in slab wood to be nearly twice that from 10-year-old trees.

Variation Patterns among Trees   Of great importance to tree breeders is the variation among trees of the same age and of the same species that are growing on the same site (Figure 12.7). Some foresters have the idea that the wood of most trees of any given species will be similar—that *Eucalyptus* wood of a given species grown in one environment will not vary much from one tree to the next. Such uniformity in wood quality among trees from seed does not exist. For all wood characteristics that have been adequately studied to date, variation among trees of the same age growing on the same site has always been found to be large (van Buijtenen et al., 1961; Thorbjornsen, 1961; Webb, 1964; Skolmen, 1972). For example, regardless of locality, if 50 loblolly pine trees of the *same age, same crown class*, and growing on the *same site* are sampled at breast height, the difference in specific gravity of the mature wood between the highest- and lowest–gravity tree will be about 0.20. This difference between the lightest and heaviest trees is equivalent to approximately 700 to 1000 lb (317 to 454 kg) dry weight of wood per 100 ft³ (2.83 m³).

There are hundreds of references on tree-to-tree variation in wood quality. The one sure thing in forestry is that wood specific gravity as well as other wood qualities, will vary greatly from tree to tree, regardless of the species or where the trees are grown. Since heritability of wood characteristics is also usually high, the two ingredients for good gains from a tree improvement program are present.

Wood Variation among Sites and Geographic Areas   Wood characteristics are the result of varying growth processes, and any factor that affects the growth pattern of a tree may also affect its wood properties. Whether trees grow on sandy or clay soils, under short or long growing seasons, or are subjected to major and differing environmental variations, some effect of the environment on wood quality is to be expected. In a number of studies involving several species,

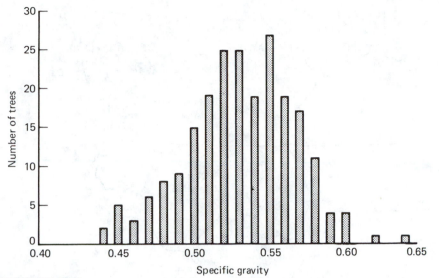

FIGURE 12.7

Within a stand of trees, there is a great variation in specific gravity. Illustrated are the gravities for a large number of *P. taeda* of the same age that are growing on similar sites and that are showing the huge differences that exist. Normally, a forest stand with trees of the same age will contain trees that vary from one another in specific gravity about 0.15 to 0.20, as is illustrated. This magnitude of variation in addition to a strong inheritance pattern enables the tree breeder to change wood specific gravity rather easily in the desired direction.

differences in specific gravity from different geographic areas have been found (Howe, 1974; Talbert and Jett, 1981). Yet it must be emphasized that although geographic differences have been observed, large individual tree-to-tree variation has been maintained in each instance. It is significant that only small differences in mean specific gravity occur among stands within a geographic area that may be growing on environments that are only moderately different.

Reports on trends of variation of wood specific gravity among natural stands have a long history. One trend often observed is that specific gravity within a species range is lower when it is inland from the coast or toward the higher latitudes or higher elevations. These trends hold for many pine species as well as for other softwoods and hardwoods. The magnitude of this trend for loblolly pine in the southern United States is shown in Figure 12.8. Note how specific gravity drops from south to north and from coast to inland; the latter differences are less than those related to latitude. These geographic differences in wood specific gravity may be relatively large and are important to wood-using organizations. Such differences are generally environmentally induced and are not highly heritable. For example, loblolly pines grown from seed from the high-wood-specific-gravity southern coastal areas do not produce higher specific gravity than

FIGURE 12.8

An illustration of variation in average specific gravity and tracheid length of loblolly pine with geographic locations is shown for the Southeast in the United States. The tendency is for lower average specific gravity and shorter tracheids in the north of the species range. Values shown are averages for plots of 23 trees each. The values shown are for specific gravity and the values in parentheses are tracheid lengths in millimeters.

that of the local trees when grown in the inland areas or in the northern latitudes. Major errors have been made by making the assumption that progeny from trees in high-specific-gravity areas will grow high-gravity wood when planted in other geographic areas. This usually does not happen.

There are numerous instances of extreme changes in wood specific gravity when trees are grown in exotic environments. One of the most outstanding examples is in the coastal region of southern Africa. There *P. caribaea* produces exceptionally low-specific-gravity wood, whereas *P. elliottii* under almost identical environments produces unusually high-specific-gravity wood. One of the greatest dangers when growing exotics is to move a species into its new environment without a previous determination of the kind of wood that will be produced (Howe, 1974). Huge acreages have been planted with no thought given about the kind of wood that would grow. There have been some very large losses when the wood of the exotic has proven to be subpar and has not produced wood similar to that produced in its indigenous environment.

Methods of Sampling for Wood Properties

A knowledge of the within-tree pattern of variation is of key importance to any studies related to wood qualities or inheritance patterns in wood. Many errors have been made by not knowing or by ignoring this pattern. The key is this: *Never compare wood of different ages* from the pith. If trees to be sampled are of different ages, then only the mature (or juvenile) woods can be compared. Even then, the comparisons are not completely valid because there is some change in specific gravity with age. For example, the breast-height juvenile-wood specific gravity of loblolly pine appears to increase with the age of the tree (Talbert and Jett, 1981). Because of its stability, it is best to use mature wood of trees for comparative purposes when such wood is available (Echols, 1959; Zobel et al., 1960b; Klem, 1966).

For many softwoods, the sampling height is of key importance. Since it is easier to sample from the ground, the breast-height level (4.5 ft or 1.4 m) is most commonly used. To be able to estimate whole tree values from such a single sampling location, a regression of breast height to total tree values must be made. Normally, about 40 trees need to be sampled to construct a regression equation suitable to predict total tree values from breast-height values (Figure 12.9). For most hardwoods, the location of the sample is not as important as it is in the softwoods because the variation from the center of the tree outward, or from the base upward, is generally much less than that of the softwoods.

It takes some skill to recognize and separate juvenile from mature wood, because there is no definite line between the two but only a period of general transition (see Figure 12.6). Often, both the juvenile and transition wood are included in the juvenile sample; therefore, the comparisons made to judge the wood quality among trees are only from known mature wood.

A proper determination of the wood characteristics of a species, race, or progeny must take into account the magnitude of the variability that is present. For trees of unknown parentage, usually a minimum of 30 trees should be sampled before a meaningful average value can be obtained. It does not require much imagination to recognize the futility of an effort to determine the specific gravity of a group of trees by sampling only one or two of them. The results of such limited sampling are often grossly misleading, as might happen when trees on the extremes of the variation pattern were those that were chosen.

Unfortunately, inadequate sampling to determine average wood values has been the rule rather than the exception. In one striking but typical example, the

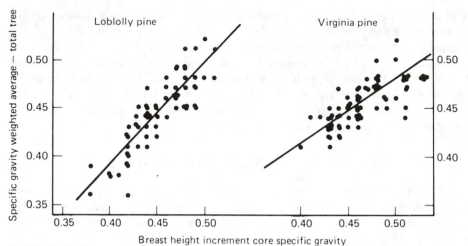

FIGURE 12.9

If samples are to be taken at any easy location, such as breast height (4.5 ft or 1.4 m), a good correlation, such as those shown for two species of southern pines, is necessary. The correlations are usually quite high; it normally takes 40 or more trees on a given site to develop a useful regression of breast height to total tree values.

owner of a large tract of land supporting a number of different species needed to select which species should be favored in a forest management program. The wood property evaluations were made on only *one tree* of each species, and *on the basis of this one sample* each species was retained or rejected for future management. When suitable tests were made by sampling 30 trees of each species, it was found that five of the eight species rated on single-tree samples of their specific gravities were not representative of the species' average specific gravity. Two of the very best species would have had to be discarded and one of the poorest kept, if the final choice of species had been based upon the single-tree sample. Such disregard of the basic nature and magnitude of variation among trees within a species would appear to be inexcusable, but it is common, nevertheless. Large sums of money are at stake, and the decisions based on inadequate data from too few trees often give results so misleading that the forestry enterprise will fail. This has happened a number of times, especially in instances in which exotic trees have been involved.

Variation among trees must, therefore, always be kept fully in mind in all situations when the woods of forest trees are sampled, regardless of whether the information is to be used for research, genetic development, or for operations. For example, if one is to assess the effects of fertilizer, site, soil, or spacing on wood properties, enough trees must be examined for each treatment to permit evaluation of individual-tree responses over and above the inherent variability of the trees. Disturbing as this may seem, there is no such thing as the *tree of average wood quality for any one stand of timber.* Often the wood from a tree of average height, diameter, basal area, or growth rate is sampled as having wood representative of a stand, but this is totally incorrect because all of these characteristics are not directly related to wood properties and environmental conditions can be important (van Buijtenen, 1969). Thus, the "average" tree might represent one of the extreme individuals for a particular wood property (see Figure 12.7 as shown).

In sampling trees for wood qualities, it is essential that a random selection be made if the mean value and the associated error terms are to be measured at some specified confidence level. Sampling can be from the whole population or it can be stratified into a desired portion of the population. For example, often only the dominant and codominant trees are sampled; the selection of trees within this group must be random. The specified confidence level to get "meaningful estimates" depends upon the needs and desires of the forester, but the general standard used is the 95% confidence level.

CONTROLLING WOOD QUALITY

There are a number of ways that wood qualities can be controlled in addition to choosing desirable species and provenances. As discussed previously, the age of harvest is a major tool because of its effects on the percentage of juvenile wood— as the stand becomes older, the proportion of juvenile wood decreases rapidly, as is shown in Table 12.6. The most effective way to control or change wood qualities in softwoods is to change rotation age. Similarly, plantation spacing will have a marked effect upon the proportion of juvenile wood.

Manipulation of tree form, whether through forest genetics or forest management, is a most powerful tool to improve wood qualities. Actual genetic manipulation of wood has proven to be very successful with high inheritance and good gains resulting from selection. Anything that alters growth or growth patterns, such as fertilizers, may change wood properties; there is a special section later in this chapter that is relevant to the effect of a growth rate on wood qualities.

For simplification, each of the factors mentioned later will be discussed primarily with respect to how they relate to specific gravity and, occasionally, to tracheid length. It is of importance to reemphasize that specific gravity is not a simple trait, but is a complex of wood properties involving cell wall thickness, cell size, summerwood percentage and other factors.

TABLE 12.6

The Relative Amount of Juvenile Wood and Mature Wood in the Boles of *P. taeda* Trees of Different Ages

Age of Tree (years)	Percentage Juvenile Wood	
	Dry Weight	Volume
15	76	85
25	50	55
45	15	19

Tree Form

The value of improving wood through improvement of tree form is very often overlooked. Manipulation of tree form is one of the easiest and quickest ways to improve wood, because it can be done both genetically and silviculturally and because gains can be considerable and rapid.

The easiest way to improve wood quality is to develop straighter trees that also have smaller limbs which grow at right angles to the tree bole. The main effect of these improvements is to reduce the percentage of reaction wood that accompanies trees that grow out of the vertical and have large or acute-angled limbs (von Wedel et al., 1968).

Whenever a tree leans, a kind of wood different from the normal is formed whose function is to help straighten the tree. This wood is collectively called *reaction wood* (Scurfield, 1973). When a softwood leans from the vertical, a kind of reaction wood called *compression wood* is formed (Low, 1964). Such wood, which develops on the underside of the bole of a leaning tree, has abnormal longitudinal shrinkage when dried, short tracheids, "fissures" in the thickened cell walls, and a higher than normal lignin content. In hardwoods, *tension wood* is formed on the upper side of the leaning tree. It has an unusually low lignin content and short fibers and often has gelatinous fibers that form planes of weakness in boards containing it. Compression and tension wood never occur in the same tree—the former is found only in softwoods and the latter in hardwoods. Because of its major effects on the anatomy and chemistry of the wood produced, reaction wood is of great importance in the manufacture of both solid-wood products and for the use of wood as fibers, energy, or a source of chemicals. It is possible for over 50% of the merchantable volume to consist of compression wood in crooked loblolly pine trees (Zobel and Haught, 1962); about 5% of the volume of straight trees is compression wood.

Branch size has a major effect on the volume of wood in knots and associated abnormal wood around knots. It has an obvious effect on the quality and size of knots as they affect the strength of sawn boards and plywood. Wood in and around knots sometimes has high resin content and low cellulose content; in some ways it is similar to reaction wood. Pulp yields and quality are low, but the wood has a high energy content. In one study, 7% of the merchantable volume of small, 12-year-old normal loblolly pine was knots and associated abnormal wood; large-limbed trees had 14% of such wood (Von Wedel et al., 1968).

Since tree straightness is moderately inherited, it is possible to improve wood quality by breeding for this characteristic (Shelbourne et al., 1969). Good forest management techniques such as control of stocking, thinning, and pruning, are silvicultural tools that are also available to improve tree form. Breeding for straightness and small limbs will be beneficial for improving wood qualities for fiber products, and selection for these traits is essential in an improvement program aimed at production of high-quality solid-wood products. Value improvement from straighter trees has been shown for both yield and quality of lumber and of plywood. One example of the effect of tree straightness and limb size has been reported by Blair et al. (1974). They found that straightness improved both the yield and quality of pulp produced, whereas the greatest effect of limb size was on tear factor, a key quality in many kinds of paper.

Genetics of Wood

One effective way of changing wood is to breed strains of trees with the desired properties. Most wood properties are from moderately to strongly inherited, enabling a rapid change in the desired direction. Some selected summary publications about inheritance of wood in varied species are listed in Table 12.7.

TABLE 12.7
A List Of Some Publications Indicating Inheritance of Wood Qualities for Several Forest Tree Species

Author	Date	Publication	Contents
Akachuku	1983	*For. Sci.*	Genetic control of wood in *Gmelina*
Armstrong and Funk	1979	*Wood and Fiber* 12(2):112–120	Genetic variation in wood of *Fraxinus americana*
Bendtsen	1978	*For. Prod. Jour.* 28(10):61–72	General, on improved wood
Burdon and Harris	1973	*N.Z. Jour. For. Sci.* 3(3):286–303	Wood density in four clones of *P. radiata*
Chudnoff and Geary	1973	*Turrialba* **23**(3):359–362	Heritability of wood density in *Swietenia macrophylla*
Dadswell and Wardrop	1959	*Appita* **12**(4):129–136	Growing trees with wood desirable for paper
Dyson	1965	EAAFRO; For. Tech. Note 16.	Wood quality and tree breeding in East Africa
Goggans	1962	Technical Report 14, North Carolina State University.	Inheritance of wood properties in loblolly pine
Harris	1965	Proc. Conf. of Sec. 41, IUFRO, Melbourne,- Australia, pp. 1–20	Heritability of wood density
Keller	1973	*Ann. Sci. Forest.* 30(1):31–62	Heritability of wood of *P. pinaster*
Kennedy	1966	*Tappi* 46(7):292–295	Heritability of several wood characteristics in clonal Norway spruce
Nicholls	1967	*Sil. Gen.* **16**(1):21–28	Wood qualities for tree breeding
van Buijtenen	1962	*Tappi* 45(7):602–605	Heritability estimates in wood density of loblolly pine
Zobel	1964	*Unasylva* **18**:89–103	Breeding for wood properties in forest trees

Actually, a whole book could be written on the importance of inheritance and variation in wood properties and their potential inclusion in a tree improvement program.

A large inheritance pattern that is useful has been found for nearly every wood property, but the emphasis has been upon specific gravity. Breeding for wood improvement is usually not a major objective; it is supplementary to breeding for growth, form, pest resistance, and adaptability. The inheritance pattern is strong enough to obtain good gains by using this approach. Wood improvement, in a sense, can be likened to "cream on the milk." The main breeding program is usually to improve the amount of "milk," and wood improvement is added to that—just as cream adds to the value of milk.

Wood Specific Gravity   Throughout this chapter the strong inheritance pattern for wood specific gravity has been emphasized. This does not need to be developed further here, other than to emphasize that the characteristic specific gravity combines a high heritability with a large variation pattern, enabling good success with a breeding program.

Wood specific gravity comes close to being the ideal characteristic to manipulate genetically because of the large tree-to-tree variation, the strong heritability, low genotype × environment interaction, and its major effects on yield and quality. For both softwoods and hardwoods, heritability of specific gravity is in the range of $h^2 = 0.5$ to 0.7 (Stonecypher and Zobel, 1966; Einspahr et al., 1967; McKinney and Nicholas, 1971; Polge, 1971; Nicholls et al., 1980; Land and Lee, 1981). Both of the ingredients for gain are present: good heritability and good selection differential. For example, one company found that moderately intensive selection for specific gravity in pine has produced a gain of 300 to 500 lb (140 to 230 kg) dry weight/100 ft^3 (2.83 m^3) of wood.

Fiber and Tracheid Length[1]   Another wood characteristic that has great variability and also a strong inheritance is cell size (Wheeler et al., 1965; Smith, 1967; Ujvari and Szönyi, 1973). Although its effect on the final product is usually much less than is specific gravity (Barefoot et al., 1970), cell length can have important effects on paper properties. This is especially true for short-fibered hardwoods as well as for the juvenile wood of some conifers, whose tracheid lengths are equal to, or are smaller than, the fibers of some hardwoods.

Tree-to-tree variations in average fiber or tracheid length are similar to specific gravity. For example, from a sample of over 300 loblolly pines of the same age growing on fairly similar sites, one tree was found with an average tracheid length of only 2.6 mm for the thirtieth ring at breast height, whereas another tree from the same area has tracheids 6.1 mm long from the thirtieth ring. Such huge

---

[1]Technically, conifers do not have fibers; they only have tracheids and ray cells. However, industrial personnel in most countries traditionally refer to both tracheids and fibers as fibers. In this book, *fibers* and *tracheids* will be referred to separately. Readers of the literature on wood properties should be aware that the terms are often used interchangeably.

differences in tracheid lengths among trees are not unusual, and to the forest geneticist they provide the variability needed in selection to change cell size. A variation pattern up the tree, ranging from shortest at the base, longest at the center, and shorter near the top, has been demonstrated by France and Mexal (1980) for *Picea engelmanii* and *Pinus contorta*. Just as for specific gravity, tracheid length drops for trees grown in the higher latitudes. Differences are as much as 1-mm average tracheid length for southern and northern stands of loblolly pine in the southeastern United States, with the shortest tracheids in the North.

Other Wood Properties   There are numerous other wood properties that might be used in a genetics program because they have reasonably strong inheritance patterns. If maximum gains are to be achieved in a breeding program, however, as few characteristics as possible should be included. Therefore, wood characteristics in addition to specific gravity and tracheid length are only included when there are special circumstances or needs.

Nearly all wood characteristics have an effect on pulp-and-paper properties. These characteristics include such things as cell wall thickness (which is closely related to specific gravity), cell lumen size, and length–width ratio of the cells. Flexibility, tensile strength, tear, burst, printability, and bendability are all important paper properties that are affected by fiber or tracheid properties that can be changed through breeding. However, most pulp-and-paper quality characteristics are so strongly controlled by wood specific gravity, with a lesser effect caused by fiber and tracheid length, that other wood characteristics are seldom used in breeding programs (Barefoot et al., 1970).

An area of wood quality breeding that has been largely ignored is the possibility of developing high-quality woods for special purposes, such as for furniture or finished cabinets. Enough has been done in this area to know that good gains can be made for quality. For example, selection for curly birch (*Betula*) has been successful (Heikinheimo, 1952). Similar gains are possible for wood such as walnut (*Juglans*) (Walters, 1951). Specialty breeding of high-quality hardwoods will likely become more important in the future, especially with tropical hardwoods and high-quality woods from the temperate zones. The anatomy of the tropical hardwoods can be very complex, and it will not be easy to develop breeding programs to improve it (Teles, 1980).

Inheritance of chemical properties of wood has been less well studied, but all indications are that the patterns may be useful. However, studies made on cellulose yield found it to be inherited in a nonadditive manner (Zobel et al., 1966; Jett et al., 1977), making a standard selection program ineffectual.

One kind of inheritance related to wood is the ability of a tree to compartmentalize diseased wood in the bole. Early studies have shown this ability to be rather strongly inherited, and if used, it will result in much less serious losses due to rot (Garrett et al., 1979; Lowerts and Kellison, 1981). Many other wood qualities, like spiral grain, are found to have from mild to strong inheritance patterns (Champion, 1929; Whyte et al., 1980).

Although less work has been done on hardwoods when compared to soft-woods, considerable information about the genetics of the woods of hardwoods is available (Zobel, 1965). Inheritance patterns for studies done are reasonably strong, but due to the complexity of wood, the results are less predictable. The feasibility of improvement of the wood of tropical species by genetic manipulation is much less well known.

Genetic Relationships among Wood Properties   Many of the different important wood qualities are genetically independent one from another; thus, one can have thick-walled tracheids and high-specific-gravity trees that are either short or long fibered. This is illustrated for different species of Mexican pines in Figure 12.10, which shows the independence of these two characteristics. Species are arranged in a decreasing order of specific gravity but show little relationship to tracheid lengths, thus indicating the independence of specific gravity and tracheid length among species. Similar results have been shown for individual trees within other species.

Some wood qualities are interdependent, such as specific gravity and wall thickness. These are often correlated, sometimes strongly. This is true of many of the morphological characteristics of cells, such as size, wall thickness, lumen size, and other factors such as the chemical characteristics of cell morphology. Thus, when one of these is changed, the other wood properties will also be affected. A prime example is compression or tension woods that are closely associated with variations in cell size, structure, and chemistry.

Because of the general independent inheritance of factors that affect the major wood properties, it is possible to "tailor make" the kind of wood desired. Thus, programs have been developed with *P. taeda* in the southern United States to produce kinds of woods that are best for newsprint, paper board, bags and boxes, writing paper, tissue paper, and other wood products. The first result from a wood improvement program will be to obtain greater amounts of wood with the desired properties that will be most useful for the desired product.

It is possible to breed strains of pine with a reduced core of low-gravity juvenile wood. This can become critical when the percentage of juvenile wood becomes more than 20% of the furnish used in the mill such as when very short rotations are used or when a high proportion of trees are used from thinning operations in young stands. Zobel et al. (1978) were able to increase the specific gravity of juvenile wood significantly in one generation of selection by using only the very highest-gravity trees from the original 1000 parents. When the 10 best families were used, at age 10 the gains in dry wood were 22.4 kg/m^3.

FIGURE 12.10

Shown are specific gravities ▭ and tracheid length ▬ of a group of Mexican pines. This illustrates the relative genetic independence of tracheid length and specific gravity among species.

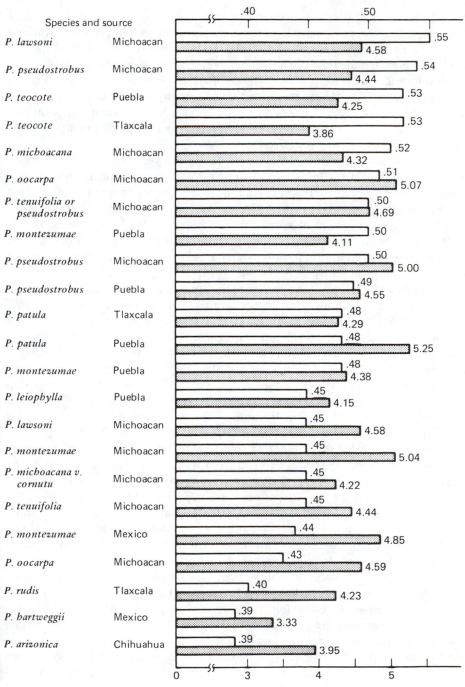

Specific gravity

Tracheid length (mm)

Effect of Growth Rate on Wood Properties

It was mentioned earlier that growth rate and wood specific gravity are by and large genetically independent, at least for some pine species. Others, like *P. radiata* in New Zealand, have a negative relationship between growth rate and specific gravity. Even when the relationship is negative, it is not strong enough to preclude the development of fast-growing high-specific-gravity trees for boards, bags or boxes or low-specific-gravity trees for tissues, quality writing papers, or newsprint. This relative genetic independence is also evident between tree form and wood qualities, and one can have straight and high-specific-gravity or straight and low-specific-gravity trees, as desired.

The relationship of growth rate to wood qualities is very important, it has been much studied, and it is confused, as is shown by the many contradictory results illustrated in the literature. It is very complicated because of the many factors that affect both wood and tree growth. As Larson (1962) stresses, anything that affects the physiology and growth of a tree can also influence the kind of wood that is formed. The literature in this area is voluminous for the temperate and exotic tree species. Only limited information is available on the relation of growth rate and wood in the tropical hardwoods (Howe, 1974).

Many foresters believe that a fast growth rate causes low specific gravity. This seems to be true generally for some genera like *Abies* and *Picea* (Stairs, 1969; Ollesen, 1976), but it is less so for many pine species in which growth rate and specific gravity are not correlated (Goggans, 1961; deGuth, 1980). Similar controversy, but on a lesser scale, exists for the hardwoods. In *Poplar,* Mutibaric (1967) found a slight decrease in wood specific gravity with increased ring width, whereas for black cherry (*Prunus serotina*), Koch (1967) found no relationship between growth and wood specific gravity. The same finding seems to be true for *Eucalyptus* (Brasil et al., 1979). Because of the confusion, it is difficult to make definitive statements about the relationship of wood to growth rate.

A major reason for the contention that growth rate is strongly and inversely correlated with specific gravity evolved from a failure to recognize within-tree variation. The wood of young wide-ringed trees has been compared frequently with the narrow-ringed wood from older trees, or the wide-ringed wood at the center of the tree has been compared with the narrow-ringed wood at some distance from the pith. Because juvenile wood usually has wide rings and low gravity, the erroneous conclusion was made that growth rate (ring width) caused the gravity differences. In fact, however, it does not matter whether rings are wide or narrow in most softwood juvenile woods; the gravity will be low. Mature wood will have higher gravity. The key point is that *if two pine trees are growing under the same environmental conditions but exhibit widely different growth rates, the faster-growing tree may have either higher or lower specific gravity than does the slow grower.* Attempts to assess specific gravity on the basis of ring width alone will lead to completely erroneous conclusions. In one study, the specific gravities of over 1000 of the very fastest-growing loblolly pines used as parents in seed orchards were compared with those of unselected trees with "average" or slower growth rates. It was found that their specific gravities were similar. The tree improver must be very critical whenever he or she is assessing the effects of growth rate on wood properties. For many of the hard pines, the pattern of little or no relationship between growth rate and wood specific gravity is well documented. This is illustrated for loblolly pine (see Figures 12.11 and 12.12). Usually, correlations between growth rate and specific gravity are very low, ranging from slightly negative to slightly positive. The lack of a strong growth rate–specific gravity correlation means that it is not necessary to sacrifice wood substance per unit volume when striving for rapid-volume production by breeding for growth and wood properties at the same time.

Results have been obtained from enough wood studies on pine so that the lack of a relationship can be accepted as a fact; that is, it is possible to have fast-growing trees with either high or low specific gravity. The forest manager has freedom to handle his or her forest to promote growth without substantially altering the specific gravity of the trees produced. In fact, a number of publications, such as those by Lowery and Schmidt (1967) on western larch or by Parker et al. (1973) on Douglas fir, show that the increased growth following thinning results in normal-specific-gravity or even increased-specific-gravity wood. One major exception is when heavy nitrogen fertilization is used to accelerate growth. Thus, both management factors and genetic factors to improve the growth rate or tree form of many conifers are independent enough from those for specific gravity that they

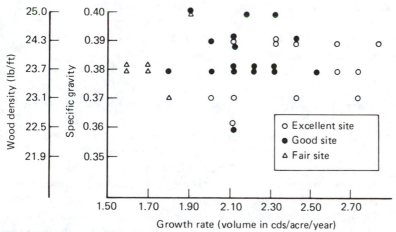

FIGURE 12.11

Shown is the relationship between growth rate and specific gravity for families growing on different sites. Each mark on the map is the average of 40 trees grown and measured for each family. Note the independence of growth rate and specific gravity.

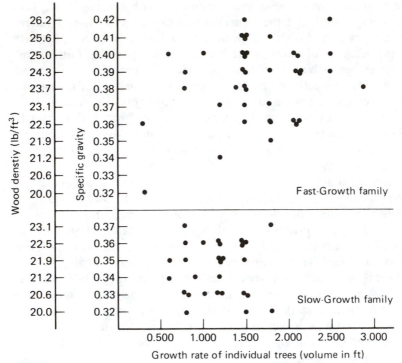

FIGURE 12.12

Growth rates and specific gravities are shown for individual trees from the fastest- and slowest-growth families in Figure 12.11. There is essentially no correlation between growth rate and specific gravity in trees, either in the fast- or slow-growth families.

give the tree improver a powerful tool to tailor-make trees for various products (Zobel, 1981).

Growth and wood-quality relationships are very complex in the hardwoods, in which a pattern is found that is directly opposite to the one frequently quoted for conifers; that is, the faster-growing trees have higher specific gravities than do slow-growing trees. This appears to be true for some of the ring-porous hardwood such as ash (*Fraxinus*) and oak (*Quercus*) but is not general for the numerous species of diffuse-porous hardwoods. In the ring-porous trees, it appears that an approximately equal volume of vessels is produced each year, regardless of the total growth during the year. Therefore, the slow-growing tree will have a greater proportion of vessels per unit volume of the annual ring, compared to the denser fibers and tracheids, and will have a low-specific-gravity wood. A fast-growing tree continues to produce the denser fibers outside the band of vessels; therefore, the wood will have a higher specific gravity. In diffuse-porous hardwoods the number of vessels formed in an annual ring is closely related to the width of the ring, and growth rate has little direct effect on wood specific gravity.

It has been found that if a pine tree is caused to grow faster because of release by thinning or by fertilization, the tracheids produced during the period of rapid growth will be somewhat shorter than those formed during a period of normal growth. This response is to a sudden or artificial environmental stimulus and usually tapers off within a few years. Within any stand, tracheid lengths are essentially uncorrelated with the inherent growth rates of individual trees, thus making it possible to have short or long tracheids, irrespective of whether the trees are genetically fast growing or slow growing. It must be stressed that for most pine species and Douglas fir, increased growth rate from environmental manipulation results in shorter tracheids, but length is independent of genetic potential for growth rate.

Sometimes the pattern is reversed for hardwoods in which the faster-growing trees have the longest fibers. Such a relationship has been shown for *Populus* by Kennedy (1957), Einspahr and Benson (1967), and several others. In actuality, the relationship of growth rate to fiber length has not been well studied for most hardwood species.

Fertilizers and Wood Qualities

As the use of supplemental nutrients becomes more widespread, there is an increasing concern about the effect that fertilization will have on wood qualities. One general reaction that has been observed and is documented for the hard pines and Douglas fir is that heavy nitrogen fertilization results in a lowering of specific gravity for about a period of 5 years (Posey, 1964) (Figure 12.13). Much more needs to be done to permit more precise evaluation of the effect of sudden and artificial changes in the nutrient environment upon wood when the growth rate has been markedly increased. Sometimes phosphorus fertilization will reduce the high-specific-gravity values that are common to trees grown under strong deficiencies of this element. The lowering of specific gravity by P reduces it to values of wood under normal growing conditions where no P deficiency exists.

One of the most desired benefits from fertilization is to make the wood of trees more uniform than when grown under normal conditions. This improved uniformity can be quite marked. In *P. taeda*, Posey (1964) found that high nitrogen fertilization reduces the gravity of inherently low-specific-gravity trees only slightly, whereas the gravity of high-density trees is greatly reduced. This is of special importance to the tree breeder because the improvement gained by breeding for high specific gravity can be partially nullified if a heavy nitrogen fertilization is used.

It appears that any environmental treatment that makes a softwood tree grow faster will result in shorter cells (Bannan, 1967). This has been explained by the rapidity of transverse divisions of the cambial initial cells that prevents develop-

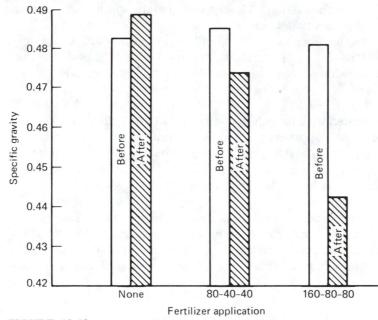

FIGURE 12.13

Wood qualities can be affected by forest management techniques. Shown are specific gravities for a stand of pine fertilized at age 16; *before* indicates the 7 years prior to fertilizing; *after* indicates the 7 years after fertilization. The high-N fertilizer was found to reduce specific gravity. It reduces individual high-gravity trees more than low-gravity ones, thus making for more uniform wood within the tree.

ment to full cell length before another division occurs. Cambial initials of slower-growing trees have sufficient time to achieve their length capabilities.

WOOD QUALITIES OF EXOTICS

Although general aspects of wood qualities have been thoroughly covered, it is essential also to discuss the wood properties of very fast-grown trees that are so common to exotic plantations in the tropical and subtropical regions. Wood qualities of exotics are often quite different from the same species in their indigenous ranges, especially in tropical plantations where young trees are harvested that contain a high proportion of juvenile wood (Zobel, 1981). In general, the wood qualities of fast-grown exotic hardwoods are acceptable, although sometimes the wood is undesirable for the product that is wanted.

Exotic Softwoods

Much of the coniferous wood produced in the future will be from fast-grown plantations of exotic species. Although the wood may be different from that from natural stands, it is not necessarily inferior; its utility is dependent upon the final product that is desired. Products such as writing papers or some tissue and newsprint can be made efficiently from the wood from young conifer plantations. For products that require good tear strength, a high proportion of juvenile wood is not desirable; therefore, wood from young exotic softwood plantations is inferior. Often tree form is very poor, resulting in a high proportion of reaction wood and large knots. When the young plantations are thinned, the combination of juvenile wood, reaction wood, large knots, and thick bark results in especially poor-quality wood, and their combined effects are that paper made from the young trees sometimes does not meet the standards for tear strength that are necessary for the world market (Dadswell and Wardrop, 1959). However, paper made from predominantly juvenile wood is often used for consumption within the country where it is grown. The strength and finishing properties of solid-wood products made from young exotic conifers with much juvenile wood often are poor (Pearson and Gilmore, 1980).

There is considerable variation among pines in their response to different tropical environments, as expressed in wood quality. Because of different responses to varied growing conditions, species such as *P. caribaea* and *P. kesiya* may have wood with essentially no summerwood when they are grown in certain environments or, in other environments, they will develop an extreme juvenile-wood core at the tree center, with very dense wood in the adjacent mature-wood zone (see Figure 12.5).

There have been a number of studies about the environmental effects on wood of tropical softwoods when they are grown as exotics; only a few can be mentioned here. For example, the wood of young *P. caribaea* was found suitable for dissolving pulps, and a series of papers by Foelkel (Foelkel et al., 1975a) for example, dealt with the value of woods of different fast-grown exotic tropical pines for sulfate pulping. Generally, these reports conclude that wood from these species is quite usable, although it is somewhat low in yield and tear strength.

Wood of young conifers is especially suitable for groundwood and thermomechanical pulping. Both processes require a heavy energy input, but the treatment of juvenile wood requires less energy usage than does the thick-walled mature wood. In the tropics some young wood, especially from thinnings, has been whole-tree-chipped. A limited amount of this kind of wood apparently can be tolerated in the mill furnish with no adverse effects, but problems sometimes result when more than 15 to 20% of the furnish is made up of whole-tree chips. Young trees have a high proportion of leaves and branches which when pulped occupy a disproportionate amount of digester space and require a heavy chemical usage although they contribute relatively little to fiber production.

Exotic Hardwoods

The undisputed leaders in fast-grown exotic hardwood plantations are various species of *Eucalyptus*. Despite the numerous trials being made and the several species planted on an operational scale, the *E. grandis–saligna* complex and *E. globulus* currently are clearly the most important. In good environments with proper care and improved genetic stock, the eucalypts can grow very rapidly and produce a most desirable wood.

There are many species and races of eucalypts that have widely diverse wood qualities (Foekel et al., 1975b). Some major eucalypt species (*E. deglupta*, for example) have wood specific gravities on the low side, whereas others, such as *E. tereticornis*, *E. citriodora*, and *E. cloeziana*, have high specific gravities that produce strong, dense wood that is especially suitable for charcoal. The two characteristics most affecting pulp qualities of the different eucalypts are specific gravity and extractives (Ferreira, 1968). Older trees frequently develop problems with deposition of phenolic substances in their wood, but young trees are reasonably free of adverse chemicals that can be very important in some species (Baklein, 1960). Eucalypt wood is often considered to be uniform wherever it is grown, but sometimes the environment in different plantations can produce somewhat different wood, and this must be watched closely (Hans and Burley, 1972; Taylor, 1973).

Although there is variability among species, it is of the utmost importance that the species of *Eucalyptus* planted most extensively (*E. grandis, E. saligna, E. regnans, E. globulus E. viminalis*) have wood in the midrange of specific gravity, and that they are suitable for many products. The wood and product characteristics of this group of eucalypts are similar to the soft hardwoods in the southeastern United States, such as *Liquidambar, Platanus,* and *Acer,* and they have wide utility.

The wood qualities of the eucalypts often vary considerably among individual trees within a species. The opportunity to change wood qualities in the desired direction through genetic manipulation is good (Rudman et al., 1969; Davidson, 1972; Doran, 1974). Wood from the eucalypts can be manipulated quite effectively in the desired direction by changes in environment through silvicultural treatment combined with breeding.

A complete treatment of wood from fast-grown exotic plantations could include numerous other hardwoods, but only a few will be mentioned here. One species with the greatest potential for tropical plantations is *Gmelina arborea;* this species grows very rapidly and is used in plantations in several tropical areas, most notably in the Amazon Basin in Brazil. Although its wood and pulping qualities have been extensively studied by many investigators, little wood-quality data have been published (Palmer, 1973). The wood of *Gmelina* is very good for most pulping operations and final products, and fast growth does not seem to have an adverse effect. Many companies have test-pulped it during the past 5 years, and reports say that it is a good wood, easy to work with, and is suitable for a number of products. *Gmelina* needs improvement in uniformity of growth and form, but it is easy to manipulate and can be handled using vegetative propagation.

Although some other fast-grown hardwoods other than *Eucalyptus* and *Gmelina* are planted as exotics, none are grown on a large enough scale to be of major importance on the world market yet. In the subtropical and temperate regions, short-rotation hardwoods are becoming more important and, although much information is now available about wood from these species, they cannot be considered as fast-grown when compared to the more tropical species (Jett and Zobel, 1974).

The current trend that will affect the usage of trees from exotic hardwood plantations is the move toward more utility of hardwoods for energy and chemical products (Goldstein, 1980). Currently, energy usage is receiving intense interest, along with the use of wood for chemicals. It appears that these two uses will become more important in the future. The ultimate effect of using wood for other than standard fiber and solid-wood products cannot be predicted, but if it occurs to any great extent, it will have a major impact on fast-grown exotic plantation culture and on tree improvement techniques.

Firewood accounts for about one half the timber consumed on a worldwide basis, and because many of the indigenous forests have been depleted, an increasing plantation culture of fast-grown exotics is being developed specifically for use as firewood. Much of the current emphasis on planting fast-grown exotic hardwoods in the tropical and subtropical areas is related to energy usage. There is an increasing interest in developing trees that are specially adapted to grow on areas close to habitations and that have wood that is especially desirable for firewood. On a worldwide basis, this generally means to breed for resistance to drought and eroded soils and for high specific gravity.

There will be a greatly increased need for genetic manipulation of wood in the tropics. Recently, methods have been developed to use the mixed tropical hardwoods as a source of pulp (Gomez and Mondragon, 1974). As a result, increased areas of tropical hardwoods will be cleared to be planted with exotics or indigenous species. Wood qualities and wood information are little known for the latter and must be included in tree improvement programs.

LITERATURE CITED

Artuz-Siegel, E. A., Wangaard, F. F., and Tamalang, F. N. 1968. Relationships between fiber characteristics and pulp-sheet properties in Philippine hardwoods. *Tappi* **51**(6):261–267.

Baklein, A. 1960. The effect of extractives on black liquor from eucalypt pulping. *Appita* **14**(1):5–15.

Bannan, M. W. 1967. Anticlinal divisions and cell length in conifer cambium. *For. Prod. Jour.* **17**(6):63–69.

Barefoot, A. C., Hitchings, R. G., Ellwood, E. L., and Wilson, E. 1970. "The Relationship between Loblolly Pine Fiber Morphology and Kraft Paper Properties," Tech. Bull. 202, North Carolina Agricultural Experiment Station, North Carolina State University, Raleigh.

Barker, R. G. 1974. Papermaking properties of young hardwoods. *Tappi* **57**(8):107–111.

Bendtsen, B. A. 1978. Properties of wood from improved and intensively managed trees. *For. Prod. Jour.* **28**(10):61–72.

Blair, R. L., Zobel, B. J., Franklin, E. C., Djerf, A. C., and Mendel, J. M. 1974. The effect of tree form and rust infection on wood and pulp properties of loblolly pine. *Tappi* **57**(7):46–50.

Brasil, M. A., Veiga, R. A., and Millo, H. 1979. Densidade básica de madeira de *Eucalyptus grandis* aos 3 años de idade. *IPEF* **19**:63–76.

Champion, H. G. 1929. More about spiral grain in conifers. *Indian Forester* **55**(2):57–58.

Dadswell, H. E., and Wardrop, A. B. 1959. Growing trees with wood properties desirable for paper manufacture. *Appita* **12**(4):129–136.

Davidson, J. 1972. "Natural Variation in *Eucalyptus deglupta* and Its Effect on Choice of Criteria for Selection in a Tree Improvement Program." Proc. 3rd Meet., Res. Comm. of Aust. For. Council, Mt. Gambier, Australia.

de Guth, E. B. 1980. "Relationship between Wood Density and Tree Diameter in *Pinus elliottii* of Missiones, Argentina." Div. 5, IUFRO Meeting, Oxford, England.

Doran, J. C. 1974. "Genetic Variation in Wood Density of *Eucalyptus regnans.*" 4th Meet., Res. Working Group, Aust. For. Council, Melbourne, Australia, pp. 99–101.

Echols, R. M. 1959. Estimation of pulp yield and quality of living trees from paired-core samples. *Tappi* **42**(1):875–877.

Einspahr, D. W., and Benson, M. K. 1967. Geographic variation of quaking aspen in Wisconsin and upper Michigan. *Sil. Gen.* **16**(3):89–120.

Einspahr, D. W., Benson, M. K., and Peckham, J. R. 1967. "Variation and Heritability of Wood and Growth Characteristics of Five-Year-Old Quaking Aspen," Inst. Paper Chem. Notes No. 1.

Einspahr, D. W., van Buijtenen, J. P., and Peckham, J. R. 1969. Pulping characteristics of ten-year loblolly pine selected for extreme wood specific gravity. *Sil. Gen.* **18**(3):57–61.

Ferreira, M. 1970. Estudo da variacão da densidade basica da madeira de *Eucalyptus alba* e *Eucalyptus saligna* Smith. Instituto de Pesquisas e Estudos Florestas (Piracicaba, Brazil) Report No. 1, pp. 83–96.

Foelkel, C. E., Barrichelo, L. E., do Amaral, A. C. B., and do Valle, C. F. 1975a.

Variation in wood characteristics and sulphate pulp properties from *P. oocarpa* with aging of forest stands. *IPEF* **10**:81–87.

Foelkel, C. E. B., Barrichelo, L. E. G., and Milaney, A. F. 1975b. Study of the *Eucalyptus* sp. wood characteristics and their unbleached sulphate pulps. *IPEF* **10**:17–37.

France, R. F., and Mexal, J. G. 1980. Morphological variation in tracheids in the bolewood of mature *Picea engelmannii* and *Pinus contorta. Can. Jour. For. Res.* **10**(4):573–578.

Garrett, P. W., Randell, W. K., Shigo, A. L., and Shortle, W. C. 1979. "Inheritance of Compartmentalization of Wounds in Sweetgum (*Liquidambar styraciflua*) and Eastern Cottonwood (*Populus deltoides*)," Forest Service Research Paper. NE No. 443, Northeastern Forest Experiment Station, Upper Darby, Pa.

Goggans, J. F. 1961. "The Interplay of Environment and Heredity as Factors Controlling Wood Properties in Conifers with Special Emphasis on their Effects on Specific Gravity," Technical Report No. 11, School of Forestry, North Carolina State University, Raleigh.

Goldstein, I. S. 1980. "Chemicals from Biomass: Present Status. Alternate Feedstocks for Petrochemicals." Proc. American Chemical Society, San Francisco.

Gomez, C. H., and Mondragon, I. 1974. The pulping of Colombian hardwoods for linerboard. *Tappi* **57**(5):140–142.

Hans, A. S., and Burley, J. 1972. Wood quality of eight *Eucalyptus* species in Zambia. *Sep. Exp.* **29**:1378–1380.

Harris, J. M. 1965. "The Heritability of Wood Density," IUFRO Section 41, Melbourne, Australia.

Heikinheimo, O. 1952. Kokemuksia visakoevun kasvatuksesta [Experiments in growing curly birch]. *Comm. Inst. For. Fenn.* **39**(5):26.

Higgins, H. G., Young, J., Balodis, V., Phillips, F. H., and Colley, J. 1973. The density and structure of hardwoods in relation to paper surface characteristics and other properties. *Tappi* **56**(8):127–131.

Howe, J. P. 1974. Relationship of climate to the specific gravity of four Costa Rican hardwoods. *Wood Fiber* **5**(4):347–352.

Jett, J. B., Weir, R. J., and Barker, J. A. 1977. "The Inheritance of Cellulose in Loblolly pine," TAPPI For. Biol. Comm. Meet., Madison, Wisc.

Jett, J. B., and Zobel, B. J. 1974. Wood and pulping properties of young hardwoods. *Tappi* **58**(1):92–96.

Kennedy, R. W. 1957. Fibre length of fast- and slow-grown black cottonwood. *For. Chron.* **33**:46–50.

Kirk, D. B., Breeman, L. G., and Zobel, B. J. 1972. A pulping evaluation of loblolly pine wood. *Tappi* **55**(11):1600–1604.

Klem, G. S. 1966. Increment cores as a basis for determining a number of properties of *Picea abies*. *Norsk Skogler* **12**(11/12):448–449.

Koch, C. B. 1967. Specific gravity as affected by rate of growth within sprout clumps of black cherry. *Jour. For.* **65**(3):200–202.

Kozlowski, T. T. 1971. *Growth and Development of Trees,* Vol. II. Academic Press, New York.

Land, S. B., and Lee, J. C. 1981. Variation in sycamore wood specific gravity. *Wood Sci.* **13**(3):166–170.

Larson, P. R. 1962. A biological approach to wood quality. *Tappi* **45**(6):443–448.

Larson, P. R. 1969. "Wood Formation and the Concept of Wood Quality," Bull. 74, Yale University School of Forestry, New Haven.

Low, A. J. 1964. Compression wood in conifers: A review of the literature. *For. Abs.* **25**(3,4):1–13.

Lowerts, G. A., and Kellison, R. C. 1981. "Genetically Controlled Resistance to Discoloration and Decay in Wounded Trees of Yellow-Poplar." M.S. thesis, School of Forest Resources, North Carolina State University, Raleigh.

Lowery, D. P., and Schmidt, W. C. 1967. "Effect of Thinning on the Specific Gravity of Western Larch Crop Trees." U.S. Forest Service Research Paper INT-70.

Marts, R. O. 1949. Effect of crown reduction on taper and density in longleaf pine. *South. Lumberman* **179**:206–209.

McKinney, M. D., and Nicholas, D. D. 1971. Genetic differences in wood traits among half-century-old families of Douglas-fir. *Wood Fiber* **2**(4):347–355.

Meylan, B. A. 1968. Cause of high longitudinal shrinkage in wood. *For. Prod. Jour.* **18**(4):75–78.

Mork, E. 1928. Die Qualität des Fichtenholzes unter besonderer Rucksicht-nahme auf Schlief -und Papierholz. *Papierfabrikant* **26**:741–747.

Mutibaric, J. 1967. Correlation between ring width and wood density in Euramerican Poplars. *Sumarstro* **20**(112) :39–46.

Nicholls, J. W. P., Morris, J. D., and Pederick, L. A. 1980. Heritability estimates of density characteristics in juvenile *Pinus radiata* wood. *Sil. Gen.* **29**(2):54–61.

Ollesen, P. O. 1976. The interrelation between basic density and ring width of Norway spruce. *Det Forstlige Försögrvaesen Danmark* No. 281, **34**(4):341–359.

Palmer, E. R. 1973. *Gmelina arborea* as a potential source of hardwood pulp. *Trop. Sci.* **15**(3):243–260.

Parker, M. L., Hunt, K., Warren, W. G., and Kennedy, R. W. 1973. "Effect of Thinning and Fertilization on Intra-ring Characteristics and Kraft Pulp Yields of Douglas-Fir." 8th Cell. Conf., Syracuse, N.Y.

Paul, B. H. 1957. "Juvenile Wood in Conifers," Report No. 2094, U.S. Forest Service, For. Prod. Lab., Madison, Wisc.

Pearson, R. G., and Gilmore, R. C. 1980. Effect of fast growth rate on the mechanical properties of loblolly pine. *For. Prod. Jour.* **30**(5):47–54.

Polge, H. 1971. Inheritance of specific gravity in four-year-old seedlings of silver fir. *Ann. Sci. Forest.* **28**(2):185–194.

Posey, C. E. 1964. "The Effects of Fertilization upon Wood Properties of Loblolly Pine (*Pinus taeda*). Proc. 8th South. Conf. For. Tree Impr., Savannah, Ga., pp. 126–130.

Rudman, P., Higgs, M., Davidson, J., and Malajczuk, N. 1969. "Breeding Eucalypts for Wood Properties." 2nd World Consul. on For. Tree Breed., Washington, D.C.

Scurfield, G. 1973. Reaction wood: Its structure and function. *Science* **179**:637–656.

Semke, L. K., and Corbi, J. C. 1974. Sources of less-coarse pine fiber for southern bleached printing papers. *Tappi* **57**(11):113–117.

Shelbourne, C. J. A., Zobel, B. J., and Stonecypher, R. W. 1969. The inheritance of compression wood and its genetic and phenotypic correlations with six other traits in five-year-old loblolly pine. *Sil. Gen.* **18**:43–47.

Skolmen, R. G. 1972. "Specific Gravity Variation in Robusta *Eucalyptus* grown in Hawaii," U.S. For. Service Research Paper PSW No. 78.

Smith, W. J. 1967. The heritability of fibre characteristics and its application to wood quality improvement in forest trees. *Sil. Gen.* **16**(2):41–50.

Stairs, G. R. 1969. "Seed Source and Growth Rate Effects on Wood Quality in Norway Spruce." Proc. 11th Comm. For. Tree Breed. in Canada, Ottawa, pp. 231–236.

Stern, K. 1963. Einfluss der Höhe am Stamm Verteilung der Raumdechte des Holzes in Fichtenbeständen [Influence of height of stem on the distribution of wood density in spruce stands]. *Holzforschung* **17**(1):6–12.

Stonecypher, R. W., and Zobel, B. J. 1966. Inheritance of specific gravity in five-year-old seedlings of loblolly pine. *Tappi* **49**(7):303–305.

Talbert, J. T., and Jett, J. B. 1981. Regional specific gravity values for plantation grown loblolly pine in the southeastern United States. *For. Sci.* **27**(4):801–807.

Taylor, F. W. 1969. The effect of ray tissue on the specific gravity of wood. *Wood Fiber* **1**(2):142–145.

Taylor, F. W. 1973. Variations in the anatomical properties of South African grown *Eucalyptus grandis*. *Appita* **27**(3):171–178.

Taylor, F. W. 1977. Variation in specific gravity and fiber length of selected hardwoods. *For. Sci.* **23**(2):190–194.

Teles, A. A. 1980. Sen Aprovertamento para Polpa a Papel Estudo de Madeiras da Amazonia Visaudo [Anatomical study of the woods of *Virola sebifera* and *Pseudobombax tomentosum* with a view to their use for pulp and paper]. *Bras. Flor.* **42**:25–34.

Thorbjornsen, E. 1961. Variation in density and fiber length in wood of yellow poplar. *Tappi* **44**(3):192–195.

Ujvari, E., and Szönyi, L. 1973. Expectable gain breeding long fibre Norway spruce. *Kulonlenyomat* **69**(2):93–99.

van Buijtenen, J. P. 1962. Heritability estimates of wood density in loblolly pine. *Tappi* **45**(7):602–605.

van Buijtenen, J. P. 1964. Anatomical factors influencing wood specific gravity of slash pines and the implications for the development of a high quality pulpwood. *Tappi* **47**(7):401–404.

van Buijtenen, J. P. 1969. The impact of state–industry cooperative programs on tree planting. *For. Farmer* **29**(2):14, 20, 22.

van Buijtenen, J. P., Zobel, B. J., and Joranson, P. N. 1961. Variation of some wood and pulp properties in an even-aged loblolly pine stand. *Tappi* **44**(2):141–143.

von Wedel, K. W., Zobel, B. J., and Shelbourne, C. J. A. 1968. Prevalence and effects of knots in young loblolly pine. *For. Prod. Jour.* **18**(9):97–103.

Wakeley, P. C. 1969. Effects of evolution on southern pine wood. *For. Prod. Jour.* **19**(2):16–20.

Walters, C. S. 1951. Figured walnut propagated by grafting. *Jour. For.* **49**(12):917.

Webb, C. D. 1964. "Natural Variation in Specific Gravity, Fiber Length and Interlocked Grain of Sweetgum (*Liquidambar styraciflua*) within Trees, among Trees and among Geographic Areas in the South Atlantic States." Ph.D. thesis, North Carolina State University, Raleigh.

Wheeler, E. Y., Zobel, B. J., and Weeks, D. L. 1965. Tracheid length and diameter variation in the bole of loblolly pine. *Tappi* **49**(11):484–490.

Whyte, A. G. D., Wiggins, P. C., and Wong, T. W. 1980. "A Survey of Spiral Grain in *P. caribaea* v. *hondurensis* in Fiji and its effects," IUFRO, Div. 5., Conf., Oxford, England.

Zobel, B. J. 1961. Inheritance of wood properties in conifers. *Sil. Gen.* **10**(3):65–70.

Zobel, B. J. 1965. "Inheritance of Fiber Characteristics and Specific Gravity in Hardwoods—A Review," IUFRO Meet. in Melbourne, Australia.

Zobel, B. J. 1981. Wood quality from fast-grown plantations. *Tappi* **64**(1):71–74.

Zobel, B. J., Campinos, E., Jr., and Ikemori, Y. K. 1983. Selecting and breeding for desirable wood. *Tappi* **66**:70–73.

Zobel, B. J., and Haught, A. 1962. "Effect of Bole Straightness on Compression Wood of Loblolly Pine," Technical Report No. 15, North Carolina State University, Raleigh.

Zobel, B. J., Thorbjornsen, E., and Henson, F. 1960a. Geographic, site and individual tree variation in wood properties of loblolly pine. *Sil. Gen.* 9(6):149–158.

Zobel, B. J., Henson, F., and Webb, C. 1960b. Estimation of certain wood properties of loblolly and slash pine trees from breast height sampling. *For. Sci.* 6(2):155–162.

Zobel, B. J., Stonecypher, R., Brown, C., and Kellison, R. C. 1966. Variation and inheritance of cellulose in the southern pines. *Tappi* 49(9):383–387.

Zobel, B. J., Kellison, R. C., Mathias, M. F., and Hatcher, A. V. 1972. "Wood Density of Southern Pines," North Carolina Agricultural Experiment Station Tech. Bull. No. 208.

Zobel, B. J., Jett, J. B., and Hutto, R. 1978. Improving wood density of short rotation southern pine. *Tappi* 61(3):41–44.

CHAPTER 13

Advanced Generations and Continued Improvement

In many tree improvement programs throughout the world, gains from first-generation activities are now being realized as seed orchards reach maturity, or through the use of vegetative propagation of genetically proven trees. Gains from advanced-generation tree improvement are potentially and considerably greater than those that are achievable in first-generation programs. In agriculture, numerous generations of breeding have resulted in improved varieties that have been altered genetically to such a degree that they bear only a faint resemblance to their ancestral relatives. Annual crop breeders have an advantage over tree breeders in that generations can be turned over at the rate of one or more per year, and many crops have been subjected to selection and genetic improvement for thousands of years both by trained and untrained breeders. The size and longevity of forest trees make them subject to breeding constraints, especially those involving time and space, which are not encountered with annual crops. Nevertheless, opportunities do exist to manage the genetic resources of important forest tree species over several generations in a way to alter their genetic potential significantly to meet the needs and desires of people.

The main objective of advanced-generation tree improvement is to maximize the gain per unit of time. To accomplish this goal, improved tree populations must be managed *from the outset* so that gains can be achieved each generation, while at the same time maintaining sufficient genetic variability to ensure continued long-term progress. Critical decisions in advanced-generation breeding include the choice of mating design and selection method, the intensity of selection applied, and proper management of inbreeding.

Breeders make genetic progress through many generations by recurrent selection (the relevant concepts were introduced in Chaper 4). Recurrent selection involves the choice of the best progeny from selected parents over successive generations. Several types of recurrent selection systems have been devised by plant and animal breeders, some of which are aimed at exploiting differences among individuals in general combining ability, whereas others have been devised to utilize specific combining ability, or to use both types of genetic effects. This chapter will be primarily concerned with improvement programs in which recurrent selection is used to improve general combining ability, although many of the methods discussed are applicable to any type of selection system. Many of the concepts covered in this chapter have been discussed elsewhere in the book, but they are mentioned together here because of their relevance to advanced-generation tree improvement.

BASE, BREEDING, AND PRODUCTION POPULATIONS

The most important concept related to advanced-generation tree improvement is a separation of populations with different functions. At any given time, three types of populations should be maintained in a breeding program, as are shown diagrammatically in Figure 13.1.

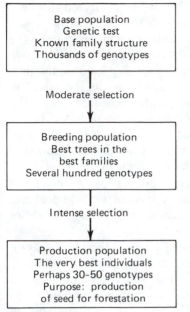

FIGURE 13.1

In most large-scale advanced-generation tree improvement programs, three different types of populations are usually maintained at a given point in time. These populations differ greatly in size and purpose as is indicated by the diagram.

The *base population* of a given generation consists of the trees from which the tree improver chooses to make selections for the next generation of breeding. In the first generation of tree improvement, the base population usually consists of trees growing in natural stands or unimproved plantations. In advanced generations, the base poulation is most often a genetic test consisting of the progeny of selected parent trees from the previous generation. The identity of at least one, and usually both, parents of all progeny growing in the genetic test is known. The base poulation most often contains hundreds to many thousands of genotypes.

The *breeding population* consists of a subset of individuals from the base population that is selected for their desirable qualities to serve as parents for the next generation of breeding. Increments of gain that result from proceeding to the next generation of improvement result from selection efforts for the breeding population. A compromise must be made between increasing the selection differential, which gives greater genetic gain but a smaller breeding population, and the need to maintain a breeding population large enough so that genetic variation is kept at a level that allows continued genetic progress in future generations. In practice, a moderate selection intensity is usually employed on individuals for the breeding population. Usually 200 or more selected trees are included in the breeding population each generation.

The third type of population in most ongoing tree improvement programs is the *production population,* which is used strictly to produce seeds or vegetative propagules for operational reforestation programs. In some instances, the production population may consist of the same set of genotypes as the breeding population, but usually the production population is a highly selected subset of the breeding population. As few as 20 to 30 genotypes may be included in the production population. These individuals represent the very best individuals in the breeding population and are the best selections that can be obtained for any given generation. They are chosen to maximize genetic gain in operational forest plantations. A small, intensively selected production population is a dead end from a breeding standpoint because of a very reduced base that would decrease progress from improvement efforts and very quickly lead to high levels of inbreeding.

Breeding and production populations are usually established in separate orchards. Design criteria for the two types of orchards generally are completely different because of their different functions. Breeding orchards are managed to expedite completion of the matings required in a generation and should be so designed. For example, if the mating design calls for control pollinations, all ramets of a single clone can be grafted next to each other to facilitate movement of the breeder among ramets. Radical cultural treatments may be applied to trees in a breeding orchard to promote the flowering required to complete the mating scheme, even if the long-term health of the trees in the orchard is affected in an adverse way. From a breeding standpoint, once the required matings are completed for a clone, the trees are no longer needed.

In contrast to management criteria and cultural treatments that are applied in breeding orchards, production orchards must be managed for long-term maximum production of genetically superior seed. A discussion of proper management techniques for production seed orchards is in Chapter 6.

The contribution of the breeding and production populations to the genetic gain that is realized in operational plantations is depicted in the hypothetical example in Figure 13.2. Percentage gain figures are not meant to represent those obtainable in any specific program or for any specific trait, but they do serve to illustrate the means by which tree improvement programs can make significant amounts of gain in both the short and long term. In Figure 13.2, gains on the order of 10% are shown to occur with each cycle of selection and breeding in the breeding population. An additional 15% gain is made by further selection each generation within the breeding population for the production population. Therefore, in the first generation, combined total gain is on the order of 25%. It is important to understand, however, that only the 10% gain through selection for the breeding population is accumulated from generation to generation. For example, the total amount of gain achieved through two generations of selection in the breeding population is 20%. The additional 15% gain that comes from establishment of an intensively selected production population is a gain that may be obtained each generation, but it is not passed to the next generation of improvement. Therefore, the total gain in second-generation production seed orchards is 35% in the theoretical example. As stated previously and as shown in Figure 13.2, the production population is a dead end for long-term improvement purposes and is used only for operational production of improved propagules.

In some tree improvement programs, the breeding and production functions may be served by the same populations (Namkoong, 1979). This situation usually results because of cost restraints that make it impossible to maintain both types of populations in one program. Although costs are less in such situations, gains in either the short term, long term, or both, may be severely compromised. When many hundreds of genotypes are included in both populations, prospects for

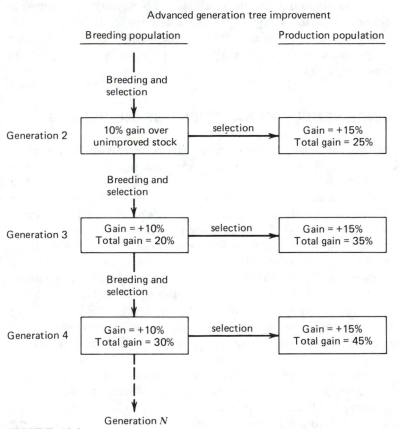

FIGURE 13.2

Genetic gain from tree improvement in a given generation arises from selection for the breeding population and further selection in this group of genotypes for the production population. The extra gain achieved by selection for the production population is not passed along to the next generation.

excellent long-term gains exist, but gains in the short term from operational production of propagules will be reduced because of a small selection differential. Alternatively, if the breeder uses a large selection differential and reduces the breeding and production population to a restricted number of genotypes, gains in the short term will be maximized, but gains in the long term will be limited by a reduced genetic base. For these reasons, breeding and production populations are kept separate in most lage-scale, long-term breeding programs.

The number of individuals that should be included in the breeding population depends upon the objectives of the program and the genetic structure of the population (Burley and Namkoong, 1980). Most tree improvement workers recommend that 200 or more trees be included in breeding populations (Libby, 1973; Schoenike, 1975; Talbert, 1979; van Buijtenen and Lowe, 1979). It has been recommended that up to 10,000 individuals should be saved if the purpose is gene pool conservation (Toda, 1965). Although it is beyond the scope of this book, it is possible to calculate the number of trees that need to be saved in order to assure a high probability of conserving alleles at various gene frequencies (Burley and Namkoong, 1980).

AVOIDANCE VERSUS MANAGEMENT OF INBREEDING

Inbreeding and its possible effect on tree improvement programs has been a major source of concern for tree breeders since tree improvement efforts were begun. Aspects of inbreeding and related matings were covered in Chapter 2. Most inbreeding studies have dealt with selfing, and in general, they have shown substantial depression in growth and also in seed yield (Orr-Ewing, 1965; Franklin, 1970; Andersson et al., 1974). There have been few investigations concerning depression at inbreeding levels less than selfing. One study with slash pine in which the effects of matings between half sibs, full sibs, and selfs were investigated showed that significant and predictable amounts of growth depression could be expected with increasing levels of inbreeding in this species (Gansel, 1971). Although many more research data dealing with low levels of inbreeding are needed, it is obvious from the selfing studies that have been conducted that tree improvers must be concerned with the potential buildup of inbreeding in their improved populations.

Inbreeding becomes a complex problem as tree improvement programs move into advanced generations. In first-generation tree improvement programs in which each parent tree has been selected from a different natural stand, or when trees have been selected from plantations established from a very broad genetic base, the chance that selected trees are related to one another is small. In the case of first-generation clonal seed orchards, all that must be done to avoid significant amounts of selfing is to separate the ramets of a clone at a sufficient distance in the seed orchard so that they do not pollinate each other. No related matings will occur in a breeding program involving control pollination unless selfs are made. In advanced generations, however, possibilities arise for matings of half sibs, full sibs, parents and offspring, or perhaps even more distant kinds of relatives. Avoidance of related matings in clonal production seed orchards becomes more difficult when several forms of relatives are included in the select population.

Related matings will eventually occur in a closed breeding population regardless of the mating design used. Even the most extensive crossing scheme available to breeders will result in the number of unrelated families being reduced by one half in each generation (see Chapter 8). The number of unrelated families is further reduced by family selection, which is used to some extent in almost all advanced-generation tree improvement programs. In some instances, opportunities may exist to delay the onset of inbreeding in the population by the infusion of new unrelated selections into the breeding program when relatedness threatens to become a problem. However, as programs proceed through several generations and genetic gains increase, it is unlikely that any new genotypes in unimproved populations will be found that could be introduced into the breeding population without reducing the amount of gain that has been achieved. It might be possible to introduce improved but unrelated material from programs in other regions, but eventually even these sources of new unrelated material will be depleted. Therefore, it is a fact that inbreeding cannot be avoided in the long term in an advanced-generation tree improvement program. As a result, the tree improver must manage the population in such a way so as to minimize deleterious effects of inbreeding or to benefit from its occurrence.

Several strategies are available to manage inbreeding in tree improvement programs. One alternative is to allow inbreeding to build up slowly in breeding and production populations. Results of a long-term experiment with corn, involving over 70 generations of selection and breeding, have shown that substantial gains have been made and are still possible at inbreeding levels that are much higher than those obtained with one generation of selfing (Dudley, 1977). Inbreeding occurred slowly in the corn population, and the effects of selection and repeated recombination of alleles through breeding served to overcome, at least to an extent, the potentially deleterious effects of inbreeding. Whether this would hold true with forest trees is not yet proven, but in advanced generations the potential does exist to select and breed relatives that are only distantly related and to allow a slow buildup of inbreeding. If high levels of inbreeding are approached slowly, rather than quickly as with selfing, inbreeding depression may be less of a problem than selfing studies would indicate.

Another strategy receiving considerable attention as tree improvement programs proceed to advanced generations is that of multiple breeding populations, which were mentioned in Chapter 8. With this procedure, a large breeding population, usually consisting of several hundred individuals, is broken into a number of much smaller populations or breeding groups. The breeding strategy calls for matings to occur only between individuals who belong to the same breeding group. Because breeding groups consist of a small number of individuals, perhaps on the order of 25 or less, relatedness and subsequent inbreeding resulting from related matings will occur quickly within breeding groups. However, if only the best individual in each breeding group were chosen for use in a production seed orchard, parents in the orchard would be unrelated to one another and seed produced would be totally outcrossed. Use of multiple breeding populations to avoid inbreeding in production seed orchards has become known as *sublining*.

Multiple population breeding has been proposed in forest tree breeding for reasons other than avoidance of inbreeding in production seed orchards, including gene conservation (see Chapter 15), and for situations in which uncertainties exist regarding the desired goals of the tree improvement program (Namkoong, 1979; Burley and Namkoong, 1980). In the latter instance, the breeder could plan for each of the small populations to have somewhat different breeding goals and could perhaps have improved material available to meet whatever new future demands the market might present. In forest tree improvement, the current major operational use of multiple breeding populations has been for the purpose of inbreeding avoidance in production seed orchards (van Buijtenen and Lowe, 1979; McKeand and Beineke, 1980; van Buijtenen, 1981).

It is clear that inbreeding is a phenomenon that must be managed but cannot be avoided in long-term tree improvement programs. Whereas the appropriate strategy to use to manage inbreeding will vary with species, region, and program, it should be recognized from the outset of a program that inbreeding will eventually become a problem and that a strategy must be developed to cope with this complexity.

MATING AND TESTING

One of the first jobs of the tree breeder who is involved in a long-term tree improvement program is to choose a mating design. Once trees are mated and seed are produced, progeny must be established in field tests in a way that they will provide the breeder with the information needed to manage the tree improvement program precisely and accurately. As stated in Chapter 8, there may be several objectives to breeding and testing. In advanced-generation breeding programs, the two most important of these are nearly always an estimation of parental breeding value (or general combining ability), along with the establishment of a base population for the next generation of selection. If methods are available to utilize specific combining ability, as with vegetative propagation or the mass production of specific crosses, then determination of specific combining abilities will also be a major objective.

The most appropriate type of mating design depends upon several factors; these include the objectives of testing, types of gene actions, the urgency of time, and the financial resources that are available to the breeder. The mating design that is chosen should ideally be one that can be repeated efficiently generation after generation, that leads to a slow rate of inbreeding, and that can be completed in a timely manner.

If the program is not well funded or is secondary in importance, simple mating designs such as open-pollinated schemes or polymix matings may be employed.

Such designs have a severe limitation because male parentage is unknown, and selection from tests will lead to unknown amounts of relatedness. However, these designs have been used in agricultural crops with reasonable success. If this type of system is used, the breeder should make an effort to include a large number of parents in the program so that the rate of inbreeding can be kept at a low level.

In more sophisticated, large-scale programs, mating designs such as some form of diallel or factorial design are usually employed. Because estimation of the general combining ability of parents is a common objective, designs that require several crosses per parent are normally preferred. Such designs as small disconnected diallels or disconnected factorials, which can be completed in a short period of time, and groups of crosses planted in the field together are being favored in many advanced-generation programs (Talbert, 1979; Goddard, 1980; van Buijtenen, 1981). These designs allow estimation of general combining abilities as well as specific combining abilities for the crosses made and serve well in maintaining genetic diversity in the progeny population. Tester mating designs that use only a few parents that are mated to every other member of the population may give excellent estimates of general combining abilities. However, they are generally not favored in advanced-generation programs because genetic diversity among progeny is restricted by the number of parents used as testers.

Mating designs and field-testing designs cannot be considered separate in a tree improvement program. There must be a "marriage" of mating designs and field-testing designs if maximum information is to be obtained from the breeding program. Therefore, the field-testing design that should be used in advanced-generation testing depends to an extent on the type of mating design that is employed. If the mating design calls for small 4×4 factorials, resulting in 16 crosses per factorial, then all 16 crosses should be planted together in the field if maximum genetic information is to be obtained. Tests would therefore consist of some multiple of 16 seedlots in addition to check lots. Information gathered from first-generation testing programs should be considered in designing an advanced-generation field-testing plan. For example, knowledge about the extent of genotype × environment interaction in first-generation tests-can be used to determine the appropriate number of environments in which each set of advanced-generation crosses should be tested. If sufficient data are available, it may be possible to define "optimal" test environments that best describe genotypic performance throughout a breeding region, and therefore reduce the number of environments needed for testing (Allen et al., 1978). The considerations for types and distribution of plots, which were discussed in Chapter 8, apply in advanced-generation as well as in first-generation tests.

ADVANCED-GENERATION SELECTION

Selection is the means by which genetic gain is made in a breeding program. A number of different selection procedures or methods are available to tree breeders, the choice of which will influence the gain obtained both in the short and long terms. A summary of different selection methods was given in Chapter 4, and general discussion of them will not be repeated here. This section will concentrate on aspects of selection that are specific to advanced-generation tree improvement.

Selection in advanced-generation programs differs from that in most first-generation programs in several ways. In the first generation, the base population from which selections are made is most often a series of mature, natural stands, or unimproved plantations in which the parentage of the trees is unknown. As a result, breeders are usually limited to using individual or mass selection as a selection method. This is not always the case; in some situations, the initial group of parent trees may be chosen essentially at random (no selection) and their offspring established in genetic tests. Following this, clonal orchards may be established using the genetically best parents based on progeny performance, or seedling or clonal orchards may be established using the best progeny in the best families. If genetically improved planting stock is needed immediately, however, the usual procedure is to use individual selection and to establish the selected genotypes in a clonal seed orchard or in an area that is managed for production of vegetative propagules.

In advanced generations, selections are nearly always made from genetic tests. However, this may not always be the case; in programs with limited finances the only alternative may be to practice individual selection in production plantations from seed orchard stock. This type of selection method in advanced generations is usually not practiced because of fears that some selections will be related through common seed orchard parentage, which would cause inbreeding to proceed at unacceptably rapid rates. However, individual selection can result in gain for heritable characteristics; this is indicated by the achievements of primitive corn breeders who for many hundreds of years collected seeds for the next year's crop from the phenotypically superior plants in their fields. If selections are to be made in production plantations from seed orchard seed, Namkoong (1979) suggests that stands be planted separately by mother trees so that some control of ancestry can be maintained.

Most programs that proceed to advanced-generation selection and breeding will establish genetic tests from selected parents. This is much preferred, because of the greater genetic gain that can be achieved by utilizing some form of family selection and because the breeder can maintain more complete control of ancestry. The discussions that follow will be about selection in genetic tests, and how it differs from that of first-generation selection in stands of unknown parentage.

Selection in a Finite Population

An important aspect of selection in genetic tests is that the base population is finite in size with a limit to the number of individuals that can be screened. This differs from the situation that is often encountered in first-generation programs in which extensive areas of natural stands or plantations can be searched and in which a very large population exists from which to choose selections. Finite population size can have a real impact on the number of traits used as criteria for selection and the selection intensity that the breeder can apply for each trait. If a very large population exists, the breeder can select for many traits and use a high-selection intensity for each trait. This is the case, because if enough acres of natural stands or plantations are searched, a suitable number of trees will be found that meet the minimum criteria. In a finite population like a genetic test, there are only a certain number of individuals that can be screened to find selections, and it may be impossible to find suitable numbers of select trees that meet very intense selection criteria for a large number of traits. For example, it might be desired to select 200 individuals as parents for the next generation from a genetic test base population of 10,000 trees. If height was the only criterion for selection, then the breeder would select the 200 best trees, perhaps using a family and within-family selection method for height. If high wood specific gravity were added as a criterion for selection and given equal weight as height, then 200 trees would still be chosen. However, unless wood specific gravity was perfectly correlated with height (which it is not), the 200 best trees for height would not be the same as the 200 best trees for wood specific gravity. Therefore, the breeder would choose the 200 trees with the best *combinations* of wood specific gravities and heights. This means a lower selection intensity must be used for both height and wood specific gravity than would be the case if selection was practiced on either trait alone. When more traits are added as selection criteria, selection intensity for any one trait will decrease even more. The preceding situation makes it of the utmost importance that the breeder decide carefully and critically which set of traits should be used as selection criteria and that traits of minimal importance be given little or no consideration in the selection scheme. Additions of more traits will result in less gain in any individual trait anytime that a certain number of individuals must be selected from a finite genetic-test population.

Early or Juvenile Selection

The importance of time in the developmental aspects of tree improvement has been emphasized many times in this book. Due to the long rotations of forest tree crops and the extended period before most tree species reach reproductive maturity, generation intervals in forest trees are usually measured in decades rather than in growing seasons. As a result of the importance of time, tree breeders are constantly searching for ways to reduce the number of years required for a generation of selection and breeding. The goal of an advanced-generation tree breeding program is to maximize gain achieved per unit of time (van Buijtenen, 1981). This means that anything that can be done to reduce the generation interval without severely reducing genetic gain per generation can result in a more efficient improvement program.

One way to reduce the generation interval in advanced-generation tree improvement is to make selections in genetic tests long before trees have reached rotation age. Selections can be made most efficiently in young tests if performance of trees in the test at rotation age can be accurately predicted from their performance at younger ages. Phrased another way, there must be a high genetic correlation between rotation age and a younger age if early selection is to be successful. These correlations are known as juvenile–mature correlations. This most important relationship has already been mentioned several times in this book. It will be covered again here because of its extreme importance in advanced-generation breeding.

Unfortunately, juvenile–mature correlations in forest trees are nearly always less than 1. This has been demonstrated for growth in coniferous species such as loblolly pine (Wakeley, 1971; LaFarge, 1972), ponderosa pine (Steinhoff, 1974; Namkoong and Conkle, 1976), Douglas fir (Namkoong et al., 1972), western white pine (Steinhoff, 1974), slash pine (Squillace and Gansel, 1974), and various hardwood species such as hybrid poplars (Wilkinson, 1973) and black walnut (McKeand et al., 1979). As might be expected, correlations of growth performance at very young ages, say age 3 or less, with performance at mature ages are very poor, but correlations improve progressively as the assessment time becomes closer to the mature age.

Because of imperfect juvenile–mature correlations, tree breeders are faced with the dilemma of deciding how much gain per generation should be sacrificed in the hope of improving gain per unit time by selecting at a juvenile age and thereby shortening the generation interval. Selection at a young age in the hope of improving performance at rotation age is a form of indirect selection or selecting for one characteristic (performance at a juvenile age) to improve another trait (performance at a mature age). Genetic formulas are available to calculate the gain that can be expected from indirect selection (Falconer, 1981).

In Chapter 4 it was shown that gain from mass selection can be expressed as

$$G = ih^2\sigma_p$$

where

$$
\begin{aligned}
G &= \text{gain} \\
i &= \text{selection intensity} \\
h^2 &= \text{heritability} \\
\sigma_p &= \text{phenotypic standard deviation}
\end{aligned}
$$

Gain from early selection, or correlated gain (CG), may be calculated in a similar manner:

$$CG = ih_j h_m r_G \sigma_{p_m}$$

where

$$
\begin{aligned}
CG &= \text{correlated gain} \\
i &= \text{selection intensity} \\
h_j &= \text{square root of the heritability of the juvenile trait} \\
h_m &= \text{square root of the heritability of the mature trait}
\end{aligned}
$$

r_G = genetic correlation between the juvenile and mature trait

σ_{p_m} = phenotypic standard deviation of the mature trait

Gain per unit time that is achieved by early selection can be calculated simply as

$$G_T = \frac{ih_j h_m r_G \sigma_{p_m}}{T}$$

where

$$G_T = \text{gain per unit time}$$
$$T = \text{generation interval}$$

and other symbols are as defined previously. The appropriate age to select would be the one that maximizes gain per unit time.

The optimum age at which to select is currently a topic of considerable debate among tree breeders. The debate revolves mainly around the magnitude of the various components in the gain per unit time formula illustrated previously. For example, the degree to which heritabilities may change with age is somewhat uncertain. There is some evidence from several conifers that heritability for growth may increase as genetic-test plantations increase with age and that heritability may culminate at about the midlife of the stand (Namkoong et al., 1972; Namkoong and Conkle, 1976; Franklin, 1979). An increasing heritability with time would increase optimum rotation age because of the appearance of the square roots of juvenile and mature heritabilities (h_j, h_m) in the numerator of the gain per unit time formula. Likewise, there is debate about the magnitude of genetic correlations (r_G) between the performance at juvenile ages and at maturity (Franklin, 1979; Lambeth, 1980).

Despite the many uncertainties, most tree-breeding organizations feel that selections should be made at relatively young ages in genetic tests. For example, utilizing growth data from numerous species in the family Pinaceae, Lambeth (1980) used a formula similar to the gain per unit time formula given previously to calculate that optimum rotation age was between 6 and 8 years for rotation ages of 20 to 40 years. This is approximately the age at which second-generation selections are currently being made in southern pine improvement programs. A 6-year-old second-generation loblolly pine selection is shown in Figure 13.3.

Anytime juvenile–mature genetic correlations are less than 1, selection at a juvenile age will result in a group of select trees that is *not* the best group of select trees at rotation age, even though they were the best group at the optimal selection age. It is important that the tree improver realize that just because the select group of trees deteriorates somewhat in quality with time, selection efforts have not failed. The deterioration is simply an aspect of imperfect juvenile–mature correlations and is to be expected when trees are selected before rotation age.

FIGURE 13.3

Selections for the next generation of improvement are often made at young ages from genetic tests, well before rotation age. Shown is a 6-year-old second-generation loblolly pine selection. Note its superiority. Selection at young ages decreases the generation interval and offsets imperfect juvenile–mature correlations, thus maximizing genetic gain per unit time.

Selection Methods

Nearly all advanced-generation tree improvement programs use some form of family and individual within-family selection system to rank trees as candidates for selection. This selection method results in a greater expected genetic gain, especially for low heritability traits (Falconer, 1981), and it makes maximum use of information derived from genetic tests. For traits that are highly inherited, such as wood specific gravity, individual or mass selection can result in nearly as much gain as selection on family and individual values. Traits with lower heritabilities, such as growth rate, greatly benefit from family and within-family selection. Mass selection is never more efficient in making genetic progress than combined family and within-family selection. Use of high-speed computers to rank individuals based upon their family and individual scores greatly facilitates use of this method.

Although use of family and individual information to rank phenotypes for characteristics of interest allows the breeder to make excellent genetic gain, it also makes it mandatory that care be taken to maintain a sufficient genetic base for long-term improvement. As mentioned in Chapter 4, family selection involves exclusion of entire families from the breeding population and can result in rapid rates of inbreeding in the improved population. Ultimately, there is a trade-off that involves greater short-term genetic gain that can be achieved by only using individuals in the very best families and restrictions in long-term genetic gain that result from a much reduced genetic base. Intensive selection on family performance can result in loss of desirable alleles from individuals in poorer families; this can also hamper long-term progress from selection. In the very long run, selection on phenotypic (individual) values alone may result in greater genetic gain because of the reduced genetic variation that results from selection on family values (Sirkkomaa and Lingstrom, 1981). However, selection systems that include selection on family performance do result in greater short-term gain and are almost always used in tree improvement because adequate levels of genetic gain are needed in each cycle to make tree improvement programs profitable. A viable alternative may be to restrict the number of individuals selected per family so that more families can be included in the program (Roberds et al., 1980). Regardless of the selection method used, very intensive selection in finite populations can reduce long-term gains that can be achieved because of a reduced genetic base, and the breeder must maintain a sufficient number of individuals in the selected population to ensure continued progress.

Advanced-generation tree improvement programs almost always involve selection for several traits simultaneously. Although the selection index approach that was first mentioned in Chapter 4 has been used only sparingly in forestry, it is receiving increasing emphasis in advanced-generation programs. The selection index system combines information on all traits of interest into a single index that enables the breeder to assign a total score to each individual. It involves attaching economic weights to all traits under consideration. In its simplest form, the selection index may combine a set of individual (phenotypic) values that is weighted by relative economic values to produce a single number for each individual. The most complex index combines family and individual information on all traits into one index. Multiple-regression techniques are used to derive appropriate coefficients for family and individual values, with the coefficients depending on economic weights, the heritabilities of each trait, and the correlations between traits. The major restriction on use of selection indexes in forest tree improvement has been the determination of the appropriate economic weights, that is, the value of a unit improvement in one trait relative to a unit improvement in other traits. Inappropriate economic weights can seriously reduce the efficiency of selection indexes.

Despite these complications, selection indexes have been proposed for use in several tree species. Selection indexes were proposed for roguing seed orchards of undesirable parents by Namkoong (1965). Other examples of the use of the index selection system are with aspen (van Buijtenen and van Horn, 1969), maritime pine (Baradat et al., 1970), and loblolly pine (Bridgwater and Stonecypher, 1979). The use of selection indexes will increase in the future as more data become available on the relative economic values of important traits to product yield and quality.

ACCELERATED BREEDING

There are essentially two ways that the generation interval can be shortened in tree improvement programs. One way is through early or juvenile selection, which has been discussed previously. The other is through *accelerated breeding,* or reducing the time required to mate parent trees and collect seed from them once they are selected. The extended period of time before trees reach reproductive maturity is a serious difficulty in tree improvement because it increases generation intervals and therefore decreases gain per unit of time. The time required to complete breeding, once selections have been made, has absolutely no impact on the gain achieved per generation of improvement, but the steps taken to decrease the time required to breed selected trees will increase gain per unit of time. As a result, much effort is being directed in tree improvement research toward enhancing flowering in young trees. These techniques are included under the general rubric *accelerated breeding.*

The impact of accelerated breeding on the generation interval in a tree improvement program can be seen in the hypothetical example in Figure 13.4. The program on the left is one employing traditional breeding methods, whereas the one on the right utilizes accelerated breeding techniques. In both programs, selections are made at 8 years of age in genetic tests and are grafted into breeding orchards. In the traditional program, however, 8 years elapse before grafts begin to produce flowers. In the accelerated program, flowering is stimulated to begin 4 years after grafting by establishing the breeding orchard in a greenhouse in which the environment can be controlled precisely. In both instances, it is assumed that 4 years elapse following the initiation of flowering before breeding is completed and seed are collected for the next generation of testing. The total number of years required to complete one generation of selection and breeding in the traditional program is 8 + 8 + 4 = 20 years, whereas only 8 + 4 + 4 = 16 years are required in the accelerated program, a time saving of 4 years. Percentage-wise, the generation interval has been reduced by 20%. Because selections were made at 8 years of age in both programs, the 20% reduction in the generation interval translates directly into a 20% increase in genetic gain per unit time in the tree improvement program.

Much of the research related to accelerated breeding has been involved with increasing understanding about the basic physiological processes involved with

Time Savings with Accelerated Breeding

Year	Traditional breeding	Accelerated breeding
1	Plant test	Plant test
	(8 years)	(8 years)
8	Select, graft outdoors	Select, graft indoors
		(4 years)
12	(8 years)	Breed, collect seed
		(4 years)
16	Breed, collect seed	Next cycle
	(4 years)	Generation interval = 16 years
20	Next cycle	Time savings = 4 years = 20%
	Generation interval = 20 years	

FIGURE 13.4

Accelerated breeding techniques can substantially reduce generation intervals and thereby increase genetic gain per unit time. In the hypothetical example shown, an accelerated breeding program has reduced generation interval from 20 to 16 years, or 20%, which translates directly into an extra 20% gain per unit time.

maturation and in reproduction in trees and how these processes can be manipulated to promote flowering. Two excellent review articles pertaining to reproduction in conifers are those by Puritch (1971) and Lee (1979). Several factors, including drought stress and fertilization during the period of the year when reproduction structures are initiated, have been shown to increase female flowering in pines. Similar results with fertilization have been found for Douglas fir (Ebell, 1972) and the true firs (Eis, 1970). The plant hormone *gibberellic acid* has been found to promote flowering in conifers if it is applied at appropriate times of the year (Greenwood, 1981; Brix and Portlock, 1982).

Because of the greater environmental control possible, breeding seed orchards are increasingly being established in containers indoors in greenhouses, where environmental factors such as moisture, temperature, day length, and fertility can be readily manipulated to promote flowering (Figure 13.5). For example, one

FIGURE 13.5

Generation intervals can often by reduced by establishing breeding orchards indoors in greenhouses. The loblolly pine breeding orchard shown will reduce generation interval by several years. Use of greenhouses to reduce generation interval is successful because of the precise control possible for environmental factors that promote flowering.

prerequisite for production of strobili in loblolly pine appears to be a resting bud (not elongating) during midsummer, the time of the year that flower primordia are initiated in this species (Greenwood, 1978, 1980). Outdoors, very young loblolly pine grafts tend to grow through the summer and not form a resting bud. As a result, flowering is minimal. In the greenhouse, a resting bud can be induced by midsummer, and flowering on young grafts indoors has been good (Greenwood, 1978). Prescriptions for management of a containerized indoor loblolly pine-breeding orchard have been given by Greenwood et al. (1979).

One aspect of an aggressive breeding program that fits into any accelerated breeding scheme is *proper management of outdoor conventional breeding orchards*. Several years can be gained each generation simply by taking care to establish breeding orchards on the proper site, at the proper spacing, and to subject the young, growing trees to intensive, proper management that will promote flowering. Such things as proper pollination techniques and care in seed insect control will also be of tremendous benefit to an early completion of the breeding of selected trees. "Tending to business" in conventional breeding orchards can not be overemphasized as being a vital part of any ongoing tree improvement program.

LITERATURE CITED

Allen, F. L., Comstock, R. E., and Rasmussen, D. C. 1978. Optimal environments for yield testing. *Crop Sci.* **18**:747–751.

Andersson, F., Jansson, R., and Lindgren, D. 1974. Some results of second-generation crosses involving inbreeding in Norway spruce (*Picea abies*). *Sil. Gen.* **23**:34–42.

Baradat, P., Illey, G., Mauge, J. P., and Mendibourne, P. 1970. *Sélection pour plusieurs charactieres sur indice.* Programmes de calcul, AFOCEL-Compte-rendu d'activité, pp. 46–70.

Bridgwater, F. E., and Stonecypher, R. W. 1979. "Index Selection for Volume and Straightness in a Loblolly Pine Population." Proc. 15th Sou. For. Tree Impr. Conf., Starkville, Miss., pp. 132–139.

Brix, H., and Portlock, F. T. 1982. Flowering response of western hemlock seedlings to gibberellin and water stress treatments. *Can. Jour. For. Res.* **12**:76–82.

Burley, J., and Namkoong, G. 1980. "Conservation of Forest Genetic Resources." 11th Commonwealth Forestry Conference, Trinidad.

Dudley, J. W. 1977. "Seventy-Six Generations of Selection for Oil and Protein Percentage in Maize." In Proc. Int. Conf. Quant. Gen. (E. Pollak, O. Kempthorne, and T. B. Bailey, Jr., eds.), pp. 459–473. Iowa State University Press, Ames.

Ebell, L. F. 1972. Cone production and stem-growth response of Douglas fir to rate and frequency of nitrogen fertilization. *Can. Jour. For. Res.* **2**:327–338.

Eis, J. 1970. Reproduction and reproductive irregularities of *Abies lasiocarpa* and *Abies grandis*. *Can. Jour. Bot.* **48**:141–143.

Falconer, D. S. 1981. *Introduction to Quantitative Genetics.* Longman, Inc., New York.

Franklin, E. C. 1970. "Survey of Mutant Forms and Inbreeding Depression in Species of the Family Pinaceae," Forest Service Research Paper SE-61, Southeastern Forest Experiment Station.

Franklin, E. C. 1979. Model relating levels of genetic variance to stand development of four North American conifers. *Sil. Gen.* **28**:207–212.

Gansel, C. R. 1971. "Effects of Several Levels of Inbreeding on Growth and Oleoresin Yield in Slash Pine." Proc. 11th Sou. For. Tree Impr. Conf., Atlanta, Ga., pp. 173–177.

Goddard, R. E. 1980. The University of Florida Cooperative Forest Genetics Research Program. In *Research Needs in Tree Breeding* (R. P. Guries and H. C. Kang, eds.), Proc. 15th N. Am. Quant. For. Gen. Group Workshop, Coeur D'Alene, Idaho, pp. 31–42.

Greenwood, M. S. 1978. Flowering induced on your loblolly pine grafts by out-of-phase dormancy. *Science* **201**:443–444.

Greenwood, M. S. 1980. Reproductive development in loblolly pine. I. The early development of male and female strobili in relation to the long shoot growth behavior. *Am. Jour. Bot.* **67**:1414–1422.

Greenwood, M. S. 1981. Reproductive development of loblolly pine. II. The effect of age, gibberellin plus water stress and out-of-phase dormancy on long shoot growth behavior. *Am. Jour. Bot.* **68**:1184–1190.

Greenwood, M. S., O'Gwynn, C. H., and Wallace, P. G. 1979. "Management of an Indoor Potted Loblolly Pine Breeding Orchard." Proc. 15th Sou. For. Tree Impr. Conf., Starkville, Miss., pp. 94–98.

LaFarge, T. 1972. "Relationships among Third, Fifth, and Fifteenth-Year Measurements in a Study of Stand Variation of Loblolly Pine in Georgia." Proc. IUFRO Working Party on Progeny Testing, Macon, Ga., pp. 7–16.

Lambeth, C. C. 1980. Juvenile-mature correlations in Pinaceae, and their implications of early selection. *For. Sci.* **26**:571–580.

Lee, K. J. 1979. "Factors Affecting Cone Initiation in Pines: A Review," Research Report 15, Inst. of For. Gen., Off. of For., Sowan, Korea.

Libby, W. J. 1973. Domestication strategies for forest trees. *Can. Jour. For. Res.* **3**:265–276.

McKeand, S. E., and Beineke, W. F. 1980. Sublining for half-sib breeding populations of forest trees. *Sil. Gen.* **29**:14–17.

McKeand, S. E., Beineke, W. K., and Todhunter, M. N. 1979. "Selection Age for Black Walnut Progeny Tests." Proc. 1st Cent. States Tree Imp. Conf., Madison, Wisc., pp. 68–73.

Namkoong, G. 1965. "Family Indices for Seed Orchard Selection." Proc. Second Gen. Workshop, SAF and Lake States For. Tree Impr. Conf., Madison, Wisc., pp. 7–12.

Namkoong, G. 1979. "Introduction to Quantitative Genetics in Forestry," USDA, U.S. Forest Service Tech. Bull. No. 1588, Washington, D.C.

Namkoong, G., and Conkle, M. T. 1976. Time trends in genetic control of height growth in ponderosa pine. *For. Sci.* **22**:2–12.

Namkoong, G., Usanis, R. A., and Silen, R. R. 1972. Age-related variation in genetic control of height growth in Douglas-fir. *Theor. Appl. Genet.* **42**:151–159.

Orr-Ewing, A. L. 1965. Inbreeding and single-crossing in Douglas-fir. *For. Sci.* **11**:279–290.

Puritoh, G. S. 1972. "Cone Production in Conifers: A Review of the Literature and Evaluation of Research Needs," Pacific For. Res. Center Inf. Report BC-X-5, Canadian Forestry Service, Victoria, British Columbia.

Roberds, J. H., Namkoong, G., and Kang, H. 1980. Family losses following truncation selection in populations of half-sib families. *Sil. Gen.* **29**:104–107.

Schoenike, R. E. 1975. Tree improvement and the conservation of gene resouces. In *Forest Tree Improvement: The Third Decade* (B.A. Thielges, ed.), 24th Ann. For. Symp., Louisiana State University, Baton Rouge, La., pp. 119–139.

Sirkkomaa, S., and Lingstrom, V. B. 1981. Simulation of response to selection for body weight in rainbow trout. *Acta Agric. Scand.* **31**:426–431.

Squillace, A. E., and Gansel, G. R. 1974. Juvenile-mature correlations in slash pine. *For. Sci.* **20**:225–229.

Steinhoff, R. J. 1974. "Juvenile-Mature Correlations in Ponderosa and Western White Pines." IUFRO Joint Meet. of Working Parties on Population and Ecological Genetics, Breeding Theory, and Progeny Testing, pp. 243–250. Royal College Forestry, Stockholm, Sweden.

Talbert, J. T. 1979. An advanced-generation breeding plan for the North Carolina State University–Industry Pine Tree Improvement Cooperative. *Sil. Gen.* **28**:72–75.

Toda, R. 1965. "Preservation of Gene Pool in Forest Tree Populations." Proc. IUFRO Sect. 22 Special Meeting, Zagreb, Yugoslavia.

van Buijtenen, J. P. 1981. "Advanced-Generation Seed Orchards," 18th Can. For. Tree Imp. Assoc. Meeting.

van Buijtenen, J. P., and Lowe, W. J. 1979. "The Use of Breeding Groups in Advanced-Generation Breeding." Proc. 15th Sou. For. Tree Impr. Conf., Starkville, Miss., pp. 59–65.

van Buijtenen, J. P., and van Horn, W. M. 1969. "A Selection Index for Aspen Based upon Genetic Principles," Inst. Paper Chem. Prog. Report 6, Lake States Aspen Genetics and Tree Improvement Group.

Wakeley, P. C. 1971. Relation of thirtieth year to earlier dimensions of southern pines. *For. Sci.* **17**:200–209.

Wilkinson, R. C. 1973. "Inheritance and Correlation of Growth Characteristics in Hybrid Poplar Clones." Proc. 20th Northeast. For. Tree Imp. Conf., Durham, N.H., pp. 121–130.

CHAPTER 14

Gain and Economics
of Tree Improvement

There is great need to make forestland more productive. Increased productivity is especially important for the tropical forest regions where there could be a catastrophe by A.D. 2000 if action if not taken to reverse the trend of reduced forest production (Holden, 1980). The urgency for action in the tropics and the need to capture the timber production potential with desired quality products from tropical forest lands have been stressed by Johnson (1976). Dozens of recent publications have dealt with the needs and methods to increase productivity, for example, the one by Anderson (1978), who feels that production in the southern United States can be doubled if all aspects of forest management (he recognizes that tree improvement is an important component) are used. Anderson makes an additional point about the need for wood-quality improvement. The concepts of how changing forestry activities will affect wood quality has been emphasized by Baskerville (1977).

It is now generally recognized that one major way to increase yield and quality from forestland is through tree improvement. However, if this aspect of silviculture is to be fully utilized, gains and improvements must be quantified and subjected to benefit–cost analyses. Tree improvement efforts become academic unless their use will enhance the value of the forest and its products. Tree improvement benefits may take several forms, such as greater adaptability, increased volume production, better quality of the wood produced, and other forms that will result in an improved final product. Shortening the optimal rotation age to obtain the desired product and developing greater uniformity in the trees produced are major potential benefits. It is clear that these are all interrelated and cannot be assessed independently.

An economic analysis of tree improvement is complex. There are so many variables, such as time, species, location, costs, inflation rates, interest rates, markets, and the like that it may appear to be nearly impossible to make comparative cost estimates. Yet, economic analyses must be attempted in order to justify the maintenance or expansion of tree improvement activities. To do this, a number of assumptions must be made that are based upon experience, opinion, and often judgment about unknown costs and returns. The difficulty is compounded by the major factor of forestry—TIME. It is most difficult to predict values, costs, and economic parameters for the future. Therefore, most economic analyses that have been made are general in nature or are developed to allow options of choosing among several levels of the unknown factors, such as costs, interest rates, or inflation rates. Although imprecise, these generalized results serve a useful purpose.

It is noteworthy that all well-designed analyses have shown that tree improvement is an economically worthwhile endeavor. This general result was stated clearly by Carlisle and Teich (1970a) with their summary sentence. "Even when subjected to this type of objective, rigorous scrutiny, tree improvement is shown not only to be economically beneficial but to be the best of a number of options."

Determination of gain is somewhat simpler than is a full-fledged economic analysis. There are difficulties with test designs and age that make gains from the use of genetics difficult to assess accurately. However, good "ball-park" figures have been obtained that are most useful as a guide. Gain estimates have often been made that are highly unreliable because they are based on immature stands of trees. Estimates made on data from young trees, especially those expressing percentages of improvement, are not only poor but they can be misleading. There are several reasons for this. The most important one is that, as the trees grow in size, the base on which the percentages are calculated changes; therefore, the percentage usually decreases with age.

A cataloging of specific economic returns or gain values in this book would serve little purpose. Only broad general trends will be mentioned, with an occasional specific reference for illustration.

GAIN

General

The question always arises about how much gain is necessary to justify continued intensification of tree improvement operations.

Increments of improvement and the costs of achieving each increments are not usually related in a linear way. Initial gains are generally obtained relatively easily, but additional gains become more difficult and more expensive to attain. The fact is that *optimum* gains are usually sought, rather than *maximum* gains. The former is the gain that is possible before the additional cost for each extra unit of improvement rises beyond the value of the improvement that could be obtained. The concept that optimum gains should be the objective of intensive management and that maximum gains are not usually economically attractive is often overlooked in forestry. It is seldom optimum to produce the maximum, according to Gruenfeld (1975). This is true for genetic gain just as it is for economics, and breeding programs should be governed by this concept.

Assessment of genetic gain is filled with numerous pitfalls. The time when the gain is assessed is all-important as are methodology, species, and location. In addition, gains are restricted by the "rule of limiting factors." For example, regardless of the genetic quality that has been developed, certain factors in the environment, such as moisture or nutrients, may be in short supply, so that the genetic improvement cannot be fully expressed.

As mentioned in earlier chapters, gain in a selection program is a function of selection differential and heritability. This appears to be simple and straightforward when it is applied only to a single characteristic. However, in a breeding program, a number of characteristics are involved simultaneously, each of which has differing heritabilities and selection differentials. In tree improvement, the interest is in the total gain from all characteristics. To express this, it becomes necessary to use some type of index to weigh the gains and importance of each characteristic as part of an overall assessment. Too often a program of multiple-trait improvement is used, but gains are expressed in terms of only one trait. To be correct, total gain should be the total value determined by the gain in each trait, weighted by its economic value. As described in Chapter 13, less gain will be obtained for any one single trait as additional traits are included in the tree improvement effort, but the optimum for all traits combined is what is being sought.

The time necessary to achieve the gain is of prime importance economically. As we expressed in earlier chapters, "time is money," and one of the greatest benefits from a tree improvement program is to reduce the time it takes to grow a desired product. Often, gains from tree improvement are expressed as total gains, but as was stressed in Chapter 13, the real economic standard should be *gain per unit time*.

How genetic gains should best be expressed is a real puzzle. Any measure that is used can be wrong or misleading; this applies to both percentage and to unit values. The safest measure of gain is a monetary one; that is, what value in currency does the gain represent? Although this is the best, it still has multiple pitfalls because the monetary value of a unit of gain, or the cost to obtain it, often is not known. For example, how much value is represented by a reduction of disease infection by 15%, or what is the value of trees that are straightened enough so that three times the number are in a straightness category that could be used for the highest-value products? When specific gravity is increased from 0.48 to 0.50, what is the monetary effect on the forestry and manufacturing operations? An estimate of the value from increased volume or from producing more dry-wood weight per unit area can be obtained reasonably accurately, but such value determinations are most difficult for most tree quality factors.

Assessment of Gains from Juvenile Measurements

Great care is needed in assessing and predicting genetic gains. The data are always needed immediately, and great pressures are sometimes placed upon tree breeders and orchard managers to make predictions and projections, even though data are incomplete or the tests are not sufficiently mature. The mature performance of some characteristics cannot be readily assessed from young trees; for example, it was stated earlier that because of differences in growth curves, growth characteristics usually cannot be reliably determined in much less than half-rotation age. Yet, because of the urgent need for guiding information, gains are often calculated and large programs initiated that are based only on the performance of young trees. Much of the genetic-gain data in the literature is totally unreliable because it is based upon an assessment that was made too early. On the other hand, some characteristics can be assessed at an early date with considerable accuracy. For example, those that are related to survival can be assessed at a young age unless growing conditions were especially favorable in the early life of the stand. Wood specific gravity of older trees can usually be estimated rather well from young trees. In contrast, the trees must be considerably older before tracheid length can be predicted with reasonable accuracy. Another important characteristic that can be assessed early is straightness of tree bole.

The importance of juvenile–mature correlations and the reliability of prediction of later performance from early assessment vary by species and characteristics. Most characteristics related to growth require extended periods to obtain reasonably accurate information, although early predictions may be somewhat more accurate on families compared to those for individual trees. The predictability from young ages is often better for provenances; for example, usually a high-elevation provenance is slower growing than one from lower elevations, no matter what its age. The concept of predicting performance at older ages from young ages always has been important but is much more so now with the advent of more use of vegetative propagation in which juvenile material is needed for easy rooting. A number of so-called "superseedling" studies have been established in which outstanding nursery seedlings were selected for possible increased yield at rotation age. Results have been inconclusive. For example, Sweet and Wareing (1966) have reported that selected large seedlings of radiata pine did not grow faster, and Brown et al. (1961) found the same situation for loblolly pine. However, Hatchell et al. (1972) found greater growth of selected loblolly pine seedlings after 10 years and strongly recommended the method, as did Nienstaedt (1981), for "super" spruce. For spruce, Barneoud et al. (1979) also found a drop-off and warned that care was needed in the use of the superseedling method. Robinson and van Buijtenen (1979) concluded that, despite a drop in growth, meaningful gains were still evident in 15-year-old loblolly pine. Several researchers have emphasized that the crown and bole form of the selected superseedlings were rough, that is, they were similar in many ways to classical "wolf trees."

All the studies made on the relation of juvenile-to-mature growth cannot be cited, but many tests indicate that the first year, or nursery performance, is not a good predictor of later growth. This was even true for fast-growing willow on a 4-year rotation. Barrett and Alberti (1972) stated that "it is concluded that the evaluation or selection based on growth performance during the first year is ineffective." On the other hand, Mohn and Randall (1971) obtained gains on fast-growing poplar by predicting third- and fourth-year height based upon short-term assessments. The authors have observed situations in which early selection of *Populus* has led to some erroneous conclusions about growth potential as early as the third year. The uncertainty was summarized by Wilkinson (1974) who mentioned that for hybrid poplars, 1- to 15-year correlations may be low enough so that selection intensities may need to be reduced to preserve the best clones at year 15. It is clear that a widespread, wholesale use of superseedlings as a major base for selection should be done only with the greatest caution and after complete testing has indicated the juvenile–mature relationships.

Another problem involving age is related to tree size and change of percentage superiority as the tests become older. For example, it is not acceptable to predict commercial volume growth based on volumes of trees that are 2.5 to 5.0 cm in diameter. Such predictions are frequently made, and the predicted volume gains can be worse than useless. Tables for small-tree volumes are often inaccurate. Tree form and relative bark percentage often change greatly with tree age. What is most serious is that a small difference in volume of small trees will translate into a large percentage value that is not at all representative of differences that will be found for the mature trees. For example, it is easy to find families of trees that are 1 m tall at 1 year of age, whereas other families in the same test are only 0.5 m in

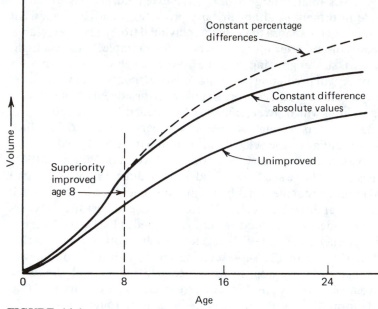

FIGURE 14.1

Improvement projections are quite different, depending on whether absolute gain values or percentage gain values are used. Shown is a hypothetical volume improvement projection from measurement of year 8, showing the kinds of differences in projected volumes whether based on absolute differences or percentage differences at year 8.

height. Thus, the best families are 100% superior to the others. Yet at age 20, one family might be 20 m tall, whereas the other is 15 m tall. The 100% superiority has dropped to 33% superiority, even though the family differences are still very outstanding (Figure 14.1). Percentage differences based on the performance at a young age should never be extrapolated to older trees because the base upon which the percentage is calculated becomes larger, and thus the percentage tends to drop. Obviously, any genetic gains reported as percentage values must be viewed with some caution, and the age at which the percentages were calculated must be taken into account.

The Importance of Plot and Test Design

Data from research plots should not be used to predict production under operational forest conditions. Most forest genetics research plots are designed to show only comparative performances among sources or families. Usually a commercial check is included to give some indication of the relative performance of the test material in relation to what was being planted operationally prior to genetic improvement. Such information is good and useful, but great care must be taken in transferring absolute-growth data from tests to operational plantations and making economic projections therefrom. Growth rate data from row plots designed to test parents in a seed orchard cannot be transferred directly to unit area growth in a plantation. It is not unusual to have a 10 to 15% reduction between the growth rates from the test planting compared to the planting made under operational conditions.

There are many reasons for not using test data for predictions in operations. A primary problem relates to differential competition. In row or pure family block plots, competition effects will be different than those in plantations from mixed seed from a seed orchard. The clones within a seed orchard do not produce equal amounts of seed; yet progeny tests use data from equal numbers per family. Often the test results include inferior clones that will be rogued from the seed orchard following testing. Another problem is that frequently the clonal mix is not equal in the seed orchard; this always occurs following roguing. Frequently, the test site may not be representative of operational planting sites. Also, the care in planting and management of genetic tests is better than that given to operational plantings. Sometimes the check lots are not representative of the unimproved planting. When all these difficulties are considered, some of which will lead to an overestimate and others to an underestimate of genetic gain when the improved trees are used in commercial plantings, it behooves the tree improver to be very careful in making production predictions. Certainly the comparison figures from genetic tests are needed and valuable, but extrapolations from them to operational planting normally should not be done.

The only true test of the gain from seed orchards over commercial stock is to collect seed directly from the seed orchard after it has reached full productivity and to test the improved stock against the unimproved in paired plantings that are designed to minimize environmental effects but still be handled in an operational manner. If realized gains from rogued seed orchards are desired, then the tree improver must test after the orchard has been fully rogued. This means that many years must elapse before tests of gains from seed orchards can be started. In their eagerness, many seed orchard managers use their first commercial seed and establish tests of gains from seed orchards with it. In a young orchard, only a few clones produce pollen or seed; a "rule of thumb" used by many pine orchard managers is that in young orchards 80% of the seed is produced by 20% of the clones. If one or two of the heaviest producers are especially good or especially bad, the assessment of orchard gains can be very wrong. Unequal genetic contribution to the seed crop can be as extreme as that found by Bergman (1969), who determined that over half the genetic makeup of the seed in one seed orchard was from one clone. This is an extreme case for a very young orchard; the balance improves as the orchards approach maturity. Furthermore, the chief source of pollen within young orchards may be from nearby unimproved stands.

A special situation exists when seed are collected and planted by mother tree or clone. In that case, testing should be by family blocks and if done correctly, quite accurate predictions can be made of the orchard stock performance. It is especially difficult to make yield projections when survival is not accurately assessed.

Large errors can occcur if an assumption of full survival is made so that yield is calculated on the basis of individual tree performance multipled by an assumed number of trees representing full stocking. Choice of site and silvicultural treatments, such as site preparation or competition control, are most important if meaningful results are to be obtained.

Realized Gain

Measurement of genetic gains through tests planted in an operational manner allows the breeder to calculate actual or realized gains. In Chapter 4, it was shown that anticipated genetic gain from mass selection could be calculated as the product of the selection differential (S) times the narrow-sense heritability (h^2). Symbolically, gain $= h^2S$. Narrow-sense heritability was defined as a ratio of the-additive genetic variance in a population to total phenotypic variance $(h^2 = \sigma_A^2/\sigma_P^2)$. However, the gain predicted by the formula is only an estimate of the anticipated gain, and often is not the same as what will actually be obtained. Establishment and measurement of studies designed specifically to estimate the realized gain will allow the tree improver to show what has actually been achieved through selection and breeding.

Methods to Obtain Gain

Although there are exceptions, gains from tree improvement can actually be put into three broad categories. Generally these are based upon genetic testing and breeding, which will enable continued progress over time and later generations. The basic categories, all of which were mentioned in earlier chapters, are the following.

1. **Mass selection followed by testing**   Most of tree improvement to date has relied on the selection of desired trees, testing these to determine the best genotypes, and on open pollination among these to produce seed for operational planting. Gains from one generation of selection using this method have been remarkably good, especially for those characteristics having reasonably high heritabilities (i.e., they are under considerable additive genetic control), and where the total variation and thus selection differentials are suitably large. Because of segregation and recombination that takes place when sexual propagation is used, the progeny will not be exactly the same as, or as good as, the parents if there is substantial nonadditive genetic variance. Nonetheless, the gains will represent a movement of the population average in the desired direction.

2. **Phenotypic selection followed by vegetative propagation and testing**   The use of vegetative propagation is becoming increasingly important in forestry. When it is suitable, gains are good because individuals with good additive and nonadditive genetic values are reproduced intact. In characteristics with a large proportion of nonadditive variance, vegetative propagation is particu-larly important because one can capture all this portion of the genetic quality

of the donor parent in the propagule. The one action in tree improvement that will increase genetic gain more than any other is to learn how to use vegetative propagation more widely and effectively.

3. **Making special crosses**   When characteristics are primarily controlled by nonadditive genetics, special crossing designs can be used to capture the desired characteristics by exploiting specific combinations of parents. When tests indicate that two parents produce outstanding offspring, this information can be used by making the required specific crosses to produce the desired seedlots in quantities that are sufficient for reforestation. This must be done by control crosses or by some system of controlled mass pollination. Male sterile plants, or those that do not produce selfs, could be effectively used in the special crossing programs involving two-clone seed orchards. Gains from this approach can be great if there is substantial nonadditive genetic variance; it is a method that will become more widely used to develop trees for problem areas or for special needs in those species in which vegetative propagation is not economically feasible. Currently, production of mass quantities of seeds by special crossing is difficult and expensive, but as technologies develop, more control-pollinated progenies will be used.

Magnitude of Gains

It is of value to mention various categories of characteristics and their responses to genetic manipulation. Genetic improvement can be obtained by choosing among populations or among individual trees within populations. As discussed in detail earlier, characteristics related to adaptability are often more strongly associated with populations, as compared to most other economic characteristics that are more closely related to individual-tree inheritance. It is of the utmost importance that for most species, many of the important tree characteristics are not strongly related genetically; therefore, it is possible to work toward "tailor-making" a population containing a package of the desired characteristics.

Gains that have been achieved for several categories of characteristics obtained by manipulation of individual trees are listed next. Large lists of references could be given for each, but only a couple representative studies will be cited.

1. **Growth characteristics**   Genetic manipulation of growth characteristics generally will achieve only modest gain through seed regeneration, although the gains certainly are worthwhile. Most growth assessments have been based on height, but when volume growth has actually been determined, gains have been increased (Jeffers, 1969; Talbert, 1982). In a summary of gains from tree improvement, Nikles (1970) reported a net economic yield of 30% on 15-year-old plantations based upon growth of *Pinus elliottii* and *P. radiata* in Australia. A 43% volume gain from the best 10% of longleaf pine families studied was found by Snyder (1973). The genetics of growth is complex and usually has a strong nonadditive and environmental component in addition to an additive component. Also, there is often a strong genotype × environment interaction for growth that restricts the gains possible

when one population is destined for use over wide areas. Thus, gains in growth using a seed orchard program will be modest but economically worthwhile. Gains in growth can be considerably greater when special-breeding or vegetative-propagation methods are used.

2.   **Tree straightness and limb-quality characteristics**   Straightness is fairly strongly inherited and responds well to either a sexual or vegetative propagation system. Because of the strength of its inheritance, it is often possible to make enough gain in straightness in the first generation of intensive selection so that selection for this trait can be relaxed in subsequent generations (Goddard and Strickland, 1964; Campbell, 1965; Ehrenberg, 1970). Limb characteristics are much more strongly influenced by the environment than is straightness, but there is enough genetic control to improve limb qualities. Gains will usually be less than for straightness (Shelbourne and Stonecypher, 1971). Because of the effect on the final product and its inheritance pattern, emphasis in a tree improvement program should generally be made on bole straightness, but limb qualities should not be ignored. They have a major effect on some products, and gains can be made even though they may be modest.

3.   **Wood qualities**   Although there are some exceptions, most wood properties are under a strong additive genetic control, and gain using a seed orchard program is good. Although many wood qualities show a reasonably strong inheritance, the two that are most studied (specific gravity and fiber length) show great gain from genetic manipulation (see Chapter 12). It now appears that the chemical property of cellulose yield per unit dry wood is also under some degree of genetic control, but because of its large nonadditive component genetic gains will only be obtained if vegetative-propagation or special-breeding systems are used. The encouraging thing is that the inheritance of a wood property such as specific gravity is so strong that gains can be made even within the framework of desired growth and form qualities.

Genetic manipulation of wood yields good results; this has been attested to by numerous authors. In a recent report, Jett and Talbert (1982) determined for 12-year-old loblolly pine trees that with only one generation of light selection in which one in two individuals were saved for higher wood specific gravity, improvement in dry-wood weight was increased about 10 kg/m^3 of wood.

Numerous calculations have been made on gains in wood qualities, particularly for specific gravity (Dadswell et al., 1961; Zobel et al., 1972). In an intensive analysis of the value of including wood in a tree improvement program, van Buijtenen et al. (1975) reported that breeding for wood specific gravity was almost always desirable for making linerboard paper. There is no doubt about the value of gains from including wood in a breeding program.

4.   **Pest resistance**   Sometimes, huge gains can be made from breeding for pest resistance, and other times the gains achieved are little or none. Results depend upon the genetic variation of the host as well as of the pathogen and

their interactions with the environment. Gains in pest resistance from the use of genetics can sometimes be so spectacular that they make forests profitable out of an otherwise nonprofitable situation. Although pest resistance breeding sometimes does not work, overall genetic gains from this breeding activity have been one of the brightest spots in tree improvement. The complexities, successes, and failures of breeding for pest resistance were reported in Chapter 9.

5. **Adverse environments**   Although much of the genetic variability related to adaptability to adverse environments is a function of a provenance or population, there still is considerable tree-to-tree variability that can be exploited. For example, occasional trees from warm areas are cold hardy, or those from moist areas are drought hardy. Using the individual trees in developing a desired tolerant strain can be difficult and time consuming, but sometimes there is no other satisfactory way to get the job done. For example, none of the sources of cold-hardy eucalypts have proper growth and form for the Gulf Coast area of the southern United States. Therefore, vigorous high-quality trees from large cold-susceptible sources are being selected and propagated to develop the needed cold-tolerant strain (Hunt and Zobel, 1978). The gain that is possible from using individuals which are especially tolerant to adverse environmental conditions has not been sufficiently exploited. Good gains not obtainable by other means are possible if the proper effort is expended.

6. **Miscellaneous**   There are many additional aspects of tree improvement in which useful and sometimes large genetic gains can be made. Things such as fruitfulness show huge differences among clones, as do graft incompatibility and rooting ability. These strongly inherited differences will be exploited more often as tree improvement activities become more sophisticated. Many other physiological characteristics, such as resin production, show moderate to strong inheritance patterns. For example, tall oil production from pines has been doubled by using the genetically high yielders as parents (Franklin et al., 1970). Growth patterns and response to light and other environmental factors have been extensively studied (Ekberg et al., 1976; Eriksson et al., 1978).

The genetic gains from forest tree breeding have been larger and easier to obtain than most forest geneticists originally expected. The large variability in forest trees and considerable additive variance have made the easy mass selection schemes used in first-generation programs very successful. Of the many studies reporting specific gain values, only one example is listed to give the reader some idea of the magnitude of gains that have been achieved.

In a recent summary paper of the gains from one generation of tree improvement for loblolly pine using the comparison tree selection system in natural stands, Talbert (1982) made a number of volume gain calculations based upon data from genetic tests up to 12 years of age; much of this was projected through growth and yield models to age 25. As an example, the following percentage improvements were predicted for age 25 for seed orchard seed over commercial check seed.

	Stand Volume (%)	Stand Value (%)[a]
Unrogued seed orchard	6.4	18.0
Rogued seed orchard	12.7	32.0

[a]Includes added value resulting from increased tree size.

Talbert (1982) especially warns about the change in percentage improvement with age in unthinned stands and states the following: "One of the reasons for decreasing percentage gains in volume with increasing age is an increasing volume figure on which the percentage is based. Another reason is increased mortality in stands planted with seed orchard stock because of more intense tree-to-tree competition resulting from more rapid stand development." It would seem apparent that the situation would be helped by a thinning at the correct time.

There is no longer any doubt that substantial gains can be made in most desired characteristics in forest trees in one generation of selection. Changes in the genetic composition of forest stands are of such a magnitude that there will be a large impact on forest management strategies (Zobel, 1979). It is important that the tree breeder and forest manager realize that gain figures for volume, stem straightness, or any other characteristic must be interpreted in relation to their effects on stand development and associated management activities. Gains for any trait, expressed in biological or economic terms, can and will change dramatically during the life of the stand and with different forest management and silvicultural practices.

Summary of Gains

Where do all of the many facets and intricacies of gain leave the tree improver and how does one use the available information? Some especially important considerations related to gain can be summarized as follows.

1. The first consideration is the warning to use care to interpret all reports in light of the product requirements, tree ages, test conditions, species, and areas involved. Do not blindly apply gain figures that may be available for other species, other areas, or other-aged forests because the result is often similar to comparing apples with oranges. This most important warning can be summarized as follows: Be critical and analytical of any gains that are reported and that are being considered for use.

2. Do not make estimates of gains for older operational forests based on those from very young tests. This is especially true for growth characteristics in which growth patterns change with age. Some quality, pest resistance, and adaptive characteristics can be assessed with confidence at young ages, but

even these should be interpreted with caution. The literature is filled with data that are misleading or plainly wrong because they have been based upon small, young, or otherwise unsatisfactory studies and applied to operational programs.

3. Try to convert all gains to monetary units. Be cautious of gains reported on a percentage basis; this caution needs to be exercised also when actual unit values of volume, quality, or adaptability are available.

4. Adaptability assessments are difficult to make. It is the extreme situation or sequence, the 1 day in 1 year or the 1 day in 10 years that can cause trouble. It is necessary that the material being tested is sufficiently exposed to the extremes of environments and unusual environmental sequences that might occur so that one can have confidence in their performance.

5. Unit values of gain for growth characteristics are easy to determine, but those for some quality characteristics are most difficult to quantify. Despite this difficulty, the value of quality improvements is often obvious; an example is producing trees with straighter stems than those found in unimproved populations.

6. A calculation of total gain from a tree improvement program is most complex and ultimately should be based on some form of an aggregate economic index.

7. Be realistic in the types of gain being sought. The objective is usually to make optimal gains rather than maximal gains. Much harm has been done to tree improvement efforts by promising gains larger than can be realistically obtained. No one gets upset when gains are achieved that are greater than a modest gain prediction, but the converse can be fatal to a tree improvement program.

8. Use the best possible methods to capture genetic variability. More refined and precise techniques are becoming available, and these should not be ignored. For example, enough information is now available to use special seed production methods like two-clone orchards or mass pollination. Vegetative propagation methods are rapidly emerging that take advantage of both the additive and nonadditive genetic variance, thus enabling larger gains to be obtained in a shorter period of time. Special developmental breeding programs are needed to reap the optimum benefit from vegetative propagation.

ECONOMICS OF TREE IMPROVEMENT

The ultimate value of tree improvement depends on how much it helps forest management; put more directly, how greatly does tree improvement increase the returns from the forestry operation? Despite the great complexities of assessing the true value of a series of complex actions and investments that often do not give a return for a number of years, numerous studies about the economic value of tree improvement activities have been conducted. At first, these assessments were concerned with costs and predicted or hoped-for gains. Recently, more realistic figures are becoming available that can be used to develop meaningful economic analyses.

As is common for all aspects of tree improvement, *time* is the most formidable hurdle in making suitable economic analyses. Even after realistic costs and genetic gains are known, there is always the uncertainty of interest rates, future markets, inflation, and general economic conditions. For most forest products, added uncertainties relate to competition from materials that can be used to replace wood products; for example, the fear that plastic might replace paper products was once prevalent, and it actually has occurred to some extent with large milk containers and grocery bags. Not many years ago, foresters were told that aluminum, steel, and concrete would greatly reduce the use of wood products in the building industry. This happened to some extent but has slowed because of the high cost of energy needed to manufacture the alternative materials.

A major deterrent to tree improvement programs and their ultimate use is product uncertainty. A rarely recognized but important aspect of this is the consideration of fashion, especially in solid-wood products for use under highly aesthetic situations, such as room finish or furniture. Fashions in wood products can vary just as they do for clothes. One example is curly birch in Europe; this unique type of wood is inherited strongly enough to make good genetic gains relatively rapidly (Johnsson, 1950). Curly birch was for many years a very rare, high-value product. As more trees with this type of wood are produced, it will become more common, and the price of this specialty product will decrease. A change in wood value occurred in the furniture industry in the United States with black cherry (*Prunus serotina*). This species was a favorite for furniture for a number of years, but after it became scarce, fashion shifted to other woods such as ash and white pecan, and black cherry was no longer in vogue. Because of this change, there have been questions recently about the wisdom of pursuing cherry-breeding projects that had been started earlier.

Future product needs and values are critical to any economic assessment of the value of tree improvement. Because they are so difficult to predict, all economic studies that have been made or will be can be challenged. Despite these imperfections, the studies already made have indicated a clear and repetitive picture of the economic worth of tree improvement. In one of the earliest studies on the economics of tree improvement, Davis (1967a) ended his assessment with the following cautious but optimistic statement.

Recognizing the many assumptions used in making the cost and required yield increases that are given, it is still fairly clear that investments in seed from commercial seed orchards are sufficiently low that the minimum expectation of increased yield will justify them. In short, from the stumpage grower's point of view, it certainly appears that current investments in loblolly pine seed orchards are well within the "ball park" with respect to financial justification. If the upper expectations of gains materialize, they should prove to be excellent investments.

This summary, based on some initial costs and very tenuous returns, sounds very similar to the one made by Carlisle and Teich (1970b) later, when more reliable data were available.

It is understandable that, because of the lack of information, many of the early economic analyses as well as some current ones refer to the economics of tree improvement only in general terms. However, some studies do deal with specific situations. One of the earliest was by Perry and Wang (1958) who calculated the added value that would accrue to the forest enterprise when seed with different degrees of genetic improvement was used for planting. Actual costs and genetic gains were not available at that time, but the calculations present enabled one to obtain a good idea about the intensity of genetic effort needed to justify certain costs. Part of the analysis related to the amount of money that could be spent in producing a kilogram of seed that would yield a desired amount of gain. A more recent and in-depth report by Weir (1975) related to the financial problems that would result from the loss of improved growth in operational plantations if insects destroyed the developing seed in the seed orchard. His calculations indicated that loss of seed from seed orchard insects would have a larger impact on the forest economy than the current losses from the southern pine beetle that is considered to be one of the most destructive pests in that region. His article clearly touched on a fact that is often overlooked, that is, "invisible" losses are occurring in forestry because improved seedlings are not being used in plantation establishment.

Economic Value of Tree Improvement

Many economists like to base their assessment on how much genetic gain is needed to justify a given program. This is a relatively easy way to illustrate economic aspects of tree improvement, and one can relate it to the genetic gains actually being achieved. One example is Davis (1976b) who calculated that a 2.5 to 4.0% volume gain over the stock currently used for planting would be necessary to justify a tree improvement program. He emphasized the difficulty of including quality improvement in such a calculation in addition to the more easily assessed volume gains. Based on Canadian experiences, Carlisle and Teich (1970b) came to a similar conclusion, stating that an increase in yield from 2 to 5% would offset the added costs of a tree improvement program. They added that if quality characteristics, such as wood, were considered and that if tree improvement and silviculture were properly coordinated, mill profitability would be profoundly affected. The interesting point about the 2 to 5% required to obtain an economic return reported by the aforementioned authors is that these percentage values are very small when they are compared with the 6 to 12% volume gains that Talbert (1982) reported, or the 12 to 14% volume gains, 5% improvement in wood specific gravity, and 10% in rust resistance that Porterfield et al. (1975) reported. In some recent work on the tropical pines, for which very little economic information has been available, Ledig and Whitmore (1981) show volume gains of 12 to 21%, depending upon the selection intensity used. However, these gains have not been translated to economic values.

A number of other studies have reported various aspects of the economics of tree improvement. Many, like the one by Lundgren and King (1966), involving *Pinus banksiana,* concluded that tree improvement is a good investment. Another study by Reilly and Nikles (1977) who based their analyses on actual costs and returns for improved volume and tree straightness, found a 14% internal rate of return from an unrogued seed orchard. This increased to 19% after roguing. Similarly, Row and Dutrow (1975) showed a 12.4% rate of return on tree improvement, whereas Carlisle and Rauter (1978) indicated an internal rate of return that varied from 6 to 21%, depending on which variables were used. Some authors are very enthusiastic about the potentials of genetic improvement. As an example, Carlisle and Teich (1970a) have stated that "it would be difficult to find a cheaper way of increasing yield" for white spruce. This was determined even though the model they used was weighted toward minimal benefits.

For fusiform rust, Holley and Vale (1977) showed that 18 to 28 million dollars are lost annually due to poorer yields and lower wood quality resulting from fusiform rust infection on southern pines. From this, it is evident that large economic gains will be achieved if reasonable rust resistance can be obtained. There have been numerous studies on gains from improving wood characteristics (e.g., Dadswell et al., 1961; van Buijtenen, 1973; Zobel, 1975; Jett and Talbert, 1982). Space does not permit citing more of the numerous studies and tree characteristics that have been assessed. Overall, it is quite evident that the greatest returns are obtained by working on a few rather than on many characteristics at one time.

Tree improvement is more economically successful with some species than with others. Dutrow (1974) emphasized this point, indicating that much larger genetic gains are required from oak than from birch in order to obtain a reasonable return on tree improvement because of the limitations in seed production. Similar comments about the relative profitability of oak and birch have been made by Marquis (1973).

In addition to the specific studies, a number of general reports indicate both the suitability and good financial returns from tree improvement. Examples of the more general types of published papers are those by Zobel (1966), Dutrow and Row (1976), Porterfield (1977), and Stonecypher (1982). Good profitability is predicted and special emphasis is indicated to the effect that initially large expenditures that are often required can be economically justified in the long run.

Most studies have been based on the shorter-rotation, faster-growing conifers. For hardwoods, Marquis (1973) suggested that tree improvement programs will be most profitable for species that are heavy seed producers, have rapid growth, and high-value products. The economic situation is less clear-cut under other conditions. For example, Ledig and Porterfield (1981) have emphasized that discount rate and length of rotation are particularly crucial variables in determining the economic worth of tree improvement. They reported on the gains needed to justify genetic application in slow-growing species with long rotations. With rotation ages up to 50 years, a volume improvement as low as 6.3% will yield an 8% return on the tree improvement investment. However, Ledig and Porterfield have emphasized that it is very difficult to show profitable gains from tree improvement when rotations from 80 to 120 years are used.

Great differences in profitability are also present, depending on the techniques and methods used. For example, van Buijtenen and Saitta (1972) have reported that mass selection is more cost-effective than progeny testing for the establishment of first-generation orchards and that the latter should be used primarily as a basis for selection for advanced generations. Most authors emphasize, as do Teich and Carlisle (1977), that much more genetic and economic information is needed and that the strategies chosen are critical. Often the mistake is made of underestimating the side benefits of the tree improvement program on tree quality or wood quality or on the cost of harvest or manufacture (Nantiyal, 1970). Predicting future markets, costs, interest rates, inflation, and the like is especially difficult. It appears that the best approach in the future is to use some linear programming technique, as Teich and Carlisle (1977) suggested, or a systems analysis approach like the one van Buijtenen and Saitta (1972) used.

The information necessary to make economic analyses of tree improvement and the methodology to do it best are badly needed. Unfortunately, use of suitable economic techniques and the required input data will likely continue to lag behind the need for results to justify another operation or more intensification of tree improvement. It is essential that the tree improver forms a team with the economist to develop the most effective program and methods of analysis.

Reporting Economic Gains

It is essential that the value of tree improvement is presented in a realistic and meaninful manner. This also requires teamwork between an economist who is knowledgeable about tree improvement and a tree improver who can appreciate and understand simple economic analyses. Of the many ways to present the value of a tree improvement program that have been attempted, one of the best is to calculate the added value that a given amount of genetically improved seed will contribute to the forestry organization. A number of organizations have used this method successfully; it is illustrated in Table 14.1. One can obtain from Table 14.1 an indication of the *present* value of added income in the future that would be obtained at the time of harvest for two contrasting cases with case 1 representing a good situation; case 2 is less favorable.

However, the values in Table 14.1 alone are quite meaningless without a comparison of the present value of a pound of improved seed with the cost of producing that seed. It can be shown by the following ratio.

$$\frac{\text{Present value of improved seed}}{\text{Present value of the cost of producing a pound of seed}} \qquad \frac{B}{C}$$

If the B/C ratio is greater than 1.0, then (because 8% interest was used in Table 14.1) the return will be greater than 8%. As an example, take the $48/lb in the

TABLE 14.1

Present Value of Additional Wood Obtained from 1 lb of Seed Orchard Seed for Several Stumpage Values, Two Growth Rates, and Two Combinations of Nursery Production, Genetic Gain, and Plantation Stocking

Case 1	Case 2
1. 1 lb of seed produces 9000 plantable seedlings	1. 1 lb of seed produces 7000 plantable seedlings
2. 500 seedlings are planted per acre (1 lb of seed plants 18 acres)	2. 800 seedlings are planted per acre (1 lb of seed plants 8.8 acres)
3. Rotation age = 25 years	3. Rotation age = 25 years
4. Genetic gain = 15%	4. Genetic gain = 10%
5. Interest rate = 8%	5. Interest rate = 8%

Stumpage Value ($/cord at time of harvest)	Base Growth (cords/acre/ year)		Stumpage Value ($/cord at time of harvest)	Base Growth (cords/acre/ year)	
	1.5	2.0		1.5	2.0
10	127	197	10	48	64
15	221	296	15	72	96
24	354	473	24	115	153
30	443	591	30	144	192
40	591	788	40	192	255

table in case 2. If it costs $50/lb to grow the seed, it is not attractive as an economic investment and will not return 8%. If it costs $15/lb, then it is a very attractive economic investment for the organization concerned.

Making Economic Assessments

With all the uncertainties and unknowns, it would appear on the surface that it is not possible to make economic analyses of tree improvement. But this is not so— one uses all available data, uses the best estimates when facts are lacking, and then proceeds. The answer will certainly provide information that would not be available if the analyses had not been done. The key to such an analysis is flexibility. The methodology must be sound and general, enabling the use of new or updated data when they become available. This is the pattern followed by most organizations.

There is no value in this chapter to suggest any particular method to be followed because each organization has its own format and objectives. There are a few general points, however, that should be kept in mind when making an economic assessment of tree improvement.

1. Bring all returns and costs to a common time. The usual method is to convert everything to present value.

2. Costs are relatively easy to determine with considerable accuracy. Care is needed to use realistic costs that will be in effect after a program is operational. Costs based on research or small plots can be very misleading.

3. Results and gains are very difficult to determine accurately, and frequently they must be based on projections and common sense. But one must be careful taking yields based on research and applying them to large operations Almost always, the research results are more optimistic than can actually be obtained operationally. Frequently, research results are based on young trees that cannot give an indication of what the situation would be at rotation age.

4. Keep the system simple and understandable. The personnel who make administrative decisions demand straightforward and simple answers. Results couched in excessive statistical jargon are often ineffective. Results that are tentative because of an excessive use of qualifiers (*about, maybe, perhaps, sometimes*) are often totally ignored or rejected.

5. The results should be conservative. No one is upset if gains are greater than those that were projected. However, do not be so conservative and choose the "worst case" so that an incorrect impression is given about the economic potential of tree improvement.

6. Foresters often become "paralyzed" into taking no action because all the facts are not known. This is not acceptable, and estimations must be made to enable determination of the economic suitability of a program.

LITERATURE CITED

Anderson, G. A. 1978. Effects of intensive pine plantation mangement on southern wood supply and quality. *Tappi* **61**(2):37–46.

Barneoud, C., Brunet, A. M., and Dubois, J. M. 1979. Behavior of the highest plants in a young spruce plantation. Association Forêt-Cellulose (AFOCEL): Annales de Recherches Sylvicoles, Paris, France, pp. 383–400.

Barrett, W. H., and Alberti, F. R. 1972. Value of early selection in progeny of willows [Valor de la seliccion temprana en progenies de sauces]. *IDIA*, Supp. For. No. 7, pp. 3–8.

Baskerville, G. 1977. "Let's Call the Whole Thing Off." Proc. Symp. Inten. Cult. Northern For. Types. U.S. Forest Service Technical Report NE-29, pp. 25–30.

Bergman, A. 1969. "Evaluation of Costs and Benefits of Tree Improvement Programs." 2nd World Cons. For. Tree Breed., Washington, D.C.

Brown, C. L., Goddard R. E., and Klein, J. 1961. Selection of pine seedlings in nursery beds for certain crown characteristics. *Jour. For.* **59**(10):770–771.

Campbell, R. K. 1965. Phenotypic variation and repeatability of stem sinuosity in Douglas fir. *Northwest Sci.* **39**(2):47–59.

Carlisle, A., and Teich, A. H. 1970a. "The Costs and Benefits of Tree Improvement Programs." Petawawa Forest Experiment Station, Chalk River, Ontario, Canada, Info. Report PS-X-20.

Carlisle, A., and Teich, A. H. 1970b. "Cost and Benefit Analysis of White Spruce (*Picea glauca*) Improvement." Proc. 12th Meet. Comm. For. Tree Breed. in Canada, pp. 227–230.

Carlisle, A., and Rauter, M. 1978. "The Economics of Tree Improvement in Ontario," Tree Improvement Symp., Ontario Min. Nat. Res. OP-7, Ontario, Canada.

Dadswell, H. E., Fielding, J. M., Nichols, J. W. P., and Brown, A. J. 1961. Tree to tree variations and the gross heritability of wood characteristics of *Pinus radiata*. *Tappi* **44**:174–179.

Davis, L. S. 1967a. "Cost-Return Relationships of Tree Improvement Programs." Proc. 9th South. Conf. For. Tree Impr., Knoxville, Tenn., pp. 20–26.

Davis, L. S. 1967b. Investment in loblolly pine clonal seed orchards. *Jour. For.* **65**:882–887.

Dutrow, G. F. 1974. "Economic Analysis of Tree Improvement: A Status Report." U.S. Forest Service Technical Report 50-6.

Dutrow, G., and Row, C. 1976. "Measuring Financial Gains from Genetically Superior Trees. U.S. Forest Service Research Paper, 50-132.

Ehrenberg, C. 1970. Breeding for stem quality. *Unasylva.* **24**(23):23–31.

Ekberg, I., Dormling, I., Eriksson, G., and von Wettstein, D. 1976. Inheritance of the photoperiodic response in forest trees. In *Tree Physiology and Yield Improvement* (M. G. R. Cannel and F. T. Last, eds.), pp. 207–221. Academic Press, New York.

Eriksson, G., Ekberg, I., Dormling, I., and Matern, B. 1978. Inheritance of bud-flushing in *Picea abies. Theor. Appl. Gen.* **52**:3–19.

Franklin, E. C., Taras, M. A., and Volkman, D. A. 1970. Genetic gains in yields of oleoresin, wood extractives and tall oil. *Tappi* **53**(12):2302–2304.

Goddard, R. E., and Strickland, R. K. 1964. Crooked stem form in loblolly pine. *Sil. Gen.* **13**(5):155–157.

Gruenfeld, J. 1975. "Leaning Flagpoles." Workshop South. For. Econ. Workers, Biloxi, Miss.

Hatchell, G. E., Dorman, K. W., and Langdon, O. G. 1972. Performance of loblolly and slash pine nursery selections. *For. Sci.* **18**(4):308–313.

Holden, C. 1980. Rain forests vanishing. *Science* **208**:378.

Holley, D. L., and Veal, M. A. 1977. "Economic Impact of Fusiform Rust." Proc. Fusiform Rust Conf., Gainesville, Fla., pp. 39–50.

Hunt, R., and Zobel, B. 1978. Frost-hardy *Eucalyptus* grow well in the Southeast. *South. Jour. Appl. For.* **1**(1):6–10.

Jeffers, R. M. 1969. "Parent-Progeny Growth Correlations in White Spruce." Proc. 11th Meet. Comm. on For. Tree Breed. in Canada, Ottawa, pp. 213–221.

Jett, J. B., and Talbert, J. T. 1982. "The Place of Wood Specific Gravity in the Development of Advanced Generation Seed Orchards." *South. Jour. Appl. For.* **6**:177–180.

Johnson, N. E. 1976. Biological opportunities and risks associated with fast-growing plantations in the tropics. *Jour. For.* **74**(4):206–211.

Johnsson, H. 1950. Ankommor av masurbjörk [Offspring of curly burch]. Årsberätt. Fören. Växtföräd. *Skogsträd*, pp. 18–29.

Ledig, T. F., and Porterfield, R. L. 1981. "West Coast Tree Improvement Programs: A Break-Even Cost-Benefit Analysis." U.S. Forest Service Research Paper PSW-156.

Ledig, F. T., and Whitmore, J. L. 1981. "The Calculation of Selection Differential and Selection Intensity to Predict Gain in a Tree Improvement Program for Plantation Grown Honduras Pine in Puerto Rico," Southern Forest Experiment Station Research Paper SO-170.

Lundren, A. L., and King, J. P. 1966. Estimating financial returns from forest tree improvement programs. *Jour. For.* **28**(1):37–38.

Marquis, D. A. 1973. Factors affecting financial returns from hardwood tree improvement. *Jour. For.* **71**(2):79–83.

Mohn, C. A., and Randall, W. K. 1971. Inheritance and correlations of growth characteristics in *Populus deltoides. Sil. Gen.* **20**(5–6):182–183.

Nautiyal, J. C. 1970. "Economic Considerations in Tree Breeding." Proc. 12th Meet. Comm. For. Tree Breed. in Canada, pp. 222–223.

Nienstaedt, H. 1981. "Super spruce seedlings continue superior growth for 18 years," North Central Forest Experiment Station Research Note NC-265.

Nikles, D. G. 1970. Breeding for growth and yield. *Unasylva* **24**(2–3):9–22.

Perry, T. O. and Wang, C. W. 1958. The value of genetically superior seed. *Jour. For.* **56**:843–845.

Porterfield, R. L., Zobel, B. J., and Ledig, F. T. 1975. Evaluating the efficiency of tree improvement programs. *Sil. Gen.* **24**(2–3):33–34.

Porterfield, R. L. 1977. "Economic Evaluation of Tree Improvement Programs." 3rd World Cons. of For. Tree Breed., Canberra, Australia.

Reilly, J. J., and Nikles, D. G. 1977. "Analysing Benefits and Costs of Tree Improvement: *Pinus caribaea.*" Proc. 3rd World Cons. For. Tree Breed., Canberra, Australia, pp. 1099–1024.

Robinson, J. F., and van Buijtenen, J. P. 1979. Correlation of seed weight and nursery bed traits with 5-, 10-, and 15-year volumes in a loblolly pine progeny test. *For. Sci.* **25**(4):591–596.

Row, C., and Dutrow, G. 1975. "Measuring Genetic Gains by Projected Increases in Financial Return." Proc. 13th South. For. Tree Imp. Conf., Raleigh, N.C., pp. 17–26.

Shelbourne, C. J. A., and Stonecypher, R. W. 1971. The inheritance of bole straightness in young loblolly pine. *Sil. Gen.* **20**(5–6):151–156.

Snyder, E. B. 1973. "15-Year Gains from Parental and Early Family Selection in Longleaf Pine." Proc. 12th South. For. Tree Impr. Conf., pp. 46–49.

Stonecypher, R. W. 1982. "Potential Gain through Tree Improvement. Increasing Forest Productivity." Proc. 1981 Society of American Forestry Convention.

Sweet, G. B., and Wareing, P. F. 1966. "The Relative Growth Rates of Large and Small Seedlings in Forest Tree Species," New Zealand Forest Service Report No. 210.

Talbert, J. T. 1982. "One Generation of Loblolly Pine Tree Improvement: Results and Challenges." Proc. 18th Can. Tree Imp. Assoc. Meet., Duncan, British Columbia, Canada.

Teich, A. H., and Carlisle, A. 1977. "Analysing Benefits and Costs of Tree Breeding Programs." 3rd World Cons. For. Tree Breed., Canberra, Australia.

van Buijtenen, J. P. 1973. Innovation by systems analysis: Southern pine kraft linerboard from breeding to boxcar. *Tappi* **56**(1):121–122.

van Buijtenen, J. P., Alexander, S. D., Einspahr, D. W., Ferrie, A. E., Hart, T., Kellogg, R. M., Porterfield, R. L., and Zobel, B. J. 1975. How will tree improvement and intensive forestry affect pulp manufacture? *Tappi* **58**(9):129–134.

van Buijtenen, J. P., and Saitta, W. W. 1972. Linear programming applied to the economic analysis of forest tree improvement. *Jour. For.* **70**:164–167.

Weir, R. J. 1975. "Cone and Seed Insects—Southern Pine Beetle: A Contrasting Impact on Forest Productivity." 13th South. For. Tree Impr. Conf., Raleigh, N.C., pp. 182–192.

Wilkinson, R. C. 1974. "Realized and Estimated Efficiency of Early Selection in Hybrid Poplar Clonal Tests." Proc. 21st Northeast. For. Tree Impr. Conf., Fredericton, New Brunswick, Canada, pp. 26–35.

Zobel, B.J. 1966. "Tree Improvement and Economics: A Neglected Interrelationship." 6th World For. Conf., Madrid, Spain.

Zobel, B. J. 1975. "Our Changing Wood Resource—Its Effects on the Pulp Industry." 8th Cellulose Conf., Syracuse, N.Y., pp. 5–7.

Zobel, B. J. "Trends in Forest Management as Influenced by Tree Improvement." Proc. 15th Sou. For. Tree Imp. Conf., Starkville, Miss., p. 73–77.

Zobel, B. J., Kellison, R. C., Matthias, M. F., and Hatcher, A. V. 1972. "Wood Density of the Southern Pines," N.C. Agricultural Experiment Station Tech. Bull. No. 208.

CHAPTER 15

The Genetic Base and Gene Conservation

In the short term, the major objective of the tree improver is to manipulate the variability present in a forest tree population to produce more trees with desirable growth, form, or adaptability characteristics. In so doing, the genetic base will usually be maintained or narrowed for economically important characteristics, and the base will be maintained or broadened for those characteristics that affect adaptability. Any tree improvement program carried over many generations will eventually reach a plateau beyond which further meaningful progress is not possible unless steps are taken that ensure maintenance and enhancement of sufficient genetic variability. Therefore, for successful long-term tree improvement, it is essential to begin with *a broad genetic base* and to use a breeding program that will *conserve genetic potential* that is already in the population.

It is vital to recognize that one can breed for a narrowing of the genetic base for certain important economic characteristics such as straightness of tree bole and at the same time develop trees that have wide adaptability that is expressed through sustained good growth and good pest resistance and adaptability to adverse environments. This is possible because in most species there is considerable genetic independence between different economic characteristics and those for adaptability. Desired objectives can be achieved either by using a provenance with the needed adaptability within which selection is practiced for desired economic characteristics, or conversely, by intensively selecting for desired economic characteristics and then choosing individuals within the selected population with the desired adaptability.

Which approach one should follow depends on the needs and desires of the organization concerned and the species of interest. What is basic to both approaches is that the breeding program must be designed to conserve genes and gene complexes of value in order to maximize long-term gains in economically important traits and in adaptabiliy. Therefore, provision must be made so that potential gains in the program are not constrained by a reduction in the genetic base or by the effects of inbreeding depression that result from related matings. It is obvious that no single method of gene or gene complex conservation can be used in all circumstances. The best procedure would be to utilize all methods available for gene conservation, but cost and space limitations usually prohibit this. How soon serious problems will occur using current breeding methods depends on the intensity of selection being used, population size, and the number of generations the program has been in operation.

It was stressed in earlier chapters that all successful tree improvement programs must have an operational, or use, phase and a developmental, or research, phase. Most of this book has dealt with the operational aspects of tree improvement. This chapter will emphasize the gene conservation activities necessary to ensure a successful and ongoing long-term operational tree improvement phase.

Gene conservation as applied to the progress and efficiency of a tree breeding program is somewhat different from the objective of saving genes for some general but undefined future purpose. This chapter will be primarily concerned with gene conservation for an applied forest tree improvement program, although the more conventional gene conservation approach will also be covered.

WHAT IS GENE CONSERVATION?

It is difficult to discuss gene conservation in an orderly and rational way because the subject sometimes becomes very emotionally charged. Many persons equate gene conservation with the necessity to prevent the extinction of a species or a provenance of a species. Sometimes wild and unsubstantiated claims are made (see Zobel and Davey, 1977). Although there are rare exceptions, the problem usually faced is not the extinction of a species but rather the reduction of a species, or a part thereof, to such an extent that it reduces its genetic potential. For the plant breeder, this concern strongly relates to saving those genes and gene complexes that have current or possibly future economic value and to adaptability characteristics (Zobel, 1971, 1978). The problem is large and real, because as Kemp et al. (1976) have stated, the general objectives with which the conservationist is concerned can often only be surmised and not positively identified. The tree breeder must be cautious and conservative because once a gene pool or gene complex has been lost, it is gone forever.

When a species is truly endangered to the extent that it is facing extinction, nearly everyone agrees that all efforts should be made to save as much of it as possible and as quickly as possible. For example, Keiding (1977) has mentioned that three pine and one teak species are in danger of being lost; certainly they are being seriously threatened by depletion of genetic resources and by contamination from related species. But the usual situation is that only a portion of a species is endangered, because its gene pool is in danger of being so reduced that some genes or gene complexes will be lost (Kleinschmit, 1979). One example of such a situation was reported for *Eucalyptus* by Turnbull (1977) who stated that not one of the 450 species that exists in the genus is in danger of extinction, but a few are definitely faced with extreme genetic impoverishment. Another example is in Central America, where a few species are truly endangered and a number of species have provenances that are being seriously depleted genetically (Figure 15.1). Genetic material from this area is a primary source for planting in South America and other tropical regions. Therefore, genetic impoverishment in Central America will have very great economic as well as biological importance. In fact, concern was so great for the Central American conifers that the international cooperative CAMCORE[1] was formed (Gallegos et al., 1980; Dvorak, 1981), including companies and governments from eight nations to help conserve the gene resources of the endangered trees.

Gene conservation therefore relates to activities directed at saving gene pools for use to prevent loss of genes, gene complexes, and genotypes and, in extreme instances, to prevent extinction of whole taxonomic categories of trees. For ease of reference and accuracy, the term *gene conservation* will be used in this book rather than the term *gene preservation*. It is of importance, however, to recognize that some authors and groups use the two terms synonymously and interchangeably.

---

[1]CAMCORE stands for Central America and Mexico Coniferous Resources Cooperative.

FIGURE 15.1

There are numerous reasons why sources within species, and even some species, are in danger of losing genetic material in Central America. Shown is a woodcutter with a load of wood. Forests are being overutilized for fuel in the drier areas, especially those close to habitations.

CONSERVATION OF GENES AND GENE COMPLEXES

The literature usually refers to *gene conservation,* but this can be somewhat misleading. Almost all economically important tree characteristics are controlled by the alleles at several gene loci. There are a few instances in which an important characteristic in forest trees is under single-gene control, such as certain disease-resistance characteristics, but these are the exception. In reality, then, efforts to conserve genes usually involve conservation of whole gene complexes that make a tree economically desirable and adapted to pests and other aspects of its environment (Zobel, 1978). In this sense, the gene complex *does not* equate to the genotype of the individual; it relates to the gene action or actions for specific traits. Conservation, control, combination, and use of gene complexes is a key concept in tree improvement.

Despite the controversies associated with gene or gene complex conservation, there is one fact on which all parties agree—conservation of genetic material is of critical importance in tree improvement if maximum long-term gains are to be achieved (Burley and Namkoong, 1980). But there are great differences of opinion about what complexes should be conserved and how this can be best achieved. One major difference exists between those foresters who define gene conservation as *saving all possible genes or gene complexes within species or races* and those who espouse *saving those genes or gene complexes that will be most helpful in a long-term breeding program.* A major argument raised by the former group is, Who knows when any given gene or gene complex might become desirable? The most cautious persons will thus insist that all possible genetic materials within the population be saved, an ideal action but one that is deterred by the fact that conservation of *all* genes and gene complexes within a species is highly impractical because of space, time, and cost limitations. In fact, in a breeding program, gain is made by changing gene frequencies; the objective is to increase the gene frequencies of desirable characteristics and to reduce those that are undesirable.

A decision must therefore be made about which genes or gene complexes are desirable, and these should have the highest priority in being conserved. This is difficult indeed, especially when one takes into consideration what might be useful or needed in future generations. Sometimes conservationists deal with saving genes that are related primarily to adaptability. These are vital, but often they are not the characteristics that are endangered. When bad logging is done, which often occurs in poorly executed selective logging systems, the characteristics in danger of loss are those of economic importance, such as straightness of tree bole, small limb size, desired wood qualities, or some other factor (Figure 15.2). When

FIGURE 15.2

A type of gene loss often not recognized can result from differential, or selective, logging. Often, all the good trees of the good species are removed, and the stand is left to grow as is. The stand in the illustration is totally degraded and is almost at the point of not growing because of a series of selective loggings, even though it is on an excellent site. Two types of genetic losses occur: (1) loss of desired species and (2) loss of the best genotypes of the trees of the desired species that are left.

endangered, these characteristics are just as vital to conserve as are those that are usually emphasized and that are related to adaptability to pests and adverse environments.

The concept that every possible gene or gene complex should be saved leads to suggestions that wild stands be preserved. The method is effective but results in tying up very large areas, which is a most inefficient approach if the same gene complexes can be conserved for later use when they are incorporated by breeding into a limited number of individuals in a clone bank or holding area. Essentially every individual tree or genotype within a population is genetically different, although many share certain genes or gene complexes. The objective of a gene conservation program should be to conserve genes and gene complexes, not necessarily genotypes.

CONSERVING GENETIC RESOURCES

The need for conservation of genetic material in forest trees is evident and not arguable (Burley, 1976). The reasons for the need are sometimes debated, and the conservation methods to be followed elicit heated discussions. However, the need for saving genetic material for use and adaptation for future unseen needs is critical.

Reasons for Loss of Genetic Resources

There are numerous reasons why the need for the preservation of forest gene resources is so important and is becoming so critical so rapidly. Any action that destroys forests or destroys one part of a forest can lead to a dangerous situation. The destructive agents are many; for example, insects, diseases, cutting for fuel, logging, clearing for agriculture, urban expansion, fire, storms, and other natural disasters all take their toll. There is no single culprit. The practice of removing the indigenous forests in the southeastern United States and replacig them with pines is sometimes believed to reduce the genetic base of the hardwoods to the endangered-species level; this contention has, however, been challenged by Popovich (1980) who cites the small amount of monoculture in the South and the fact that hardwoods are always a part of conifer plantations.

The most dangerous situations occur in the forest tree populations with restricted ranges or where there are disjunct populations. This can occur for unusual ecotypes on restricted sites, or sometimes for entire, restricted endemic species. In these circumstances, not only genes or gene complexes are endangered, but whole unique populations and species can be totally eradicated. Some forest tree species or provenances have been reduced to a few hundred survivors.

The problem of conservation has become especially critical in the dry tropical areas where the only economically acceptable source of energy is firewood. Wood cutters systematically take the closest trees and move out from their villages, leaving no forests behind (see Figure 15.1). Because these areas are usually hot and dry, forest regeneration is not good. An attempt is often made to farm the denuded areas, which destroys any regeneration. If farming is not done, grazing, usually along with fire, is almost always used, and these two factors wipe out any new trees that might become established. Goats and sheep can be among the worst enemies that forests have. They frequently prevent reestablishment of trees in cutover areas, and thus they create a real problem in gene conservation (Zobel, 1967).

Another problem, which is locally recognized but often not considered, is the destruction of forests by shifting agriculture. Especially in the tropical regions, whenever roads are built through the forests, an army of settlers usually follows. The forests are cut, farmed for a short time, and then are abandoned when the residual nutrients in the soil are gone or after erosion has removed the topsoil. Loss to shifting agriculture is serious indeed in many tropical areas and is considered a most important reason for forest destruction.

Any type of large-scale destruction, such as fires, insects, diseases, or storms, sometimes causes great concern about gene loss. However, the key to the seriousness of this concern is what happens after the disaster. If a new forest becomes established, there is little danger of gene loss, but if the destruction is so great that trees do not become reestablished, severe gene loss can occur (Figure 15.3). For example, Zobel has seen fires in Central America so large (up to 200,000 ha) and so severe that little natural tree regeneration could become established. This type of wholesale destruction is most common in more droughty areas and is a major cause of loss of genetic resources.

Examples of losses large enough to threaten loss of genetic potential due to insects or disease are numerous. Perhaps the most notable case is the loss of the American chestnut (*Castanea dentata*) from eastern North American hardwood forests due to the chestnut blight that was caused by the introduced fungus *Endothia prasitica*. Once a widespread and extremely important species economically, the American chestnut has been almost totally lost except for occasional sprouts that are usually killed before reaching productive maturity or a very rare tree that grows outside the species range. The gene complement of this species has been reduced to such a dangerous degree that, for all practical purposes, it has been lost.

Another example is the current plight of Fraser fir (*Abies balsamea*). Fraser fir is a species that is restricted naturally to a few mountaintops in the southern Appalachian Mountains in the eastern United States. It has considerable economic importance as a high-quality Christmas tree. Natural stands of Fraser fir are currently being threatened by the balsam woolly aphid (*Adelges piceae*). Often, entire natural stands of mature fir are destroyed by this pest. Reproduction by seed is abundant, but if these stands are attacked and destroyed before reaching reproductive maturity, the genetic resources of the stand could be lost. This would be severe from biological, economic, and aesthetic standpoints.

The logging activities of human beings for lumber or pulp can be destructive to the gene base. If all trees are removed, and this action is followed by frequent fires or grazing that prevents regeneration, gene loss can be severe (see Figure 15.3). Another less recognized loss is by selective logging when only a few high-value

FIGURE 15.3

Many species of pines planted in the tropics have their origin in Central America. Loss of gene complexes from this area has special significance for tropical forestry. Shown is an aerial view of a logging operation that totally removed all *P. caribaea.* After logging, regeneration was good, but a subsequent fire killed all the young trees. No further seed source is available. Shown are the roads and skid trails.

species, which occur only sparsely on each hectare, are removed (see Figure 15.2). Regeneration is restricted and severe genetic losses result, even though the forest appears to be intact. The lack of regeneration results partially from a lack of a suitable seedbed for the desired species and from competition from less desired species that prevents the reestablishment of the species that had been selectively logged. There are a few temperate as well as numerous tropical species in which gene complex loss resulting from so-called selective logging has, or will soon, become critical.

Methodology of Conservation

There is a great diversity of opinion about which is the best method to use to preserve forest gene resources. Broadly speaking, conservation efforts follow one or two general approaches: *in situ,* which means preservation of trees and stands in natural populations, and *ex situ,* which refers to saving the genes or gene complexes under artificial conditions, or at least not in their native stands.

In situ Conservation   *In situ* conservation appears to be very popular; it certainly has the greatest appeal and is best understood by persons not involved in land management. *In situ* conservation is applied by setting aside and preserving stands of the desired species or complexes to prevent further losses, usually from the activities of humans. One of its major benefits is the conservation of ecosystems. There is no other effective method to achieve conservation of ecosystems than through the "*in situ* approach."

The philosophy of a complete "hands off" strategy with respect to the stand that is to be conserved has great appeal but will fall short of the desired objective of maintaining the genetic structure of the current stand because of the dynamic changes that take place in all forest stands. If natural stands are left solely "to nature" and are not managed, the species content and gene complex distribution will change while the stands pass through one stage of succession to another. For example, many pine forest types have been established as the result of some past, usually dramatic environmental change caused by fires, tornadoes, hurricanes, or clearing for farming. If the current pine stands in the southeastern United States are left to grow and are not managed, the subclimax pine forests are often gradually replaced by hardwoods, and over time the pine component will disappear or be greatly reduced. Thus, if it is desired to preserve and maintain the genes and gene complexes already present in the pine stands, they must be managed to halt the natural succession to a high hardwood component. A *major need for in situ conservation of genetic resources,* which is either not understood generally or is ignored, *is to manage the preserved stand in order to maintain the desired genetic composition.*

One of the least understood aspects of managing stands for *in situ* conservation is the size needed. The general attitude seems to be that "if a few trees are good, a lot of trees are better," and most *in situ* conservation areas are larger than needed. Large-sized *in situ* conservation stands are not harmful from the genetic standpoint, but they impose an excessive strain and drain on the economics of the organization underwriting the conservation effort and often on society in general, because potentially useful forest products are less available. It is not necessary to conserve hundreds of thousands of individuals that contain the same desired gene complexes; at most, a few thousand such trees are usually sufficient. If these individuals are conserved and managed so that they can regenerate themselves, there is no need to set aside many thousands of hectares for gene conservation.

The public, as well as some foresters, has become confused between *conservation of forests for ecological and aesthetic reasons* and those for *gene conservation.* The former usually requires large acreages, but this should not be interpreted as the need for equally large areas for gene conservation per se. For example, there are moves at the present time to save huge sections of the tropical forests in the Amazon as a "gene conservation measure." The large areas may be desired for ecological conservation but are not needed for gene conservation; a few well-chosen forest stands of moderate size will serve the gene conservation purpose well. It is amazing how many people who want to preserve forests for one reason or another incorrectly use gene conservation as the reason why this should be done. This practice hinders the true gene conservation efforts and often causes unwarranted anatagonism against the gene conservation effort. It usually is not necessary to set aside huge areas of natural forests to achieve the gene conservation objective.

One common problem with *in situ* conservation is that often the wrong forest populations are saved. Instead of conserving the truly endangered populations, stands are often chosen that are growing at high elevations or on other unique poor-site types, because they are less expensive, easier to obtain, or are the desired size. In so doing, the objective of conserving the populations occurring on more typical sites is defeated. Great care must be used in the selection of stands for *in situ* conservation.

One conservation difficulty in some areas arises from contamination of the desired genetic material from related sources that are planted near the area to be conserved. As plantation forestry becomes more common, the danger from such contamination increases. It has been said, for example, that in several countries in western Europe, there are no truly indigenous Scots pines because of a complete mixture between the indigenous and imported sources of this species. Sometimes it is desired to save thinned stands of exotics to preserve their genetic composition as desirable land races. This is most difficult because other undesired sources of exotics planted nearby may affect the stand that is to be saved.

From the botanical standpoint of preserving pure races, gene mixing from contamination of geographic sources is undesirable. For a breeder interested in producing the best yields in the given environment, it can be of minor concern and can represent an unusual type of land race; if the newly created genotypes are superior in the environment of interest, they should be used by the breeder. For an operational program, there is no particular virtue in purity of the local source as such, so long as the new genotypes have the combination of gene complexes that make them superior for the growth and products desired. The "new" trees may have good gene complexes, but the breeder will often need to stabilize them before they can be used in an operational program.

***Ex situ* Conservation** There are many ways to save desired genes or gene complexes that either contain genetic material that is the same as natural populations or for specific uses. The most common of these is through reproduction by the conventional vegetative propagation techniques of grafting, rooting cuttings, or air layering (Longman, 1976). Foresters have a special opportunity, in that given genotypes (or gene complexes) can essentially be conserved "forever" through vegetative means. Rather than savings hundreds or thousands of acres of trees, a few vegetative propagules of the desired trees can be established, maintained, and crossed when and as desired. Although the objective is gene conservation for special uses, seed orchards effectively maintain and increase the frequency of desired genetic characteristics (Zobel, 1971).

As breeding programs develop, it becomes possible to incorporate many genetic qualities into a few individuals through control pollinations. Saving of the gene base can thus be accomplished through selection and breeding of a limited number of individuals, and new combinations can be obtained by crossing among them. This method of packaging, saving by propagation, and outcrossing of genes and gene complexes is actually what is accomplished by clone banks in applied tree improvement programs. As pressures for forestland use and contamination of natural stands of trees become more widespread, the packaging method will undoubtedly become more important as a method of gene conservation.

Many other *ex situ* methods are available. For example, seed storage methods are very good for some species, and preservation of genotypes, genes, and gene complexes can be done in this way. Ultimately, the seed will lose viability and will need to be replaced. Seed of some species can be kept under proper storage conditions for very many years, but for other species that have transient viability, gene conservation by seed will not work. One danger with seed storage is that mutations may occur in the stored seed so that trees grown from the stored seed will have a genetic component that is a little different from the original population. Similarly, pollen can be stored for a long period of time. But pollen represents only half the desired material, and suitable females need to be available upon which the pollen of the conserved genes can be used.

Some of the more recent methodologies, such as tissue culture, have great potential for *ex situ* gene conservation. As this method becomes more operational, it will be possible to "store" the genetic potential of large numbers of genotypes in a very small area. This is a hope for the future.

Because of the difficulties and costs of *in situ* gene conservation, it is likely that *ex situ* methods, especially those related to packaging the desired genes, preserving by vegetative propagation, or seed and pollen storage followed by multiplying by crossing, will be the most widely used methods of gene conservation in forest trees in the future.

CONSERVATION OF PROVENANCES WITHIN A SPECIES

Too often, discussions about gene conservation leave the impression that the only need is to conserve genes within a single gene pool. However, most important forest tree species have one to several geographic races or provenances that possess rather large and important genetic characteristics that are unique to each. Therefore, the first major job of gene complex conservation for the plant breeder is to save the unique characteristics of geographic races (Zobel et al., 1976). Differences among provenances are primarily caused by a few major, differing gene complexes that give the source a unique advantage for growth and survival in a special environment. Other gene complexes may be common to most or all races within the species.

An important decision relative to gene conservation is immediately faced for species with a wide distribution—does one conserve the characteristics of the outlier provenances, each of which has undergone natural selection in extreme and different environments, or should there be a concentration on populations within the center of the species range? The latter contain a broad group of gene complexes that can be used widely throughout most of the species range but that are not especially adapted for the environmentally extreme fringe areas. Usually, some adaptability to the fringe areas is found in individuals from the center of the species range, but such trees usually occur in low frequencies. It would be ideal to conserve key complexes of all provenances, including the fringe ones, but this is not feasible with species that have a wide geographic distribution and many geographic sources, unless there is well-coordinated and outstanding cooperation among a number of organizations.

Although geographic differences have been widely studied in some species, there often is a considerable lack of knowledge about the gene complex differences that exist among provenances within a species. It is important to recognize that the best sources of genetic material for a breeding program, emphasizing characteristics related to adaptability, come from existing gene complexes that are relatively common to entire populations or provenances, rather than from occasional individuals within a population that may have a desired genetic structure. Conserving gene complexes for adaptability by saving poulations requires a different approach than that for conserving characteristics that are unique to individuals within a population.

A decision to conserve the gene complexes of different provenances raises some interesting questions. Should the trees within the provenance that are to be conserved be obtained randomly, or should they be selected to save only the phenotypes that have the most desirable growth and form for a breeding program? The fact is that there is little indication or reason to believe that the best-formed trees within a provenance will not have the same adaptability as others that typify the provenance. This is true because the lack of genetic correlation between adaptability and morphological chracteristics appears to be the rule rather than the exception. A too common misconception is that because a tree is a scrawny, poorly formed individual, it will magically carry alleles for pest resistance or other types of adaptabilities that are not carried by better-formed, faster-growing trees. Intensive studies have shown few close genetic relationships between adaptability and tree form in forest trees. There are, of course, notable exceptions to this statement, such as prostrate tree form in areas of continuous wind, or foliage characteristics that enable a tree to withstand drought stress better.

The number of trees required to minimize losses of alleles has been debated, and methodology has been developed. Although the mathematics involved is beyond the scope of this book, it has been developed in the article by Namkoong et al. (1980). They propose that a sample of 50 individuals would provide security against gene loses and would preserve genetic variances. They also state that losses and potentials for future breeding would be greatly reduced in populations in which 20 or fewer individuals are used.

From a tree-breeding standpoint, it becomes clear that if funds and land resources are limiting (as they usually are), only trees with the most desirable phenotypes should be used as sources of seed when collecting for conservation of gene complexes among geographic races. If possible, several hundred trees (200 to 400) should be used as the basic number for the collection base, and individuals that are to be conserved should preferably be from separate areas to avoid relatedness. If so many races exist that 200 to 400 trees cannot be represented from each, then fewer individuals should be saved from the races of lesser value or importance. This method will preserve the most outstanding gene complexes that make different provenances of special value.

The initial reaction to conservation of the adaptive characteristics of different geographic races is that it is an impossible job. But the stakes are high, and the need to save and utilize the material in breeding programs is great. It is difficult to set an upper limit on the number of geographic races and trees within races that should be saved, because tree breeding will always be working on the lower limit of the ideal.

CONSERVATION OF CHARACTERISTICS OF INDIVIDUAL TREES

To many people, gene conservation encompasses the preservation of as many indivdiual genotypes as possible. A quesion that is often raised is whether to save trees with currently undesired economic characteristics, such as those with crooked boles, pest infection, slow growth, or undesired wood. It would appear logical to assume that trees with currently unwanted characteristics will not carry any greater potential for combating future pest or other adaptability problems that might arise than will those trees possessing good form and growth.

However, there can be a danger in rejecting all trees having inferior characteristics. The definition of what constitutes desirable or undesirable characteristics is determined by current usage and economic standards. Market requirements change, and what is the *despised* today can become the *wanted* of tomorrow. A good example of such a reversal was the acceptance of knotty-pine lumber. For many years, boards with multiple knots had low value, but then the market desires changed, and knotty pine came into considerable demand as a specialty finish in houses. Another real-life example of this reversal has occurred for wood specific gravity of pines for use in paper manufacture. When the tree improvement program in the southern United States began 30 years ago, the major paper products from southern pines were kraft bags and corrugated boxes, and all emphasis was on trees with high-specific-gravity wood that gave good yields and made paper with good tear strength. The seed orchards established usually contained only high-specific-gravity-wood parents. Fortunately, the select trees meeting all grading criteria but which had lower wood specific gravities (amounting to 40 to 60% of the trees graded) were not destroyed but were preserved (as gene complexes) in clone banks. When tissue, newsprint, and quality printing papers suddenly became of major importance from southern pine wood, it was possible to develop a seed supply immediately, using parents that would produce progeny with low-gravity wood that was highly desired for these end products.

It is essential to look into the future as much as possible and to be careful not to make a judgment about what should be preserved that is based solely on the currently high economic worth. It might even be justifiably argued that the currently increasing pressures toward urban and amenity forests would dictate saving deformed and slow-growing phenotypes for their ornamental or protective value.

Obviously, then, a concern of any plant breeder involved in a long-term improvement program is the fate of genes that are currently neutral, or are of no economic consequence, but that may become important in the future. A prime example would be genes that confer resistance to an insect or pathogen that is unknown or that currently poses no threat but that could become a major pest in later generations. A prime example of a new pest is the pine woolly aphid (*Pineus pini*), which was introduced into Rhodesia (Zimbabwe) in 1962 and spread rapidly, attacking nearly all the hard pines. Great variation among individual trees in resistance to this pest was found (Barnes et al., 1976). Trees under stress were the most susceptible. The authors stated that "these provenance trials illustrate the principle of adaptation to plantation and local climatic conditions as insurance against possible catastrophe when a new injurious organism is introduced."

These "neutral" genes are usually not selected against in breeding efforts, but they can become lost in the improved population due to chance occurrences of genetic drift resulting from small population size. Obviously, maintenance of a large breeding population will increase the chance that so-called neutral genes will be present when needed. An alternative strategy, which uses genetic drift to *increase* the frequency of neutral genes, is the multiple small-population concept of van Buijtenen and Lowe (1979), Burley and Namkoong (1980), and other researchers. When a large breeding population of several hundred genotypes is divided into numerous smaller groups, perhaps of 20 genotypes or less, the effects of genetic drift will be different for each group. In some groups, neutral alleles will be totally lost by chance, because of the small population size. In other groups, however, drift may operate to increase the frequency of the neutral allele, so that if it ever does become of economic importance, plant material may be immediately available that contains that allele in high frequencies.

In a tree breeding program, *a decision has to be made* about what is, or may be, the most important genotypes to conserve. Every genotype cannot be saved. Trees selected should be those having characteristics known or foreseen to be of major importance in a breeding program (Figure 15.4). Each group of trees conserved will represent a small population sample containing desired or potentially desirable gene complexes that, when combined with other groups, will ensure conserving the bulk of the gene complexes that are essential to an ongoing tree improvement program.

THE SPECIAL SITUATION FOR TROPICAL REGIONS

Although gene conservation is vital to all tree improvement programs, it is especially critical when exotic forestry is being practiced (Brazier et al., 1976; Brune and Melchior, 1976; Kemp et al., 1976). Both variation among prove-

FIGURE 15.4

When possible, trees with the best qualities for a breeding program should be saved, such as this beautiful
Pinus tecunumanii from Guatemala. This species is
under great stress and is rapidly disappearing. To conserve this species is a major goal of the CAMCORE
Cooperative. (Photo courtesy W. Mittak, Germany.)

nances and conservation of genetic variability within provenances are involved;
this must all be combined with the land race concept (Brune and Zobel, 1981).
Recognition of the importance of maintaining a proper genetic base and preserving the best for a given environment in exotic plantations has been delayed too
long, although there has been a recent upsurge of interest, especially in tropical
species. Roche (1979) states that the most pressing conservation problems are in
the humid tropics. He makes a plea to expand conservation efforts also to those
species that are of value for usage other than for standard forest products of
boards or fiber. For example, the book by Burley and Styles (1976) is devoted to
various aspects of gene conservation and breeding of tropical forest trees.

In some ways, the most pressing need in the tropics is for gene conservation in the indigenous forests (Myers, 1976; Roche, 1979). This was emphasized by King (1979) who reported that 65% of the land in the tropical areas of the world supports fragile ecosystems; people who live in these regions total 630 million, or 35% of the total population of the developing countries. The need for food and fuel has led to land utilization practices that in turn cause degradation of the fragile ecosystems with a resultant depletion of gene pools. The situation is aggravated because many of the species in the tropical forests are very sensitive to ecological changes. Further, only a limited amount of work has been done on some species; their biology, reproduction, and silvicultural care are poorly understood. Add to this the large numbers of species per hectare in some tropical hardwood forests, and one sees that the job of gene conservation becomes most formidable. A number of authors, such as Wood (1976), feel that one of the best ways to preserve the tropical forests is to remove pressures for their use by growing the (usually) much more productive and uniform exotic forest trees on suitable areas and leaving the rest of the tropical forestlands without intense pressures to manage them. Wood called such plantations of exotics *compensatory plantations*.

Thus, the tropical areas need strong conservation measures to assure that the species used as exotics are not impoverished (Davidson, 1977) or that the indigenous species are not lost or seriously reduced genetically by changing land use patterns. The problems are awesome but must be attacked—and soon—if the biological structures of tropical forest species are to be preserved and if forestry in the tropics is to remain productive.

THE POLITICS OF CONSERVATION

Concern about gene conservation is evident for species growing in both tropical and temperate regions. Publications dealing with gene conservation date back many years, such as the one by Kanehira (1918) who stressed the need for forest conservation in Taiwan. The Committee on Exploration, Utilization and Conservation of Plant Gene Resources of the Food and Agricultural Organization of the United Nations (FAO) has been active for many years. FAO publishes a newsletter and annual report entitled *Plant Genetic Resources* that contain some excellent information related to gene conservation. Of particular value are the forestry occasional papers entitled *Forest Genetic Resources*, which are published by FAO and which carry excellent summaries of work in progress related to forest gene pool conservation.

As population pressures grow and as the standard of living increases, forest utilization will become greater, and with it comes the potential of increased genetic loss. Gene conservation is one of the most pressing aspects of forestry, one that demands action.

Because of the nature of the problem, which often crosses national boundaries, the long-term payout, and the obvious social implications, gene conservation must necessarily be strongly and governmentally financed and directed. Although gene loss may be of specific importance to a local area, it usually has widespread implications. It is frequently international in scope, thus making the problem difficult to attack. To add to the difficulty, the areas most needing conservation activities often are in the developing countries that have neither the resources nor the leadership to undertake the preservation effort. Frequently, although the loss will occur in one country, the major use of the forest species is in other countries; examples are the pines of Central America that are used in many of the plantation programs in South America. Therefore, international organizations are the primary candidates to take the leadership. This is now being done; organizations such as FAO of the United Nations are playing a major role in the conservation effort. Other organizations, like the Commonwealth Forestry Institute in Oxford, England[2] and the Queensland Forest Service in Austrtalia, are sponsored by one nation, but they work on an international basis; they are spearheading and organizing conservation efforts. These organizations are very active and have been most successful, especially in relation to the exotic species planted in the tropical areas. But much more is needed, particularly for the tropical hardwoods. No one really knows how severe the genetic loss is in tropical forests that often contain economically specialized and relatively unknown species. It is feared that loss of genes may be great, especially when selective logging of a few species is practiced.

All tree improvers must place gene conservation high on their list of needs and activities. Private organizations must become more involved. A good example is the CAMCORE Cooperative (Central America and Mexico Coniferous Resources Cooperative) that combines private companies and governments of a number of nations in a joint conservation effort (Gallegos et al., 1980; Dvorak, 1981). If tree breeders do not do their part to conserve and broaden the gene base in forest trees, continued long-term gains in tree improvement will not be possible.

LITERATURE CITED

Barnes, R. D., Jarvis, R. F., Schweppenhauser, M. A., and Mullin, L. J. 1976. Introduction, spread and control of the pine wooly aphid *Pineus pini* in Rhodesia. *S. Afr. For. Jour. No. 96:1–11.*

Brazier, J. D., Hughes, J. F., and Tabb, C. B. 1976. Exploitation of natural tropical resources and the need for genetic and ecological conservation. *Tropical Trees,* No. 2:1–10.

Brune, A., and Melchior, G. H. 1976. Ecological and genetical factors affecting exploitation and conservation of forests in Brazil and Venezuela. *Tropical Trees,* No. 2:203–215.

---

[2]The CFI is sponsored by a consortium of 20 Commonwealth companies in addition to FAO. It is an international organization and has supplied genetic material to over 50 countries in the tropics.

Brune, A., and Zobel, B. J. 1981. Genetic base populations, gene pools and breeding populations for *Eucalyptus* in Brazil. *Sil. Gen.* **30**(4–5):146–191.

Burley, J. 1976. Genetic systems and genetic conservation of tropical pines. *Tropical Trees,* No. 2:85–100. Academic Press.

Burley, J., and Styles, B. T. 1976. *Tropical Trees—Variation, Breeding and Conservation.* Academic Press, London. Academic Press.

Burley, J., and Namkoong, G. 1980. "Conservation of Forest Genetic Resources." 11th Commonwealth For. Conf., Trinidad.

Davidson, J. 1977. "Exploration, Collection, Evaluation, Conservation and Utilization of the Gene Resources of Tropical *Eucalyptus deglupta.*" 3rd World Consul. For. Tree Breed., Canberra, Australia, pp. 75–102.

Dvorak, W. S. 1981. CAMCORE is the industry's answer to coniferous preservation in Central America and Mexico. *For. Prod. Jour.* **31**(11):10–11.

Gallegos, C. M., Zobel, B. J., and Dvorak, W. S. 1980. "The Combined Industry–University–Government Efforts to Form the Central America and Mexico Coniferous Resources Cooperative." Symp. on Fast Growth Plantations, São Pedro, São Paulo, Brazil.

Kanehira, R. 1918. The necessity of natural forest conservation. *Jour. Nat. Hist. Soc. Taiwan* **8**(36):56–66.

Keiding, H. 1977. "Exploration, Collection and Investigation of Gene Resources: Tropical Pines and Teak." 3rd World Consul. For. Tree Breed., Canberra, Australia, pp. 13–31.

Kemp, R. H., Roche, L., and Willan, R. L. 1976. Current activities and problems in the exploration and conservation of tropical forest gene resources. *Tropical Trees,* No. 2: 223–233.

King, K. F. 1979. Agroforestry and utilization of fragile ecosystems. *For. Ecol. Mgt.* **2**(3):161–168.

Kleinschmitt, J. 1979. Limitations for restriction of genetic variation. *Sil. Gen.* **28**(2–3):61–67.

Longman, K. A. 1976. Conservation and multiplication of gene resources by vegetative multiplication of tropical trees. *Tropical Trees,* No. 2:19–24.

Myers, N. 1976. An expanded approach to the problem of disappearing species. *Science* **193**:198–202.

Namkoong, G., Barnes, R. D., and Burley, J. 1980. "A Philosophy of Breeding Strategy for Tropical Forest Trees," Tropical Forestry Papers No. 16, Commonwealth For. Inst., Oxford, England.

Popovich, L. 1980. Monoculture—A bugaboo revisited. *Jour. For.* **78**(8):487–489.

Roche, L. 1979. Forestry and conservation of plants and animals in the tropics. *For. Ecol. Mgt.* **2**(2):103–122.

Turnbull, J. W. 1977. "Exploration and Conservation of Eucalypt Gene Resources." 3rd World Cons. For. Tree Breed., Canberra, Australia, Vol. 1, pp. 33–44.

van Buijtenen, J. P., and Lowe, W. J. 1979. "The Use of Breeding Groups in Advanced-Generation Breeding." 15th South. For. Tree Impr. Conf., Starkville, Miss., pp. 59–66.

Wood, P. J. 1976. The development of tropical plantations and the need for seed and genetic conservation. *Tropical Trees,* No. 2:11–18. Academic Press.

Zobel, B. J. 1967. Mexican Pines. In *Genetic Resources in Plants—Their Exploration and Conservation* (O. H. Frankel and E. Bennett, eds.), FAO Tech. Conference on Exploration, Utilization and Conservation of Plant Gene Resources, Section 6(IX), pp. 367–373, FAO, Rome.

Zobel, B. J. 1971. "Gene Preservation by Means of a Tree Improvement Program." Proc. 13th Mgt. Comm. on For. Tree Breed. in Canada, Prince George, British Columbia, pp. 13–17.

Zobel, B. J. 1978. Gene conservation—As viewed by a forest tree breeder. *For. Ecol. Mangt.* **1**:339–344.

Zobel, B. J., and Davey, C. B. 1977. A conservation miracle. *Alabama For. Prod.* **20**(5):5–6.

Zobel, D. B., McKee, A., Hoek, G. M., and Dyrness, C. T. 1976. Relationship of environment to composition, structure and diversity of forest communities of the central-western Cascades of Oregon. *Ecol. Mono.* **46**:135–156.

CHAPTER 16

Developing Tree Improvement Programs

Forestry is playing an ever more important role in the world's society and economy. Forests have become widely recognized as a most valuable renewable natural resource. The heavy emphasis in past years on development of food production in preference to fiber is changing somewhat, although food production still has priority (Zobel, 1978a; Brown, 1980).

Forests must be managed for all types of products, including recreation and conservation, and the contributions to society from forests are rapidly changing. There is an urgent need to make forestland fully productive (Armson, 1980). This need is indicated by the predicted timber shortfall by A.D. 2000 and the accelerated reduction of forestland productivity by fuelwood cutting, farming, use of forestlands for other purposes, and the poor harvesting and regeneration programs that are still being used in a number of areas (Keays, 1975; Zobel, 1980). The increasing need for greater forest production can only be met if more intensive forest management is practiced. Per capita use of forest products is one of the best barometers to gauge societal development. Many developing countries are arriving at an explosive stage of utilization because of the increasing needs for forest products.

More intensive forest management requires the application of tree improvement. Use of wood for fuel and for energy generation and for the production of organic chemicals is revolutionizing forestry operations (Goldstein, 1980; Parde, 1980; Lin, 1981). Trees with different kinds of wood and growth patterns are needed to produce the desired biomass; genetic manipulation can be used as a primary method to achieve this. For example, *Eucalyptus* trees that are to be used for charcoal production should have high-specific-gravity wood, and should grow so that they can be harvested on short rotations.

It is becoming widely recognized that it costs little more to plant improved rather than ordinary trees and that any successful forestry operation will become better by the inclusion of a tree improvement program. But good programs do not just happen; they need much planning and an investment in time and money. Time is critical because trees are long-lived organisms, and it takes time to change quality or to increase production (Zobel, 1978b).

The returns that will be obtained from tree improvement relate to the ultimate use that is made from the information obtained and whether the program has been sufficiently founded to enable continued long-term progress. Research or information about forest management is of limited value unless it is put to use in operational forestry programs. It is now generally accepted in forestry that planting requires better trees and that the only way to obtain optimal gains is to combine good silviculture with improved planting stock. Thus, the stage is set to take full advantage of tree improvement. If maximum benefits are to be obtained from tree improvement, careful planning is needed to assure the greatest efficiency, both in the short and long terms. This chapter is a summary of how to use some of the most important information discussed in the preceding chapters to establish new programs.

STARTING A TREE IMPROVEMENT PROGRAM

In Chapter 1, a greatly simplified way of viewing tree improvement was outlined, which consisted of determining variation patterns, assessing the intensity and cause of variation, packaging the variation into desirable trees, and mass-producing the improved individuals. This can only be accomplished if the species is well known to the researcher and if one has an in-depth knowledge of the species. Within this broad approach there are numerous specific decisions and actions that must be taken.

The basic requirement for starting a tree improvement program is to determine whether or not one is really needed. When programs are initiated with only vague ideas and poorly defined objectives, they are doomed to mediocrity or failure. Planning requires projection into the future as well as an estimation of current needs. In addition to the biological factors, assessments of the need for a potential tree improvement program require estimates of future markets, economic climates, utilization practices, and political conditions.

The following basic principles must be incorporated if a tree improvement program is to be successful.

1. *What is the objective?* The products needed and the urgency to produce them must be determined. Will solid-wood products be a major part of the forest production or are fiber products the primary end goal? Will forest uses such as energy or chemicals from wood be of primary concern? Answers to these questions will determine the species or seed sources as well as the emphasis and methodology best suited for the program. The most serious error made in starting a tree improvement program is to organize and to become committed to specific goals before the objectives of the program have been determined.

2. *What basic biological facts are needed?* The biology of certain species is so poorly known that intensive studies are required before a program can proceed (Figure 16.1). Such simple things as how to collect and store seed and pollen, how to grow seedlings in the nursery, and how to outplant the trees and care for the plantations must sometimes be determined. This need is especially critical for the lesser known tropical species. Foresters in the tropics often choose to work with pines and eucalypts because they know the silviculture of those species, whereas such information is often lacking for the indigenous species. Very large programs have proven to be unsuccessful because the basic information about how to grow seedlings in the nursery or how best to establish and care for plantations was not known; therefore, all the potential advantages from a tree improvement program could not be realized. When an applied tree improvement program is developed *it must be accompanied by parallel efforts to solve forest management needs.*

FIGURE 16.1

Before a tree improvement program can be successful, many basic biological facts must be known. Although well understood at present, in the past the biology of the developmental stages of pine reproductive structures had to be worked out; an early stage of conelet development is shown. Such information is missing for many tropical hardwood species.

3. *What species, or seed sources within species, should be used?* Choosing the correct species and seed sources is essential for the success of a tree improvement program. This important step is often given inadequate consideration because of the time required to make meaningful tests. Pressures sometimes dictate using results of studies from trees assessed at very young ages, and sometimes decisions on species or seed source are based upon such relatively unimportant items as cost of seed. When working with indigenous species, the initial decision should be to use the local material unless (or until) later tests show that something else is better. The problem of determining the proper species or sources is more complex when exotics, rather than indigenous species, are used. Although the steps that should be taken in deciding which exotic species or seed sources should be used were described earlier, they are summarized here.

Evaluate any existing plantations in the area for tree growth, form, and wood quality. Because of the potential change of wood quality when trees are grown in differing environments (see Chapters 3 and 12), it is especially essential to determine wood quality in the specific environment where the exotic will be planted. If there are suitable stands of exotics, and if the trees are producing seed, collections from the best trees can be used as immediate seed sources for planting. If plantations are extensive, with a broad enough genetic base, selections of the most outstanding trees can be made and a land race developed as a future source of seed.

When there is no indication about which species or source to use, rely on the judgment of people with extensive experience and consult the available literature. The importance of this preliminary step cannot be overemphasized. Species or source decisions are made too often upon a whim or upon the advice of someone with limited experience. The fate of whole programs and millions of dollars depends on such "nondecisions." When political or economic pressures require immediate planting, the best known source can be used with the understanding that land race development and further source testing will be started as soon as it is feasible to possibly replace the initial source that was used.

Regardless of whether the situation exists as described in the two preceding paragraphs, tests of the potentially best species and sources need to be started immediately. These tests should be carefully chosen; a common error is to waste energy and resources on material that has little or no potential for the environments under consideration. Once indications of good species and sources are evident, they should be planted on a scale that is large enough to allow for selections that will ultimately be used for the development of land races.

Tests should be made of sources that have proven their worth in other regions. Often, improved stock has already been developed in exotic environments and is usually available for most of the species that are used widely as exotics. If tests show that these are well adapted to the new environment, good gains may be quickly achieved by obtaining plant material from the original improved plantations for local use. It is important not to use improved stock from other regions on a large scale until the utility of the improved material in the new environment has been well determined.

4. *Always have at least one secondary species or source along with the primary one.* The secondary species or sources should be sufficiently tested so they can be used in the event that something unforeseen happens to the primary species and source. At least some commercial plantings should be made with secondary species or sources to provide the necessary experience about how to grow them and the opportunity to develop the needed land races if catastrophe should strike the primary species.

5. *Establish the necessary seed production capacity* (or facilities for vegetative propagation) in seed orchards or clonal "nurseries." Selecting and breeding the genetically improved material is only a first step; it must be possible to produce the improved trees in mass quantities at a reasonable cost.

6. *Do the testing necessary for the determination of the genetic worth of the selected trees and to form a base for long-term, advanced-generation improvement.* Tests should be made in an area and under conditions that are typical of

those used for commercial planting; there should be progeny tests to determine the value of the parents and genetic tests to develop populations that can be used for advanced-generation breeding. Accomplishing both objectives simultaneously is possible only with considerable planning.

7. *The tree improvement program must have both an operational (applied) and developmental (research) function.* For a successful tree improvement program, the operational phases must be associated with the equally important developmental phases (see Chapter 1), so that long-term as well as short-term objectives are required if a program is to be successful (Figure 16.2). It is easy to become so involved in the day-to-day chores of conducting an operational tree improvement program that new ideas are not pursued or that a reassessment is not made of the current operational program. New possibilities and technologies need to be explored.

8. *Make certain there is sufficient commitment to support and finance the project over the long term, with capable people to carry on the program so that its*

FIGURE 16.2

Testing newly developed strains of trees takes much time and effort. Shown is a 6-year-old test planting of especially wet-site-tolerant loblolly pine. These trees grew well on a site on which standard loblolly pine is ill adapted.

objectives can be fulfilled. Too often, programs are initiated, get started well, and then lose support because of a lack of knowledge about how much it costs to maintain a proper program over many years. The assurance of proper and adequate support is especially critical in the developing nations where the economic situation may change rapidly, or governmental agencies may alter their goals following changes in governmental leadership.

9. *Combine breeding for improved economic characteristics with breeding for a broad genetic base for adaptability and to develop* trees suitable for new products (Figure 16.3). Because most economically important traits are usually genetically independent, it is possible to develop them simultaneously. One of the greatest needs in forestry is to have trees that will grow on marginal sites. The genetic base that is necessary to develop trees for shorter rotations and wood for nonconventional products such as energy or chemicals must also be maintained.

FIGURE 16.3

Broadening the genetic base is a necessary part of a tree improvement program. One way to do this is to make wide crosses within a species. Shown are 3-year-old tests of wide crosses of loblolly pine; crosses are North Carolina × Texas and Louisiana × South Carolina.

METHODS TO BE USED

Advice is plentiful, varied, and often contradictory about the best approach is organizing new tree improvement program. Each person has knowledge about a special set of circumstances related to past experience, and he or she will tend to recommend what has worked best under those conditions. Advice is often positive and dogmatic about why a certain method is the only (or best) one to use in setting up the program. However, each species and each situation must be approached differently (Zobel, 1977; Nikles, 1979).

It is of the utmost importance to recognize that there is no single "correct" way to organize a tree improvement program. An intelligent decision requires a sorting among the options and acceptance of the methods that are the most correct for the set of circumstances involved. After the objectives have been set and available funding, facilities, and personnel are known, the following questions must be asked and other factors must be considered before the methodology of a program is set.

1. "Has it been tried before?" If it has, making a decision is simple. Existing programs should be studied and analyzed, always with the question in mind, "Will it work for us?" The tendency to copy ongoing successful programs without analyzing the parts can result in costly failures. A small tree improvement program cannot operate in the same way as a large one; therefore, methodologies must often be altered. In large programs, such as one established through a cooperative, partial results can be obtained by each member and pooled for the benefit of all. An organization with a small program has to be more self-sufficient and have a broader testing and crossing program than does each individual member of a cooperative.

2. "What are the advantages and problems of the proposed program?" Some program approaches should be automatically rejected as being too expensive or too difficult for the objectives and resources available. A program must be suitable biologically, or it will not result in optimal returns. No program is perfect, but the pros and cons must be considered before a decision is made. The decision with respect to the best program must take into consideration both the long-term and short-term criteria. Shortcut analyses of program alternatives can be dangerous and misleading. In the final analysis, the program must be viewed with respect to what is essential, and the other factors should be given a lower priority or deleted.

3. "Should a new method be used?" New and sometimes quite different ideas are constantly being proposed for development of tree improvement programs (Figure 16.4). Those that sound promising in theory must be evaluated for operational conditions. New methods sometimes are based on the concept of quick returns combined with minimal costs; they may even be a version of get-rich-quick schemes. New ideas should never be rejected until well-designed and well-executed studies have shown their true worth. How-

FIGURE 16.4

Many innovations must be tried in tree improvement; these are much more possible with a cooperative approach. Shown is the vacuum seed harvester developed through the combined efforts of members of the North Carolina State–Industry Tree Improvement Cooperative. Such development would not have been done by individual members.

ever, a full-fledged program should never be developed on theoretical or unproven methods. Always make the intended program flexible enough so that new ideas can be incorporated into it with no interruption of the ongoing work.

4. Accelerated breeding plans should be incorporated into all tree improvement programs. A quick turnover of generations is very important in obtaining genetic gain through a breeding program (Zobel, 1978b). The objective is not to see how much total gain can be obtained per generation but *how much gain can be achieved per unit* of time (Weir and Zobel, 1975). Although the total gain from the conventional and accelerated programs may be the same, obtaining the gain several years earlier results in planting improved trees that would not have been available otherwise during the interim period while a conventional program was being developed.

5. One must estimate future methodologies, such as vegetative propagation, and include the ability to incorporate new methods into the program so that advantage can be taken of advancements. In addition, the following must all

be considered: changes related to nonconventional forest products, how much uniformity will be needed, how much will quality needs change with improved technology, and how important will coppice regeneration become (Berger et al., 1974; Ceancio, 1977).

IMPORTANT GENERAL CONCEPTS

In addition to the previous lists of specific actions needed to start a tree improvement program, certain other important factors need to be considered. The first of these is the time required to complete the various steps of the program (Zobel, 1978b). How urgent is the need for use of the material developed in the tree improvement program? As discussed previously, time is of key importance in any economic assessment because profitability is tied to the length of time an investment is to be carried. Thus, if one can reduce rotation age for a desired product by a few years through tree improvement, returns on the planting investment will be greatly increased.

Underlying all decisions in determining the type of tree improvement program is whether the gain is worth the *risk*. The more intensive the selection program, the greater will be the gain; however, a greater reduction in the genetic base will result. A balance must be maintained between gain and an acceptable amount of risk. A determination of this balance takes great skill and experience and is often ignored during planning, or it is not seriously considered. Usually the financially oriented foresters want to maximize gain, whereas those with other orientations place priority on risk reduction. It is necessary that people with the two viewpoints work closely together to achieve a suitable compromise to develop a well-balanced tree improvement program.

The general understanding is that improved seed, and thus benefits from tree improvement, will not be available until after seed orchard seed are available. However, modest gains, especially in tree quality, adaptability, and pest resistance, can be obtained in the interim before the seed orchard comes into production. Seed can be collected from good individual trees in the forest, or seed production areas can be established. Such seed sources are particularly valuable in exotic forestry programs in which adaptability is such a highly important characteristic. Some genetic improvement is even possible by means of the proper manipulation of natural regeneration and through control of harvesting practices or by application of precommercial thinning.

Often, tree improvement programs are begun with the only objective being to obtain short-term gains in operational planting programs. When this is done, tree improvement is viewed as consisting only of selecting good trees, establishing them in seed orchards, and collecting the seed. This method will eventually lead to problems, due to the fact that the true genetic value of the parents is unknown because there are no progeny tests and there will be no possibility for advanced-generation breeding. Tree improvement programs are continuously developing, and constant effort is required to obtain additional increments of improvement. To achieve this best, the developmental phase should be started at about the same time as the operational phase, so that the results can be available to supplement and perhaps even replace the initial operational phase. As tree improvement becomes more sophisticated and moves into advanced generations, plant material must be available to take advantage of advanced-generation improvements. For example, a program using vegetative propagation will result in only limited gains unless an accompanying breeding program has been especially designed to produce material that can be used in the vegetative propagation program.

THE COOPERATIVE APPROACH

Tree improvement requires a large expenditure of effort and money, trained people, and suitable facilities. Because of this, it is usually not economical to undertake tree improvement on a small scale. It is an activity that is best suited for large corporations and governments to undertake, especially in situations where the forestland base is extensive enough to obtain a good payback (van Buijtenen, 1969). This restriction does not mean that smaller organizations cannot share in its benefits. The way to obtain more general application of tree improvement is to establish cooperatives. Tree improvement is well suited to a cooperative effort in which the members share costs and returns as well as exchange equipment, plant material, and information.

A cooperative effort becomes especially valuable when advanced generations are developed (Weir and Zobel, 1975). A few trained professionals with proper technical support can direct a very large program. Each member cannot afford its own specialists, but it can share them by means of a cooperative. The success and efficiency of the cooperative approach have been well proven, and many cooperatives have been started throughout the world. Most of them have been successful and have made great contributions. Much of the information and many of the ideas expressed in this book are based upon experiences with cooperatives. There are many rules and methods necessary to make a cooperative successful; these could fill a small book. A few of the more important ones have been outlined by Zobel (1981).

1. Enthusiasm is needed. The supporters must be sold on the need and value of a tree improvement program. Halfhearted support will doom a cooperative.

2. All members must make a full commitment. Nothing will ruin a cooperative quicker than having a member or members who fail to give their share fully. This includes not only financial support but also technical support and action in field operations. A minimum contribution of time and action is expected from each member. Also, every member must make a contribution that is of value to the other members.

3. Development and tests of improved materials should be on the lands of the cooperators. Contributions of money are not enough. Each member must feel that "this is my program." There must be pride in what the cooperative achieves, both for the overall membership and for each member. Without such pride, a cooperative will not be effective nor will it survive.

4. Information and ideas must be fully exchanged among members; a proprietary attitude cannot prevail. Exchanges should take place in written reports but also in meetings within the cooperative in which members can compare ideas and see what others are doing. In forest biology, it is not the obtaining of information that will give an organization an economic advantage; rather, it is the using of available knowledge that enables one organization to forge ahead of another.

5. The membership usually has common objectives for all members, although there may be specific ones for individual members. For example, developing seed orchards is a common objective, whereas developing a strain of trees that is especially adaptable to an unusual adverse environment may be of interest to only one or to a few of the members.

6. Strong leadership is needed to bring the members together and to keep them all headed in the same direction. This need cannot be overemphasized. A cooperative will fail unless it has at its head a person who is respected by the members, one they will follow to get the job done, and one who has the authority to get necessary action. Such a director needs to have rather broad decision-making powers. There should be responsibility to a group, such as an advisory committee composed of the cooperative's membership, but the director is the one who has the authority over the operation of the cooperative and who directs actions to achieve its objectives. Occasionally, cooperatives have been administered by committee, but this system often is less than successful.

7. It is always best to have the cooperative headquarters located at a neutral location or neutral site. Often, universities are selected; if the cooperative has its headquarters at one industry or in one governmental unit, jealousies soon appear among the members. The advantage of being located at a university is the backup available in the technical fields. Of special importance is the fact that graduate students can efficiently undertake phases of the necessary research.

8. Each member should have equal authority and responsibility and receive equal benefits. Cooperatives in which the larger organizations pay more and thus also have more influence frequently develop political problems. The cooperative must be large enough to justify its activities economically; this is also true for the operations of each member. It is important that each potential new member be carefully assessed to determine his or her sincere interest, commitment, and ability to make the needed contribution to the cooperative before he or she is accepted into it.

9. Both short- and long-term objectives are needed. No matter what the attitude is at the beginning of a program, there always comes a time within a few years when the question will be asked, "What are we getting for our money?" The program must be so designed that a continuous feedback of results to the program supporters is possible so that they can see the worth of the effort. Some short-term projects serve this purpose well.

10. Good communication is necessary. This is achieved by using the "language" that the program supporters understand. Highly scientific or complex reporting of results to cooperative administrators is self-defeating. One cannot teach the supporters a new "language"; the cooperative staff and associated scientists must learn the method of expression that is best understood by the administrators and present results in such a way that they are understandable to the administrators. This is particularly necessary with respect to the people who control finances.

11. The terms *basic* or *fundamental research* are often viewed unfavorably by forestry administrators. Yet, both types of studies are needed if programs are to be efficient and continue to make progress. Use of the word *supportive* research rather than *basic* research has been most successful in enabling the administrators to appreciate the need for basic studies and to accept them in the applied program.

The success of the cooperatives that are already established in several countries on several continents has been outstanding. Most applied tree improvement programs would just now be in the establishment phases if cooperatives had not been organized and functioning at the time when the need for tree improvement was recognized.

THE COMMITMENT

A tree improvement program must have continuity. It is a long-term program and costly. All too often a program is started, only to falter because the people involved are transferred within or out of the organization. This results in a program that flounders and loses its initiative. The first part to be lost is the developmental or research aspect. Once nonfunctional, it cannot be easily regained, and time will always be lost. Lack of continuity from fluctuating leadership or nonregular support caused by lack of commitment means loss of potential income to the organization involved. The authors feel so strongly about this that their advice to those considering starting a tree improvement program is EITHER CONDUCT THE PROGRAM CORRECTLY, WITH TOTAL SUPPORT IN MANPOWER, FACILITIES, AND EQUIPMENT, OR DO NOT DO IT AT ALL. Tree improvement does not come free; it is expensive. Without tree improvement, forestry can never come close to reaching the goal of optimum productivity.

LITERATURE CITED

Armson, K. A. 1980. "Productive Forest Land—The Factory Base." The Forest Imperative. Proc. Can. For. Cong., Toronto, pp. 41–42.

Berger, R., Simoes, J. W., and Leite, N. B. 1974. Method for economic evaluation of the improvement of *Eucalyptus* stands reserved for cutting. *IPEF,* No. 8:55–62.

Brown, L. R. 1980. Food or fuel: New competition for the world's cropland. *Interciencia* 5(6):365–372.

Ceancio, A. 1977. Il metodo dell' invecchiamento nella conversione der cedui di faggio [On the coppicing period and productivity of *Eucalyptus camaldulensis* and *E. globulus* stands at Prozzo Armerma]. *Ann. Ist. Sper. Selvi.* 8:17–96.

Goldstein, I. S. 1980. New technology for new uses of wood. *Tappi* 63(2):105–108.

Keays, J. L. 1975. Projection of world demand and supply for wood fiber to the year 2000. *Tappi* 58(1):90–95.

Lin, Feng-Bor. 1981. Economic desirability of using wood as a fuel for steam production. *For. Prod. Jour.* 31(1):31–36.

Nikles, D. G. 1979. "Forest Plantations: The Shape of the Future. The Means to Excellence—Through Genetics." Proc. Science Symp., Weyerhaeuser Co. Tech. Cent., Tacoma, Wash., pp. 87–118.

Parde, J. 1980. Forest biomass. *Comm. For. Bur.* 41(8):343–362.

van Buijtenen, J. P. 1969. The impact of state–industry cooperative programs on tree planting. *For. Farmer* 29(2):14, 20, 28.

Weir, R. J., and Zobel, B. J. 1975. "Managing Genetic Resources for the Future, a Plan for the N.C. State–Industry Cooperative Tree Improvement Program." Proc. 13th Sou. For. Tree Impr. Conf., Raleigh, N.C., pp. 73–82.

Zobel, B. J. 1977. Increasing southern pine timber production through tree improvement. *South. Jour. App. For.* 1(1):3–10.

Zobel, B. J. 1978a. The good life or subsistence—Some benefits of tree breeding. *Unasylva* 30(119/120):5–9.

Zobel, B. J. 1978b. "Progress in Breeding Forest Trees—The Problem of Time." Proc. 27th Annual Session, Nat. Poultry Breeders Roundtable, Kansas City, Mo., pp. 18–29.

Zobel, B. J. 1981. Imbalance in the world's conifer timber supply. *Tappi* 63(2):95–98.

Zobel, B. J. 1981. "Research Needs in Tree Breeding—To Make a Cooperative Work." Proc. 15th N. Amer. Quant. For. Gen. Group Workshop, Coeur D'Alene, Idaho, pp. 130–132.

Species Index

493

Scientific Name	Common Name	Reference Pages
		209–213, 215, 227, 229, 243, 262–263, 265–267, 271–272, 274–275, 286–287, 290, 292, 298, 302–304, 306, 315, 325, 336–338, 341–343, 351, 353, 355, 358, 361, 363, 369–371, 373, 379–395, 401, 403, 408–413, 426, 428, 430, 432–434, 441, 446–447, 456–458, 484–485
Pinus tecunumanii	Tecun Uman pine	473
Pinus thunbergii	Japanese black pine	290, 340
Pinus virginiana	Virginia pine	58, 115, 338, 392
Platanus sp.	sycamore	71, 101, 209, 358
Platanus occidentalis	sycamore	43, 172, 187–188, 230, 282, 292, 299, 303, 341, 406, 410
Populus sp.	poplar aspen and cottonwood	54, 71, 73–74, 79, 97, 99, 101, 162, 164, 220, 282–283, 287, 291, 299–300, 303–305, 310, 325, 335, 355, 357–358, 370, 400, 410, 426, 430, 435, 441
Populus deltoides	Eastern cottonwood	409, 457
Populus tremuloides	trembling aspen	87, 154, 408, 435
Populus trichocarpa	black cottonwood	292, 409
Prunus serotina	black cherry	112, 327, 339, 400, 410, 450
Pseudobombax tomentosum	—	411
Pseudotsuga menziesii	Douglas fir	43, 71, 73, 79, 87–88, 97, 108, 130, 139, 141, 180, 190, 196, 209–211, 214, 282, 290–291, 300, 303–305, 337–338, 341, 343, 368, 401, 410, 426, 431, 434, 455
Quercus sp.	oak	325, 339, 351, 358, 367–368, 370, 372, 401
Quercus alba	white oak	230, 339

Subject Index

Adaptability (adapted, adaptation):
to adverse environments, *see*
Resistance
to air pollution and acid rain, *see*
Resistance
definition, 80, 270
gains from, 21, 270–271, 440, 446–447
to marginal sites, 12–14, 270, 484
to pests, *see* Resistance
selection and breeding for, 218, 270,
463
Air layers, 319–320
Allelopathy, 293

Breeding, tree:
accelerated, 430–431, 487
advanced generation, 416–417, 419–
422, 428, 445, 483, 489
development, 8
general, 2–3, 5–6, 445
history, 5
intuitive, 10–11
methods, general, 5, 445, 472
objectives, 20
population, 417–418
production, 416–419
for several characteristics simultaneously, 22
systems, *see* Mating (breeding systems)

Cells:
chromosomes:
composition of, 42, 45
definition, 42
numbers of, 42–44
components of, 41–42
Cline:
definition, 83–85
examples, 84
general, 81
Clone:
degradation, 331–332
deployment, 328–331
general, 314, 316, 318–319, 328–329,
331–332
number, to use, 328–329
Competition, 443

Correlations among characteristics:
genetic, 22
juvenile-mature, 23, 330, 426–427,
440–441
Cyclophysis, 323

Ecotype:
definition, 84–85
general, 84–85
Environment:
definition, 25
limiting factors, 13–14
manipulation of, 20
Environments, adverse:
cold, 297, 447
definition, 296
gains, from breeding for, 296, 447
moisture extremes, 297
resistance to, 296–298
see also Resistance
Exotics:
definition, 80
general, 93–94, 482
gene resources, 472–475
land races, use of, *see* Land race
pests of, 99, 101, 103–104, 277–278
problems with, 83, 94, 98–101, 111,
278
rules for movement, 107–111
soils, importance of, 95–96
use and value of, 97–99
when, where, why to use, 88, 93–97,
483
where to obtain, 88–89, 99, 104–108
why wrong sources are used, 102–103
Experimental designs:
general, 245–246
plot-definition, 244–245
plot-designs, 245–247

Family, definition, 26
Forest genetics:
applied, 2
definition, 6
history, 2–3, 5
see also Forest tree improvement